Wayfinding Behavior

Wayfinding Behavior

COGNITIVE
MAPPING
AND
OTHER
SPATIAL
PROCESSES

EDITED BY
Reginald G. Golledge

THE JOHNS HOPKINS UNIVERSITY PRESS
Baltimore and London

© 1999 The Johns Hopkins University Press
All rights reserved. Published 1999
Printed in the United States of America on acid-free paper

9 8 7 6 5 4 3 2 1

The Johns Hopkins University Press
2715 North Charles Street
Baltimore, Maryland 21218-4363
www.press.jhu.edu

Library of Congress Cataloging-in-Publication Data

Wayfinding behavior : cognitive mapping and other spatial processes / edited by
    Reginald G. Golledge.
        p.   cm.
    Papers originally presented at a seminar held at the Chateau de la Bretesche in
July 1996.
    Includes bibliographical references and index.
    ISBN 0-8018-5993-X (hardcover : alk. paper)
    1. Cognitive maps (Psychology)—Congresses.   2. Human information
processing—Congresses.   3. Spatial behavior—Congresses.   4. Spatial behavior in
animals—Congresses.   I. Golledge, Reginald G., 1937–   .
    BF314.W38   1999
    153.7'52—dc21   98-25999
                                                                           CIP

A catalog record for this book is available from the British Library.

To Bill and Betty Behling

Contents

Preface

The use of maps—in forms ranging from dirt drawings to stone carvings, from rice paper scrolls to Automobile Association trip-tiks, from topographic map sheets to disposable tactile strip maps—appears to be a cultural universal (Uttal 1997). Maps both record what is known and remembered about an environment and act as wayfinding aids. In the absence of these artifacts, humans and other animals rely on internal representations or stored memories of experienced environments. It is frequently assumed that these stored memories, now commonly referred to as "cognitive maps" or "internal spatial representations," are used to guide travel. Cognitive maps are always there—they cannot be left at home, torn to pieces by fractious children, rendered apart and pieced together incorrectly such that map reading errors result in a traveler becoming lost. But they do have their problems.

The idea that animals also possess internal spatial representations resulted in Tolman's (1948) first identification of the term *cognitive map*. But, rather than state that animals—particularly the rats that took short cuts through his mazes—stored spatial information as a map, Tolman used the term metaphorically. In other words, he suggested that the animals used in his studies appeared to be able to use spatial information in a manner *as if* the places they remembered were recorded in a *maplike* manner. For decades, controversy has raged over whether animals do have cognitive maps or if they have other forms of internal spatial representations that allow them to behave as if they were being guided by a map-reading operation. After decades of research in zoology, other biological sciences, and experimental psychology, in particular, various alternatives have been posed to account for successful animal travel behavior. Many have argued that the practice of returning directly to home after a meandering search for food by many nonhuman species indicates that the species did continuous spatial updating, then returned home by a procedure well known to ocean shipping or aircraft pilots—the process of dead reckoning. Called *path integration,* this process enables a traveler constantly to update current position with respect to an origin without recording details of the path already followed. Because there is no need for a memory trace of the path, route retrace may be difficult or impossible. The need by many foragers who are responsible in part for feeding other members of their species to return home with food appears to make the short cut (or "beeline" or "crow-fly") return trip the more reasonable one. If food is consumed at the spot on which it is found, safety considerations might also consequently dictate an immediate shortest-distance return.

Flying insects and avian species appear to use landmarks, sun compasses, magnetic compasses, or celestial guides to help them with migratory and shorter-distance travel. Naturally enough, questions have arisen as to whether the landmarks used are captured as a perspectively viewed retinal image, or whether their configuration or layout is either stored and recalled in sequence as a route is followed or represented as layouts or configurations similar to a survey (or overview-based) representation of a large-scale and complex space.

Despite the existence of these two vigorous research areas, focused on nonhuman and human travel, respectively, until recently there has been little deliberate attempt to combine the two literatures. This lack of attention provided the rationale for a small seminar funded by the Borchard Foundation at the Chateau de la Bretesche in July 1996. The purpose of the meeting was to bring together researchers from both the human and nonhuman research domains who had specialized in navigation or wayfinding behavior and who were familiar with the idea of cognitive mapping and the potential role that cognitive maps might play in wayfinding behaviors. As the guests of the foundation's director, William F. Behling and his wife, Betty, at the Chateau de la Bretesche, nine contributors to this volume first presented position statements on the relationships between cognitive maps and wayfinding in humans and other species; they then engaged in spirited discussion about the nature of this problem. Attending that meeting were Reginald G. Golledge, geographer, University of California, Santa Barbara; Jack Loomis, psychologist, University of California, Santa Barbara; Gary Allen, psychologist, University of South Carolina; Lynn Nadel, psychologist and neuroscientist, University of Arizona; Mary Peterson, psychologist, University of Arizona; Tommy Gärling, psychologist, University of Göteborg; Alain Berthöz, psychologist, Laboratoire de Physiologie de la Perception et de l'Action (LPPA), CNRS Collège de France; Michel-Ange Amorim, psychologist, Laboratoire de Physiologie de la Perception et de l'Action (LPPA), CNRS Collège de France, and General Hospital at Vienna, University Clinic of Neurology; Ariane Etienne, ethologist, University of Geneva; and Catherine Thinus-Blanc, Research Center in Cognitive Neuroscience, Marseilles.

Although this small group covered a variety of interests, it was recognized that there were other areas which were not well represented at the meetings but which have contributed significantly to work on either cognitive mapping or wayfinding, particularly in nonhuman species. Invitations were thus sent to a number of scientists to join with the initial Borchard group to write chapters for this current volume. These included Eliahu Stern, geographer, Ben Gurion University, Israel, and his associate, Juval Portugali, geographer, Tel Aviv University, Israel, whose work on route choice in complex urban environments has captured the essence of human wayfinding and decision making in various settings; John Rieser, psychologist, Vanderbilt University, Tennessee, whose work on wayfinding without vision in adults and children is seminal; Thomas Collett and his associates from the Sussex Centre for Neurosciences, UK, an acknowledged leader in research on insect navigation; Wolfgang and Roswitha Wiltschko, Zoological Institute, Goethe University, Frankfurt, world-respected authorities on avian navigation; and Eric Chown, computer scientist and artificial intelligence modeler from Bowdoin College. Herbert Pick of the Institute of Child Development of the University of Minnesota was not able to complete an assigned chapter because of prior commitments but, instead, acted as a chapter referee.

The book is organized in four parts. In the initial section, emphasis is placed on human wayfinding and the cognitive maps that are said to be important in the wayfinding process. One concern emanating from an initial meeting at the Chateau de la Bretesche was for the embarrassment of terms that can be found throughout the literature essentially referring to the central concept for which we have chosen Tolman's original expression, the *cognitive map*. Kitchin (1994) has produced a collection of phrases found in the literature that discusses spatial memory and spatial representations.

They include abstract maps (Hernandez 1991); cognitive configurations (Golledge 1978b); cognitive images (Lloyd 1982); cognitive maps (Tolman 1948); cognitive representations (Downs & Stea 1973b); cognitive schemata (Lee 1968); cognitive space (Montello 1989); cognitive systems (Canter 1977); conceptual representations (Stea 1969); configurational or layout representations (Golledge, Briggs, & Demko 1969; Kirasic 1991); environmental images (Lynch 1960); imaginary maps (Trowbridge 1913); mental images (Pocock 1973); mental maps (Gould 1966; Gould & White 1974); mental representations (Gale 1982); orienting schemata (Neisser 1976); place schemata (Axia, Peron, & Baroni 1991); spatial representation (Allen, Siegel, & Rosinski 1978); spatial schemata (Lee, 1968); survey representation (Downs & Stea 1973b); topological representations (Shemyakin 1962); topological schemata (Griffin 1948); and world graphs and cognitive atlases (Lieblich & Arbib 1982). Given this diversity of names for what usually are taken to be synonymous terms, one must speculate about the concept validity of whatever term is selected. The term *cognitive map* is used throughout this book to refer to the internal spatial representation of environmental information. Its use varies between the metaphorical ("as if" the information was stored in maplike format) and a hypothetical construct.

The term *spatial representation* is also used throughout the book. This might be regarded as a shorthand notation for the organization of components of spatial knowledge or other partly investigative processes (e.g., neurophysiological structures and place cells, cell assemblies, phase sequences; Kaplan 1970). The term also can be used metaphorically involving an "as if" quality, particularly when referring to purported maplike properties of representations. It has also been used as an intervening variable in which it is interpreted as a "note" attached to an economical grouping of measured variables in a statement of functional relations between other measured variables (MacCorquodale & Meehl 1953). In this way it performs a summarizing function, and at first Tolman's (1948) use of the term *cognitive map* was seen in this way. Today many believe it is more reasonable to assume that Tolman used the term metaphorically in suggesting that cognitive maps "exist" in rats and men. This stems from a statement of Tolman's (p. 192), which says: "We believe that in the course of learning a maze, *something like* a field-map of the environment gets established in the rat's brain" (my emphasis).

Piaget (1951) made a distinction between "representation" in the sense of knowledge (which he termed "conceptual representation or cognitive representation") and representation in the sense of the re-presentation of absent realities (termed "symbolic representation or symbolization").

Symbolic representations are directly observable, are external, and are often today described by the term *spatial product* (Liben 1982). Cognitive representations are not directly observable and are often termed *internal representations*. As such it may be reasonable to consider the term *cognitive representation* as a hypothetical construct referring to hypothesized and not directly observable subjective knowledge of the environment. Regardless of how the term is used, however, until neurophysiological evidence indicates that humans or other species have specific "place cells" that are locationally specific and can be activated in a linked form to form a maplike layout, the presence of internal representations must be inferred from one or more external symbolic presentations or some form of observed behavior. Cognitive mapping is one such process that is not directly observable, but is implied to exist when an organism is capable of solving certain types of problems (such as deducing economical or shortest path routes from a currently occupied destination through unknown space back to an obscured origin). The idea of a cognitive map that can be used as an aid to navigation is most commonly attributed to humans, who are guided by internal representations in their interaction with familiar and unfamiliar physical environments.

The reader will notice some differences among terms between chapters devoted to human and nonhuman species; in the latter, the more general term *spatial representation* is often used, particularly with respect to insects and nonprimates. The discussion of the meaning of the term *cognitive map* dominated much of the informal exchange of ideas at the Chateau. There was consensus that *spatial representation* was an acceptable and general term that could probably be used across all species, while *cognitive map* should best be reserved for use with those species traditionally regarded as using cognitive processes to solve spatial and nonspatial problems. Thus, humans and primates exhibited behaviors encouraging inferences that cognitive processing of stored information had occurred in order to produce the observed behaviors. The reservation expressed by Bennett (1996) as to whether or not animals had cognitive maps was reflected in much discussion at the meetings, where it was agreed that spatial representations could be encoded in a variety of formats, depending on the species being examined. As Etienne, Maurer, Georgakopoulos, and Griffin (chapter 8) suggest, "To be credited with a cognitive map a subject therefore must be able to perform adequate new routes, that is, choose the most economical alternative path . . . under new conditions" (pp. 197–98). They imply that to possess a cognitive map, a traveler must be capable of planning and performing a goal-directed act of path selection by deducing an itinerary from the memorized spatial relationships between the goal and the traveler's current position. The traveler must also be capable of defining new routes by choosing the most economical alternative (e.g., via shortcut or detour) when faced with unfamiliar environments. These actions have to occur without the help of external guides, such as sun compass, celestial navigation, maps, or landmarks.

The difference between this form of high-level spatial information processing and the act of locomotion that helps an organism to explore and exploit an environment in an adaptive manner so as to satisfy biological needs represents an agreed-upon difference between the cognitive map of humans or other primates (and possibly some other nonhuman species such as avians), as opposed to the spatial representation (or the representation of locomotor space) more typical of other animals and insects. This distinction seems most relevant for animals occupying sedentary space (i.e., ground-based species). The Wiltschkos (chapter 10) provide fascinating experiments that indicate that avians appear to use grid maps to help guide their locomotor activities (particularly long-distance migrations, and the medium-distance trips of homing pigeons). In fact the evidence they show of the development of wayfinding ability in young pigeons seems to parallel the stages of spatial knowledge acquisition and wayfinding behavior of pre-teenage children as summarized by Siegel and White (1975).

In much of the literature concerning cognitive maps in humans, the term *survey knowledge* is used. This represents the most advanced form of geometric comprehension of the external environment and is said to consist of layout or configurational understanding similar to that obtained by overlooking a place (such as from a "bird's eye view"; Hart & Moore 1973). Gallistel (1990) provides an integrating hypothesis that suggests that our general representation of the space through which locomotion occurs owes as much to apprehension of the environment from continuous internal feedback and external reaferences (such as visual flow) resulting from locomotion as it does to the direct perception of discrete spatial cues such as might be obtained from surveying an environment from a bird's eye position.

We all anticipate that this linking of ongoing research on both human and nonhuman cognitive mapping and wayfinding research will be a stimulant to encourage researchers from many disciplines to continue interacting on this significant academic and practical problem. Other groups may not be as fortunate to have such pleasant surroundings as the Chateau de la Bretesche in which to exchange ideas and formulate plans for the future, but we are all sure that further cross-fertilization of ideas relating to spatial representation and wayfinding behavior of many species will produce positive long-term theoretical and practical results.

Acknowledgments

It is fair to say that this book would not have been written had it not been for the support of the Borchard Foundation. William and Betty Behling acted as charming and convivial hosts at the Chateau de la Bretesche, providing the stress-free and supportive environment that encouraged free exchange of ideas among the representatives from the different disciplines invited to the meetings. The sessions were conducted sitting around a magnificent banquet table in the great hall of the chateau. For the three days of meetings, neither Bill nor Betty left the meeting room except to order refreshments. Their interest in the discussion was as keen as our own, and the additional insights we obtained from them while enjoying refreshments in their beautiful garden (laid out between the imposing chateau and a surrounding lake) proved that their interest was much more than idle curiosity. It is indeed a pity that those later invited to participate in the book were not able to join us at the initial meetings. (It should be of no surprise that this book is dedicated to our charming hosts, Bill and Betty.)

There are others whose contribution to the production of this book must be acknowledged. In addition to writing their own chapters, all authors reviewed selected other chapters in the book. Jack Loomis and Tommy Gärling in particular reviewed multiple chapters and I am very grateful for their insights and comments. On the production side, Ann Haque, Sandra Jacobs, Mary MacDonald, and Howard Pommerening at various times provided significant input into the preparation of the final manuscript, the reference lists, the index, and the drawing of many figures. Their assistance is gratefully acknowledged. I also thank Ginger Berman of the Johns Hopkins University Press, whose interest in the theme of our interactions led to her instrumental role in getting the press to issue a contract for the book. We further acknowledge the help of the editors and publishers of a variety of books and journals for their permission to reproduce illustrations previously appearing in their publications. Finally, we acknowledge Peter Strupp and Princeton Editorial Associates for thorough and sympathetic editing and final production.

Wayfinding Behavior

I

Human Cognitive Maps and Wayfinding

This first section explores the strong theoretical and empirical links between cognitive maps (or the internal representation of environmental information), the cognitive mapping process itself, the internal manipulation of information in the form of spatial choice and decision making, and the directed acts of human wayfinding through simple or complex environments. The evidence is clear and overwhelming that human wayfinding is directed and motivated, and follows sets of procedural rules whose content and structure are the focus of much ongoing research. The consensus is clear: humans acquire, code, store, decode, and use cognitive information as part of their navigation and wayfinding activities. Although over the centuries they have developed numerous ways of supplementing personally stored environmental information (e.g., maps, written descriptions, and various forms of image representations), it appears that humans rely on personal cognitions to make many spatial decisions, and to guide their movement behavior. There is evidence that internal representations and their externalizations (spatial products) do not necessarily match well, and that the existence of fragmented, incomplete, or distorted cognitive maps appears to account for many behaviors that might otherwise be labeled as spatially irrational.

The purpose of this part is to examine sets of concepts deemed relevant to both human wayfinding and cognitive mapping. Two chapters are by geographers (Golledge, and Stern and Portugali) and two by psychologists (Allen and Gärling). Although some disciplinary perspectives are evident in the chapters, there is much overlap and common concern. The first two chapters, by Golledge and Allen, respectively, provide overviews and summaries of theories and concepts relevant throughout the entire book. The following chapters by Gärling, and Stern and Portugali have a tighter focus: Gärling emphasizes the sequential spatial choice processes so important to human wayfinding, and Stern and Portugali emphasize decision making in urban environments, the complex scenarios in which most humans live and interact. All four chapters contain examples of relevant research.

In the first chapter, Golledge reviews critical definitions relating to cognitive maps and wayfinding. He provides an overview of the role of cognitive mapping in human wayfinding and describes the processes of acquiring and storing spatial information about large-scale complex environments. Further, he discusses how humans record and represent environmental knowledge. The role played by landmarks and routes in anchoring knowledge and in wayfinding is examined, and the differences between path following and route-based environmental learning are explored. Errors commonly related to encoding, decoding, and internally manipulating cognized spatial data are highlighted. Wayfinding by humans in contexts other than with landmark usage is also examined,

and an elaboration of errors commonly found in human wayfinding follows. Throughout, ties are made to treatments of similar problems in later chapters that focus on the nonhuman domains of internal spatial representations and wayfinding.

In the second chapter, Allen provides insights into the nature of spatial abilities and the role they play in cognitive mapping and wayfinding procedures. He gives considerable emphasis to the concept of individual differences in spatial cognition and in behavior. Allen argues that the scientific literature in psychology and geography contains a vast number of studies concerned with spatial abilities and a growing body of research on wayfinding, although little has been done to establish the relevance of the former for the latter. Thus the question of why some individuals are better than others at wayfinding has been difficult to address. Allen suggests that a potentially informative way to think of wayfinding is to differentiate between wayfinding tasks and wayfinding means. Tasks include traveling to a previously known destination, exploration with the purpose of returning home, and traveling to a novel destination. Means include oriented search, following a continuously marked trail, piloting (between landmarks), habitual locomotion, path integration, and reference to a cognitive map.

Spatial abilities in the past have been examined from psychometric, information processing, developmental, and neuropsychological perspectives. Allen suggests that broad families of abilities involved in the identification of manipulable objects, those involved in anticipating the trajectory and speed of moving objects, and those involved in supporting oriented travel within large-scale environments summarize the dominant research themes. He also suggests that some spatial abilities are commonly found in each of these families. This implies there is considerable utility associated with the concept of interactive common resources for cognitive and perceptual-motor tasks. The result of the use of spatial abilities is support-oriented travel, but they also serve as a resource for acquiring additional environmental knowledge. Cognitive maps are considered as knowledge of places and cognitive mapping includes rules for establishing spatial relations among such places. Allen also suggests that an interactive common resources framework provides a new approach to examining individual differences in spatial cognition and behavior. This framework includes perceptual capabilities (e.g., sensitivity to displacements in the visual periphery), fundamental information-processing capabilities (e.g., visual working memory), previously acquired knowledge (e.g., cardinal directions), and motor capabilities as resources. In various combinations, these resources serve as adaptive processes (e.g., path integration ability) in support of wayfinding objectives. In summary, he suggests that this framework could potentially provide a unifying frame of reference for a great deal of research on spatial abilities, cognitive maps, and wayfinding.

Next, Gärling discusses human information processing in sequential spatial choice, which summarizes the essential acts involved in wayfinding. He begins with the premise that human locomotion in space is goal-directed. Spatial orientation and navigation are, therefore, primarily means of monitoring travel plans. Travel plans are developed prior to initiating movement. The chapter focuses on the formation of travel plans and their consequent execution. Such planning entails spatial choices that are multiattribute, sequential, and stated rather than actual. Spatial attributes influencing spatial choices entailed by travel plans are then discussed at length. In particular, he summarizes research on how people process information when solving the traveling salesman problem (i.e., finding the shortest distance between an origin and a set of destinations that might be sequentially visited). He explores the psychological bases behind the widely used linear programming method developed for solving transportation problems—particularly movement through complex networks. He details research on how time and priority are traded off against spatial attributes in sequential spatial choices. The chapter draws from a number of examples relating to Gärling's work on spatial cognition and transportation planning—a recently developing area linking cognitive psychology and transportation science. Gärling explores the multiattribute nature of spatial choice, and the difficult problem of accounting for actual versus stated choices when planning travel.

The fundamental role of distance, sequencing of actions, and the identification of spatial configurations are explored theoretically and empirically. Realizing the importance of nonspatial attributes, Gärling examines tradeoffs such as those resulting from temporal constraints or activity priority on the spatial choice and consequent wayfinding behavior of humans.

Stern and Portugali next examine the relationship between environmental cognition and decision making in urban navigation. They define urban navigation as a sequential process of decision making concerning route choice. They claim that traditionally many choice situations are described by a blackbox approach, which does not specify choice rules but rather deals only with the relationship between input and output variables. The more advanced are the models using this approach, the more they introduce a higher degree of detail to explain the interior of the so-called black-boxes. Even in these circumstances, however, a cognitive explanatory mechanism of the choice process is still missing. Their chapter presents two complementary conceptual frameworks as possible ways to solve this problem. The first is the inter-representational network (IRN), and the second is decision field theory (DFT). It is suggested that both frameworks can explain the dynamics and high variability in the choices of persons navigating in urban environments. IRN suggests how cognitive maps, which they acknowledge as the basis of human decision making, are created; DFT explains the process of decision making. They follow their discus-

sion of these frameworks with conceptual simulations of two external scenarios to demonstrate the applicability of both frameworks for explaining the choice process in urban navigation. They finish by speculating on the potential contribution the two frameworks would make as a consequence of their integration.

In this section the authors have explored human wayfinding and its various components, which range from cognitive maps to objective maps, from the nature of the decision processes involved in choosing and executing routes to the spatial choices of particular types of routes for particular trip purposes. Wayfinding functions are shown to vary with individual differences, between males and females, and according to one's spatial abilities. The authors commonly believe that *cognitive maps* are important in human wayfinding, but some treat this as a hypothetical construct or convenient fiction that provides a common idea about the internal representation of spatial information. Some treat the term as a metaphor, suggesting that if we observe human wayfinding behaviors they indicate that humans behave *as if* they have access to a maplike image of an environment. Suggestions are even made that a map equivalent *is* constructed—perhaps temporarily in working memory—so that problems such as route selection and following can be implemented. The range therefore is from hypothetical construct to analog model. But which is favored most? Your interpretation of these chapters will answer that important question.

REGINALD G. GOLLEDGE

1

Human Wayfinding and Cognitive Maps

Anyone experiencing even momentary disorientation and lack of recognition of immediate surrounds has experienced the uncertainty of being "lost." This state occurs when the wayfinding process being used to guide travel fails in some way. Failure may be due to many different circumstances. To more fully understand how to compensate for being lost, it is essential to know about human wayfinding and the cognitive and environmental factors that influence it.

The purpose of this chapter is to discuss human wayfinding and to examine the role of cognitive maps in human wayfinding activities. It examines the different processes involved in human cognitive mapping, discusses how wayfinding takes place in unobstructed or obstructed environments, in familiar or unfamiliar places, and reviews how humans have altered the natural environment to simplify the act of wayfinding.

Historically, the need for people to communicate or move necessitated the development of skills and procedures for travel. Because travel is so fundamental to human existence, a rich collection of terms and concepts has been developed. Some of these are defined herein; others are discussed in later chapters. This chapter concentrates on human travel over land surfaces. Travel through the air or over water is not considered, nor is navigation by computer, robot, or other independent, mechanical means.

DEFINITION OF TERMS

Wayfinding is the process of determining and following a *path* or *route* between an origin and a destination. It is a purposive, directed, and motivated activity. It may be *observed* as a *trace* of sensorimotor actions through an environment. The trace is called the *route*. The route results from implementing a *travel plan,* which is an a priori activity that defines the sequence of segments and turn angles that comprise the path to be followed. The travel plan encapsulates the chosen strategy for path selection. The *legibility* of a route is the ease with which it can become known, or (in the environmental sense) the ease with which relevant cues or features needed to guide movement decisions can be organized into a coherent pattern. Legibility influences the rate at which an environment can be learned (Freundschuh 1991). Repeated path-following facilitates remembering the path components and recalling them for later use (i.e., route learning). Paths or routes are represented as one-dimensional linked segments or, after integration with other paths, as *networked configurations.* The latter, along with landmarks, the spatial relations among them, and other spatial and nonspatial features of places, make up the remembered layout of an experienced environment. Route learning and route following strategies help build up cognitive maps via an integration process. Cognitive maps are the internal representation of perceived environmental features or objects and the spatial relations among them. Difficulties experienced in mentally integrating different routes and their associated features into networked structures help to explain why cognitive maps may be fragmented, distorted, and irregular.

Human travel takes place either freely in an unstructured natural environment, along paths delineated by repetitive travel, or along natural or artificial ways (e.g., streams, roads, tracks). Free-ranging pedestrian movement or path-following through natural or cultivated areas may be likened to the ground-based movement of nonhuman species. Human-guided movement of vessels on water or in the air may be likened to the swimming and flying navigation of nonhuman species. But human operated vehicles or human guided beasts of burden have few (if any) obvious parallels in the nonhuman world.

During human movement it is possible to distinguish two types of guiding processes. The first of these, used to guide vessels over large water bodies or to fly aircraft, is called *navigation.* Formally defined, to navigate means to steer or direct a ship or aircraft (Webster 1995). Colloquially it means to deliberately walk or make one's way through some space. In its present form, navigation is most frequently used to refer to the science of locating position and plotting a course for ships and aircraft.

The second process involves selecting paths from a network, and is called *pathfinding* or *wayfinding* (Bovy & Stern 1990). For successful travel, it is necessary to be able to identify origin and destination, to determine turn angles, to identify segment lengths and directions of movement, to recognize on route and distant landmarks, and to embed the route to be taken in some larger reference frame. This information is required to plot a course designed to reach a destination (previously known or unknown) or to return to a home base after wandering. If a destination is known but is not directly connected by a path, road, or track to the origin, successful travel may involve search and exploration, use of landmarks, spatial updating of one's location, recognition of segment length and sequencing, identification of a frame of reference, and mental trigonometry (triangulation, dead reckoning, and the like). Human movement is often guided by external aids (cartographic maps, charts, compasses, pedometers, and the like).

For wayfinding, when traveling over roads or tracks either physical or cognitive maps can be used. Other instruments are not generally used unless a trip is planned through an unfamiliar area. Most human wayfinding uses natural skills and abilities and memory-based spatial knowledge. Much of this chapter is an elaboration of the nature and use of these skills, abilities, and knowledge structures.

ENVIRONMENTAL KNOWING AND SPATIAL REPRESENTATIONS

Knowing an environment is a dynamic process in which the current state of information is constantly being updated, supplemented, and reassigned salience depending on the short- and long-run purposes that activate a person's thoughts and actions. Often it is assumed that subjective conceptions of the environment are quite different from objective reality. Information extracted from large-scale external environments and stored in human memory exists in some type of psychological space whose metricity may be unknown. In western cultures, however, much emphasis is placed on interpreting and using space represented as a Euclidean metric. Although this may be somewhat arbitrary, and in some cases unrealistic (see Baird, Wagner, & Noma 1982), much of human common sense and expert knowledge of space is traditionally represented as a Euclidean metric. Given this emphasis, the basic geometry of spatial representations and cognitive maps can be summarized in terms of points, lines, areas, and surfaces. Consequently, if it is assumed that training and experience help structure cognition, then it also seems reasonable to assume that as environmental learning occurs, some of the standard geometry of identifiable physical space will be included in its cognitive representation. But cognitive representations need not bear a one-to-one correspondence with their counterparts in physical space (e.g., distance A to B in objective reality may be perceived to be longer, shorter, or the same as the distance from B to A in cognitive space). Thus, symmetry and reflexivity axioms of Euclidean metrics need not hold universally in a human cognitive representation. Similarly, the axiom of triangle inequality may be violated some portion of the time in the cognitive space (Rivizzigno 1976), and directions may not be as easily specified in a psychological space dominated by a personalized (or egocentric) frame of reference (Moore & Golledge 1976a).

Piaget (1950) used the term *representation* to refer to knowledge or thought. It is often this latter meaning that is implied when people speak of the "images" humans have of their environments (e.g., Lynch 1960; Strauss 1961). Piaget (1951) also makes a distinction between representation in the sense of knowledge (what he calls "conceptual representation or cognitive representation") and representation in the sense of the re-presentation of absent realities ("symbolic representation or symbolization"). Symbolic representations are directly observable and are external. There are many different "external" representations of environments (called "spatial products" by Liben 1982). These are modified by perceptual screening and human abilities.

Until neurophysiological evidence confirms that humans have specific "place cells" that define where spatial information is stored in the brain (see Nadel, chapter 12) and identifies the means by which place cell information is integrated and used, internal representations must be inferred from one or more external symbolic representations (e.g., sketch maps of a city) or from some other forms of observable behavior (e.g., search behavior to find a specific location).

Guided by their internal representations, humans interact with physical environments. The most universal way of doing this is to travel through them. The process guiding travel is *navigation,* and the process of choosing a path is termed *route choice* (Bovy & Stern 1990).

MODES OF ACQUIRING
AN INTERNAL REPRESENTATION

There are many ways that one can learn an environment (Tellevik 1992). When the environment is new, novel, or unexperienced, possible learning strategies include (1) active search and exploration according to specific rules or heuristics; (2) a priori familiarization with secondary information sources about the environment (such as maps, sketches, written or verbal descriptions, artists' renderings, videos, photographs, photographic slides, movies, and virtual realities; MacEachren 1991); and (3) experience of the environment using controlled navigational practices, including exploration using path integration to maintain knowledge of a home base, exploration and retrace methods, exploration by boundary following, sequenced neighborhood search, and so on. Only humans appear to have regular access to communicate materials of the type listed in (2) above (but see von Frisch 1967 for a discussion of the "dance" of the honeybee which appears to pass on the equivalent of "mapped" information regarding the route to nectar locations).

In general it is accepted that the two most common ways of learning an environment are by (1) experiencing it through a travel process guided by sets of procedural rules, and (2) learning the layout either from an overlooking vantage point or via some symbolic, analog, or iconic modeling (e.g., maps or photographs). Learning that takes place at the eye level perspective (usually associated with travel experiences) and learning that takes place by examining configurations or layouts (i.e., when using maps, photographs, or overlook points) are differentiated (MacEachren 1992a; Thorndyke & Hayes-Roth 1982). In the literature, these two procedures have usually been termed *route-based knowledge* and *survey* (or layout or configurational) *knowledge*. Although many argue that route knowledge contributes most (quantitatively) to our stored representations of environments (e.g., MacEachren 1992a), others have shown that the accumulation of spatial information from sources such as planar maps provides more accurate spatial relational knowledge (Thorndyke & Hayes-Roth 1982).

Before proceeding, it is essential to differentiate the notions of learning a route versus route-based learning of the environment. Learning a route (or path following) involves recognizing an origin and a destination and identifying route segments, turn angles, and the sequence of segments and angles that make up the desired path. The major concern is learning the structure of the route and the sequence of behaviors needed to traverse it, rather than learning the environment through which the route passes. Environmental features such as landmarks are learned only insofar as they help to prime turns or distances along segments. When route learning takes place, on-route information is dominant and takes precedence over all off-route information, which is usually regarded as incidental. Routes are often learned unidirectionally in the laboratory, but in practice humans almost invariably learn routes in both the original and reverse direction (unless some barrier such as one-way travel or other constraint prevents an exact retrace). Golledge, Parnicky, and Rayner (1980) have shown that this one-dimensional route learning is the primary mode of environmental knowing by mildly or moderately retarded persons, while locationally matched low socioeconomic mentally able persons quickly develop two-dimensional layout knowledge of both routes and the surrounding environment. Also, some blind or vision-impaired people develop only a one-dimensional understanding of their environment, apparently because their orientation and mobility (O&M) training teaches only route following, not layout learning (see procedures detailed in Welsh & Blasch 1980).

Route-based environmental learning (learning an environment by experiential procedures) is perhaps the most common way used by humans (MacEachren 1992a). In the process of daily or other episodic activities, humans learn routes and become aware of features in the proximal and distant environment. These latter features may be landmarks that can be used for orientation purposes, or they may be landmarks that prime upcoming decision points (e.g., when to search for an entrance to a freeway or where a shopping center is located). Environmental learning in this way develops by integrating and overlaying specific routes, districts through which the routes pass, and on and off route landmarks. Overlay processes are used to integrate different features from different experiences into a single layout coverage (Golledge, Bell, & Dougherty 1994). In this way the segment-by-segment information learned by route following can be parlayed into a network or configurational structure (Bovy & Stern 1990). Simultaneously the linearized geometry of the path becomes modified and integrated into two- or three-dimensional geometric and trigonometric concepts related to network structures and feature locations.

Sholl (1996) suggests that travel requires humans to activate two processes that facilitate spatial knowledge acquisition—person-to-object relations that dynamically alter as movement takes place (egocentric referencing) and a more stable object-to-object knowledge representation that anchors their cognitive map. This explains why someone can become locally disoriented (poor person-to-object updating), while still comprehending the basic structure of the larger environment through which travel is taking place. Misspecifying local relations, or incorrect encoding or decoding of anchor-point geometry, can produce the distortions often found in spatial products.

Shortcutting is an example of following a route that implies not just learning the route but something more (Wagener, Wender, & Wagner 1990). The process required to allow a traveler at the end of a wandering path to turn and face the home base and consistently move toward it in as direct a manner as the local transportation network allows is usually taken as evidence that some type of survey, layout, or configurational knowledge has been achieved (usually Euclidean). Such an inference is made even stronger if at any point on the return home the individual can indicate the direction and approximate distance from the current location to a landmark, place, or choice point, experienced on or visible from the initial route. An alternate explanation is that a process similar to that found to underlie the homing behavior of nonhuman species, path integration, is responsible for homing. The relative significance of these alternatives in human and animal domains are examined in several other chapters in this book (see chapters 5, 9, and 13 in particular).

Sectoral or local regional knowledge may accrue in the vicinity of a route. Initially, therefore, knowledge of an area may develop as a series of strips or corridors surrounding specific routes. This facilitates knowledge integration if the routes are known and are overlapping. Evidence exists that integration of information learned from different routes is not automatic, and may be achieved only partially. For example, in their study of route learning in an unfamiliar environment, Golledge, Ruggles, Pellegrino, and Gale (1993) showed that when individuals learned two partially overlapping routes in the same unfamiliar area, even after several trials most were unable to point accurately from specific places on one route to specific places on the other route. On the other hand, researchers studying pointing in highly familiar environments have shown that people could point reasonably well from selected known locations to other specified known locations within the study area if configurational, survey, or layout knowledge had been achieved (Sholl 1987, 1988, 1996; Siegel 1981).

HOW HUMANS RECORD AND REPRESENT ENVIRONMENTS

Places can be real or imagined. They can have an explicit spatial component (location) or they can be spatially fuzzy or difficult to pinpoint, such as a "place" of mystery or beauty, or "the beaches" of California (Tuan 1977). Attributes of places can be objective or subjective; where one person perceives an undistinguished terrace house, another sees a monument to the birthplace of a famous person. Some places are identified by a feeling (e.g., fear) rather than by objective features.

The location of places is usually identified using a global, local, or relational frame of reference (called "absolute" location). Since the advent of the chronometer, latitude and longitude have been accepted as the most universal and accurate global locational system. This system is tied to conventionally specified compass directions. But few people really know how to use this type of coordinate system. Virtually no one knows the latitude and longitude of even the best-known features in their environment—their home. Instead, they rely on local reference systems, such as street systems, or relational forms of references such as "near the Town Hall." This fuzzier identification is called *relational* location. It is specified by spatial prepositions (see Landau & Jackendoff 1993, for an elaboration of spatial prepositions and their use in specifying location).

By using one or another of these representational forms, direction is established. Direction summarizes the relative position of two (or more) places, from each other and with respect to a common frame of reference. Although Sholl (1996), Siegel (1981), Smythe and Kennedy (1982), and others have shown that, in familiar environments, humans have a reasonably good sense of direction, the range or variance in even local distance or direction estimates can be very large (greater than 180°).

In addition to locating places within some frame of reference, humans also use a variety of "surveying" practices to locate places. These include right-angle offset (e.g., for off-route landmarks), triangulation (e.g., using two angles at either end of a common baseline to locate an unspecified point), resection (i.e., using back bearings from three known locations to estimate an unknown but occupied location), and trilateration (i.e., using three estimated distances to identify an unknown location) (see Golledge 1992 for an elaboration). Another procedure known as "projective convergence" (Siegel 1981) involves a person pointing to an unseen destination from three occupied or imagined locations; the recorded pointing angles usually produce a triangle of error whose midpoint is taken to be the probable location of the occluded destination. As Golledge, Gale, Pellegrino, and Doherty (1992) show, however, this works only if the absolute locations of the three occupied places are highly familiar. Otherwise, misspecification of angles can produce "triangles of error" so large as to be of little use in locating a place.

Today, location is mostly determined using instruments—range finders, compasses, theodolites, laser measuring devices, atomic clocks, and satellite-based global positioning systems (GPS). For recording the accuracy needed to produce maps, guide missiles, launch space vehicles, or remotely steer an exploration vehicle on Mars, instrumental accuracy is essential. But for most people, when building their own knowledge structures or representations of places and environments, instruments are not used and the basic human senses of vision, acoustics, touch, or sensorimotor experience prevail. But the use of human-based methods carries with it all the error baggage that instruments were designed to eliminate. So it can be expected that spatial representations in humans are incomplete and error prone, providing the distortions or fragmentations frequently mentioned by research on human spatial representation (Baird et al. 1982; Casey 1978; Gärling, Böök, & Ergezen 1982; Gärling, Böök, Ergezen, & Lindberg 1981a; Gärling, Böök, & Lindberg 1979, 1984, 1985, 1986a; Gärling, Böök, Lindberg, & Nilsson 1981b; Gärling, Lindberg, Carreiras, & Böök 1986b; Golledge 1987; Lloyd & Heivly 1987; Presson, DeLange, & Hazelrigg 1989; Tversky 1981; Warren & Scott 1993).

If it is assumed that specifying even a single place's location includes error, then what happens when humans have to represent multiple locations, their attributes, and connections or other spatial relations among them? And what happens when they have to use these data to guide movement?

Whether focus is on cognitive or external representations of experienced environments, the most commonly used representational form is *the map*. Maps can represent the commonly stored knowledge of individuals or societies. They function as archival and activity support. When dealing with wayfinding, maps appear to be a fundamental component of explanatory schemas. But what are they? And how can they be used?

There is emerging evidence that maps are cultural universals (Stea 1997; Uttal 1997). Maps are the attempts by humans to record the absolute and relative location of places, features, and spatial relations among phenomena. They can summarize objective physical reality or subjective worlds. Conventionally, maps used in wayfinding are two-dimensional planar representations of a segment of the earth's surface. Essentially these planar maps are analog models in which symbols are related to real features via a map legend, where common perspectives, such as North at the top, are added; where a given reference frame, such as a regular grid tied to cardinal directions, allows a coordinate system to give locational precision; and where interfeature intervals are related to real world spatial relations via a scale. Given these properties, an objective representation of an area can be developed and its validity and reliability can be determined.

Traditionally, perhaps the simplest way for humans to represent a route internally or externally is by a *strip map*. Strip maps have been in existence for thousands of years (Bell 1995; MacEachren 1986). They consist of concatenated segments of a route that need have no scale, orientation, or frame of reference. Simple rules guide their construction. Strip maps are often oriented by assuming that travel proceeds from the bottom of the page to the top of the page. The map can thus be regarded as always being "pointed in the direction of travel."

Recent discussion of human use of both strip and planar maps as orientation aids and wayfinding guides has been offered by Golledge (1995a); Levine (1982); Levine, Marchon, and Hanley (1984); MacEachren (1992b, 1995); Presson et al. (1989); Shepard and Hurwitz (1984); Tversky (1981); and Warren and Scott (1993). These authors have examined orientation effects in spatial representations such as sketch maps and verbal descriptions (see also Ferguson and Hegarty 1994), and have produced conflicting evidence about the impacts of map orientation on human behavior. In particular they have found alignment effects, orientation effects, rotation and translation effects, all of which can diminish a person's understanding of a mapped environment.

EXTERNALIZING INTERNAL REPRESENTATIONS

Internal representations of places and environments are externalized as a variety of "spatial products" (Liben 1982). These come in different forms based on using verbal, sketching, estimation, reproduction, or modeling techniques. The most relevant for wayfinding include

1. Verbal directions for connecting the location of places or from recall of experienced routes (Frank, Campari, & Formentini 1992; Moar & Carleton 1982);
2. Verbal estimates or reproductions of the distance or direction of the location of a place from other locations or places (Cratty, Peterson, Harris, & Schoner 1968; Evans & Pezdek 1980; Klatzky et al. 1990; Montello 1991a, 1991b; Pick, Montello, & Somerville 1988; Worchel 1951);
3. Pointing to places from given locations (Da Silva & Fukusima 1990; Da Silva, Ruiz, & Marques 1987; Haber, Haber, Penningroth, Novak, & Radgowski 1993; Loomis, Da Silva, Philbeck, & Fukusima 1996);
4. Recognition of sequences or orders of route segments between specific locations or places (Allen 1981, 1982; Golledge et al. 1993); and
5. Recording specific trips or behavior traces through an environment (Aitken & Prosser 1990; Denis & Cocude 1992; Denis & Denhière 1990).

Once externalized (for a summary of other methods for doing this see Golledge 1976), it is possible mathematically or statistically to check the accuracy of the externalization against the objective reality of the pertinent spatial information (Golledge & Spector 1978; Spector & Rivizzigno 1982; Tobler 1994). A spatial product may not accurately reflect a stored knowledge structure, or it may be deliberately or mistakenly distorted or misrepresented because of the inappropriateness of the mode chosen to express spatial information (Baird et al. 1982; Golledge & Hubert 1982).

COGNITIVE MAPPING AND COGNITIVE MAPS

Cognitive map, a concept coined by Tolman (1948) and now used widely in many human sciences (but often by substituting different terminology; see Kitchin 1994), is used to specify the internal representation of spatial information. In humans, other primates, and some nonhuman species, the term implies deliberate and motivated encoding of environmental information so that it can be used to determine where one is at any moment, where specific encoded objects are in surrounding space, how to get from one place to another, or how to communicate spatial knowledge to others. In many nonhuman species, the nature of stored information (also called a *spatial representation*) is such that it facilitates path following and homing, but may not enable other actions or behaviors that require route recall or layout knowledge.

Despite varied terminology used to describe them, it is commonly agreed that cognitive maps consist of points, lines, areas, and surfaces. These are learned, experienced, and recorded in quantitative and qualitative forms. When quantitatively encoded or interpreted they facilitate manipulation of information using Euclidean geometry and mental trigonometry. When qualitatively encoded they provide information on order, inclusion, exclusion, or other topological relations. The geometrical structure of knowledge thus includes points (such as landmarks and reference nodes); lines including routes, paths, and tracks; areas (for example, regions, neighborhoods, and topological containment or inclusion); and surfaces—some three dimensional characteristic of features or places such as density (Golledge 1990) (see table 1.1).

Several levels of spatial information appear to exist. At the fundamental level, humans experience and learn feature names or identities; the location of features and places; the size, magnitude, and frequency of occurrence of the features or places; and their temporal domain or times of existence (Golledge 1990, 1995b). This level is constrained by the way of knowing or the type of experience by which knowledge is accrued. It is also influenced by the legibility and familiarity of bits of information. At this fundamental level awareness is gained of the shape or pattern of spatial distributions of places or features (such as landmarks, residences,

TABLE 1.1 | **Geometric Components of Spatial Knowledge**

	Knowledge of spatial structure		
Points	Lines	Areas	Surfaces
A. Landmarks as organizing concept	A. Lines as boundaries/ edges (e.g., Zannaras' perceptual neighborhoods; Lynch's districts)	A. Areas as 2-D spatial classification devices	A. Physical topography
Landmark identity			Slope or gradients
Landmark location		Regions	Continuities or breaks (erosion)
Landmark dominance		Neighborhoods	Elevation
		Communities	
B. Landmark as navigation aid	B. Lines as routes	Urban places	B. Density (population)
Landmark as choice point	Crow-fly distances and connections	Arbitrary (political) districts	Peaks and sinks in preference surfaces
Landmark as origin or destination	Over-the-ground connectors (paths)	Fragments in cognitive maps	Shape or pattern templates
Landmarks of route orientation node	Length (total and segment)	Shape or pattern	Easy or difficult to negotiate
Landmark as regional differentiating feature	Linearity or curvature	B. Areas as cognitive concepts	
	Directionality to or from anchors	Superordinate frames (e.g., Reno versus San Diego)	
Landmarks as home bases (for path integrations or homing vectors)	Retrace constraints		
	Networks or connectivity	Containers of layouts of landmarks and points	
Landmarks as onroute choice points	Tools for experiencing learning	Uniform regions	
Landmarks as priming features influencing expectations	Methods of parsimoniously experiencing areas (e.g., search results)	Nodal regions	

shops, and towns), as well as the degree to which features are dominant or subordinate according to some spatial ordering principle (McNamara 1992; McNamara, Ratcliff, & McKoon 1984; Stevens & Coupe 1978).

THE USE OF LANDMARKS IN COGNITIVE MAPPING

Landmarks may be defined in a number of ways, such as strategic foci toward or away from which one travels, intermediate foci on courses and routes that assist spatial decision making, or significant physical, built, or culturally defined objects that stand out from their surroundings. They may be surrounded by nodes of various significance (less significant places such as street intersections or lesser known places such as elementary schools). Landmarks often act as significant primers for other features or actions; nodes often act as primers for landmarks, so that once a specific node has been perceived the expectation that a given landmark will occur is heightened (Shute 1984).

Landmarks are often noticed and remembered because of dominance of visible form, peculiarity of shape or structure, or because of sociocultural significance (Appleyard 1969, 1970). But the traditionally used concept of a landmark has two distinct components. In one sense the landmark is something capable of attracting attention, and being commonly recognized by many people (e.g., the pyramids, the Statue of Liberty, the Eiffel Tower). But some places and features accrue landmark significance in an idiosyncratic way (e.g., one's home or place of work). In other words, places or features may accrue salience for an individual at a level equivalent to the salience attached to the most widely known and recognized landmark in an area. As pointed out by Golledge and Spector (1978), for most of the 400 participants in a survey of landmarks in Columbus, Ohio, approximately half the features listed as "best known and most familiar" were places or features tied to individual activity patterns. The other half were features commonly identified by other participants as well.

Whether defined quantitatively or qualitatively, landmarks usually act as anchor points for organizing other spatial information into a layout. For example, they may be used as a centroid for spatially partitioning a region; they may have visible dominance such that surrounding features can be most easily described by relating their locations to the nearby landmarks or reference nodes. Landmarks thus may act as primary organizing features in cognitive maps by dominating a spatial classification or clustering process to facilitate environmental knowing and understanding (Couclelis, Golledge, Gale, & Tobler 1987). They may act as superordinate features in a hierarchical organization or multiple leveling of lesser known, lesser experienced, and more incompletely imaged and remembered environmental features. Golledge and Spector (1978) suggested an anchor point theory of environmental knowledge acquisition in which locations, features, path segments, or familiar districts "anchored" cognitive maps and influenced the encoding, storage, and decoding processes used when accessing stored information in a decision making context. An example of how this anchor point theory works for spatial knowledge acquisition is given in figure 1.1.

Landmarks also act as organizing features in a wayfinding context. For example, if one has an insignificant home base that does not stand out from the surrounding environment, then a nearby landmark may effectively take the place of the home base for orienting, directioning, and homing

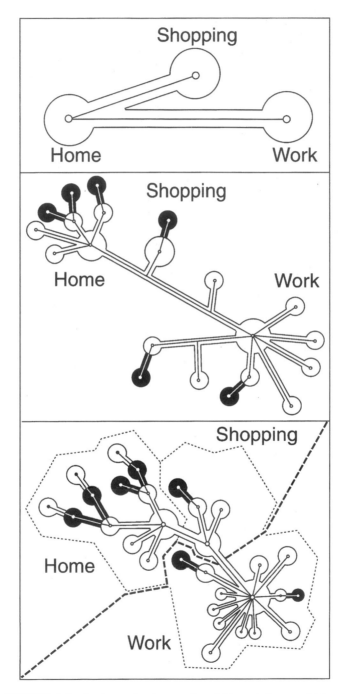

FIGURE 1.1. Anchor point theory of spatial knowledge acquisition.

behaviors. Such a substitution procedure also appears to be used by fly-ing or crawling insects (see chapters 9 and 10). Along routes, when choices have to be made at intersections (whether they be simple or complex), delib-erate choice of an environmental feature accompanied by specific memo-rization of one or more of its characteristics, can turn that choice node into a landmark that helps to specify both when to turn and selection of the next segment. Either off-route or on-route landmarks can help identify where one is along a route, can give warning of hazardous or other expectations at various places along the route, can signal where intermodal transfers or direction changes are required, or may prime the upcoming destination point. Used in these ways, landmark usage is common to both human and nonhuman navigation (see also chapters 2 and 11).

Embedded within a set of local landmarks there will be a subset that are the primary anchors for the area. These will be the best known, most familiar, most salient, and most recognizable to the greatest number of people in the region. Gale (1980) and Golledge, Richardson, Rayner, and Parnicky (1983) have shown that even in a local area such anchors can be determined by examining familiarity ratings of places obtained from rep-resentative sample populations.

Locational accuracy is more important when landmarks are used as orga-nizing principles than when used as wayfinding aids. In the former case when an anchor point is mislocated, many of the features it dominates will also be mislocated (Couclelis et al. 1987). Alternatively, when following a route, it may be sufficient to know that a critical landmark is always kept "ahead and to the left" rather than being able to locate it precisely.

THE ROLE OF ROUTES IN COGNITIVE MAPPING

Traveling through an environment is commonly recognized as the way humans most frequently acquire spatial knowledge. For centuries, human travel has been focused on specific paths and these—by focusing travel into well-known routes—provide common environmental experience for many people. In turn, this aids communication and information exchange by pro-viding communal understanding of where things are, and allows common frames of reference to be established. Routes connect places; often they overlap or cross, and consequently they can be integrated into a *network*. This structure provides a local and global referencing system that allows integration of all the one-dimensional route trajectories into a geometric mosaic. The resulting configuration embeds hierarchies of routes. For humans this might mean a structure consisting of freeways, highways, roads, streets, lanes, and alleys. Well-known and frequently traveled path segments provide linear anchors for portions of cognitive maps (Golledge 1978b). Thus internal spatial representations of what is known about surrounding networks influences the choice of routes to be followed for any given trip purpose and determines what path segments might be used when attempting a shortcut. There is little doubt that for environmental knowledge acquired from travel, reference either to physical recordings of parts of a network structure (e.g., planar or strip maps) or to the cognitively processed and stored knowledge acquired from perceiving things during wayfinding constitute the two most frequently used ways to build a cognitive map.

Wayfinding and route following contribute to the development of cognitive maps. Wayfinding initially takes place when a route is not previously known even though an origin and/or destination are known (Rieser, Guth, & Hill 1982). Route following takes place after decisions have been made as to what segments are to be traveled and what turn angles are required to connect different segments. Route (or path) following implies that a route already exists or has been planned a priori so that expectations exist about segment length, segment numbers, and turn angles (Péruch & Lapin 1993). This a priori process is known as travel planning (Gärling & Golledge 1989).

THE CONFIGURATIONAL PROPERTIES OF COGNITIVE MAPS

Given the inherent mobility of humans and the fact that things by necessity are distributed over space, one might argue that the primary purpose of a cognitive map is to facilitate place recognition and wayfinding. Lynch (1960) and others have suggested that a secondary purpose of a cognitive map is to act as an organizer of spatial experiences. It is my belief that this latter purpose is equally as universal as the former (if not more so).

Collections of locations or places combine to form configurations. These are the two- or three-dimensional arrays or layouts of environmental features known to a person. Configurations can consist of combinations of places and locations (points); routes, paths, and tracks (lines); regions and districts (areas); and natural, built, or hidden surfaces (refer back to table 1.1). Places and locations form spatial distributions; tracks, paths, and roadways form networks; landmarks and nodes form hierarchies. All these combine to represent the total knowledge structure (cognitive map).

Configurational understanding of layouts can be attained in a variety of ways, including

1. Defining boundaries around a complete or partial distribution;
2. Integrating separately learned route information into a network or a configurational whole; and
3. Overviewing from a survey point or bird's eye view.

Configurational understanding is presumed to be a level of spatial knowledge beyond route knowledge (Hart & Moore 1973; Siegel & White 1975). That is, configurations are considered to have more formal geometric (usually Euclidean) properties: they have the necessary robustness to allow trigonometric functions to be used to explain the spatial relations embedded in the configuration; they can be described by metric and nonmetric geometries and topologies; and they provide a convenient form of summarization or generalizations about experienced features, places, and connections. Whereas routes may be adequately described using only ordinal information, configurations are usually best described using metric information. This involves accurate comprehension of interpoint distances and directions, linkage and connectivity, and scale. When configurations include information with different levels of familiarity or confidence, qualitative or nonmetric features may dominate in their comprehension. These may include adjacency or nearest neighbor, regional inclusion or exclusion, directional sequencing or ordering, hierarchical structure, dominance, and subordination.

USE OF COGNITIVE MAPS

Accessing a cognitive map is designed to provide answers to questions such as the following: Where am I (Castner 1996)? Where is my home base (Castner & Anderson 1996)? Where are the phenomena for which I am searching (Hill et al. 1993)? How do I select a route between places (Passini 1980a, 1980b)? How do I return home (see chapter 5)? How can I go directly home after wandering (Fujita, Loomis, Klatzky, & Golledge 1990)? How can I know where to go next from any given point in a layout or on a route (Leiser & Zilbershatz 1989; Levine 1982)? How do I know when I'm lost (Montello & Lemberg 1995)? What strategies can be used to compensate for realized or unrealized errors in my representation of location or connectivity or course (Hill et al. 1993; Lawton 1996)? How can I determine the distance apart of locations (Montello 1991b; Da Silva et al. 1987)? What criteria are used to make decisions at choice points (Golledge, Smith, Pellegrino, Doherty, & Marshall 1985)? How is the relative direction between pairs or sets of points determined (Haber, et al. 1993; Loftus 1978; Loomis et al. 1996; Presson, DeLange, & Hazelrigg 1987; Presson et al. 1989; Shepard & Hurwitz 1984)?

Traditional answers to these questions vary considerably. For example, regarding the question on how to get back to an origin, techniques such as homing (Fujita et al. 1990), dead-reckoning or path integration (Gallistel 1990; Gallistel and Cramer 1996; Müller and Wehner 1988), spatial

updating (Golledge, Klatzky, & Loomis 1996b), route reversal or retrace
(Golledge et al. 1993), spatial search and exploration (Hill et al. 1993),
scent following (Gallistel 1990), landmark recognition (Collett 1996b),
and sun or celestial compass using (Able 1996; Alerstam 1996) may all contribute to a successful return. The question remains as to which of these
procedures are most likely to be activated in any given environment,
under any given set of conditions, and with respect to any given set of constraints. Other chapters in this book explore such questions in detail.

Wayfinding in circumstances where configurational properties dominate the cognitive map means that the greatest amount of supplementary information is available to ensure that correct route following and
path completion is achieved (but see Radvansky, Carlson-Radvansky, &
Irwin 1995). Momentary mistakes on route can be corrected by referring
to the larger-scale environment as reflected in the location or positioning of environmental features. When route knowledge alone exists, errors
are eliminated primarily by reexperiencing the route and recoding the segments and turn angles in their correct sequence. When only landmarks
are known, route selection may be inefficient and result in departure from
many optimal routes defined by different path selection procedures (such
as the shortest, fastest, most direct, and so on; Hanley & Levine 1983;
Rieser & Heiman 1982; Säisä, Svensson-Gärling, Gärling, & Lindberg
1986).

Both human and nonhuman species appear to have a dominant anchor
point for the territories over which they range, that is, their den or home
base. Numerous studies have clearly shown that such species can return
directly to their home base after irregular wanderings whose trace would
have been very difficult to have memorized (Etienne, Hurni, Maurer, &
Séguinot 1991; Etienne, Maurer, & Séguinot 1996; Gallistel 1990; Müller
& Wehner 1988). This can be taken to indicate that either the species only
need access to occasional landmarks for guidance (Bennett 1996) or that
some geometric understanding of layout must have been achieved. Some
argue that homing can be produced solely by dead reckoning or path integration activities that do not require either a record of the route chosen
or knowledge of surrounding environmental features such as landmarks (see
chapter 5 for elaboration of these ideas). Then, either by marking the environments (e.g., with scents), remembering landmarks, attempting route
retraces, or following paths, places previously visited can be revisited using
simple path selection criteria such as always moving in the direction of the
goal object, minimizing distance or time, or choosing the fastest route (see
Menzel's 1973 and 1979 work on chimpanzees).

Errors in Cognitive Mapping

Errors incurred during the encoding phase of knowledge acquisition can produce distortions in the material stored and recalled for later use. Such later use includes externalizing stored information in a spatial product such as a sketch map or diagram or providing written or verbal descriptions of routes or layout. Thus it can be assumed that the information stored in the cognitive map could produce error-prone travel or spatial behaviors if that information was used as part of a route selection and guidance process.

Given that errors can occur when encoding, internally manipulating, decoding, and representing information, it is no wonder that cognitive maps are usually assumed to be fragmented and incomplete (Baird et al. 1982; Bryant 1982; Bryant, Trersky, & Franklin 1992; Buttenfield 1986; Cadwallader 1977; Cohen, Baldwin, & Sherman 1978; Downs 1981b; Gale 1985; Lieblich & Arbib 1982; McNamara 1992; Siegel, Herman, Allen, & Kirasic 1979; Taylor & Tversky 1992a; Tversky 1981, 1992). To illustrate types of distortions, Golledge (1978a) overlaid a standard grid on a base map of Columbus, Ohio. Using software designed by Tobler (1978), he then warped the grid to fit configurations of where people thought places and features were located. The warped grids revealed "fish-eye lens" types of configurations by some people and "magnetic attractions" by other people. Spaghetti-like distortions were common among newcomers to the city, that is, those with less than six months' residence (figure 1.2).

The question then arises as to how distorted a cognitive configuration can be before it inhibits successful problem solving (e.g., wayfinding)? In particular, can one continue to get from one place to another even when the intervening knowledge structure is sparse and error ridden (Walsh & Martland 1996)? If no time constraints exist, it may well be possible for

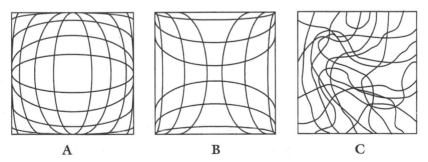

FIGURE 1.2. Distorted grids: (**A**) fisheye lens distortion; (**B**) magnetic distortion; and (**C**) irregular non-Euclidean distortion.

a person to explore different routes or wander through a layout until a familiar place or segment is happened upon (Leiser & Zilbershatz 1989). This process underlies many "adventure" type video games. Obviously as learning continues via multiple exposures, the chances of making errors in segment sequencing or direction of turn should decrease. For example, Golledge et al. (1985) recorded the wayfinding errors made by a 12-year-old boy during a five-day route learning task in a suburban California neighborhood, documenting the decline in errors while traveling and while later identifying slides of on- and off-route features, as well as by increasing the accuracy of sketch maps obtained in post-trial debriefing tasks (figure 1.3). Similar results have been obtained by Cornell and Hay (1984) and Allen (1981, 1982). For regular episodic activities such as the diurnal home-work-home trip, there is a certain urgency associated with learning a route that can be followed on successive occasions with a minimal chance of becoming lost and a minimal need to worry about making on-route decisions.

Some routes are embedded in a cognitive map by using local frames of reference and spatial relational systems, whereas others are constructed using standardized and more global frames of reference or relational systems (Warren & Scott 1993). Embedding errors then depend on attributes of the chosen reference frame and usually are scale dependent.

HUMAN WAYFINDING

Wayfinding refers to the cognitive and behavioral abilities of humans and nonhuman species to find a way from an origin to a destination. The way may involve following existing paths or tracks or free-ranging movement. The latter is quite common among nonhuman species, but human wayfinding is usually confined to well-defined paths, constructed surfaces, or designated oceanic or atmospheric corridors. There are a number of terms that are commonly used to identify channels of regular human movement, including tracks, paths, sidewalks, footpaths, alleys, streets, highways, freeways, railroads, trolley tracks, monorails, and air and ship navigation lanes. Nonhuman species may develop trails and paths, and may follow particular routes, as on long-distance migrations on land, water, or air (Able 1996). But, for the most part, it is the human species that now concentrates its travel on constructed surfaces or via designated lanes or corridors.

In this chapter a route has been defined as a trace in the environment of the planned or traveled sequence of path segments and turn angles that are followed in order to get from an origin to a specific destination. The process of defining a route and finding one's way along it depends on the

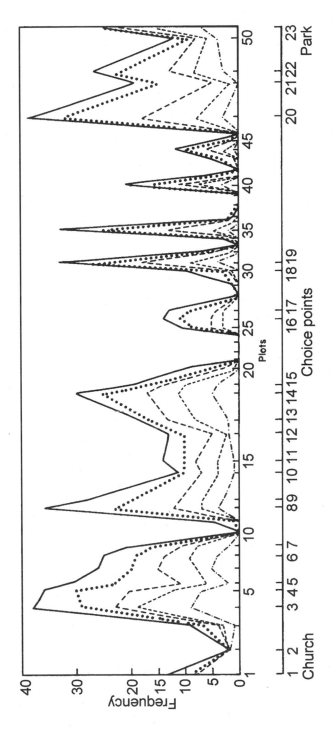

FIGURE 1.3. Graph of multiple trials of a 12-year-old boy learning a route. Change in errors at places along a route. Cumulative frequency of cue recognition over successive days by plot and choice point. —•—, day 1; – – –, day 2; — — — day 3; •••••, day 4; ——— , day 5. (Reproduced from Golledge et al. 1985, by permission of Academic Press Ltd.)

purpose(s) (single or multiple) of the trip, and whether it involves single or multiple stops. Route selection takes place within the constraints of specific criteria.

KEEPING TRACK OF A HOME BASE

Human and other species keep track of home base by methods such as

1. Remembering an outbound route so that it can be retraced (Klatzky et al. 1990; Welsh & Blasch 1980);
2. Remembering the relative location of the home base within a given layout and using layout features such as landmarks to help return home by direct or indirect means (Collett 1996a; Cornell, Heth, & Alberts 1994; Etienne et al. 1996);
3. Using dead reckoning or path integration procedures which involve integrating velocity to constantly update a bearing and distance to the home base from the currently occupied location (Fujita, Klatzky, Loomis, & Golledge 1993; Müller & Wehner 1988; Poucet 1993); and
4. Being able to obtain a bird's eye or overlook view of an environment from a prominent place or by using a map such that a home base can be defined in a layout context (Golledge, Dougherty, & Bell 1995; Matthews 1992; Thorndyke & Hayes-Roth 1982).

As the major anchor point for most human and nonhuman travel, home bases are equally as important in anchoring human or other cognitive maps as they are in providing foci for the route vectors that typify food search or other activities in the spatial representations of nonhuman species. The bees, ants, wasps, gerbils, golden hamsters, and other insect and rodent groups examined in later chapters of this book all rely on keeping track of home base in order to know where they are and, consequently, what path should be followed. Although some credit is given to local landmark use for recognition of current location, the home base provides the anchor to which such landmarks are referred. For humans, home-base-generated trips dominate the activity schedules of most individuals and aggregates of people in neighborhoods, communities, and regions, with more than 50 percent of daily trips being home-base-generated. The need to comprehend the location of a home base and to understand its position in an internal or external network configuration or its location according to a given reference frame is paramount.

Route Selection

Many studies of human activity (e.g., trip generation and commuting studies in transportation research) make the assumption that regardless of the complexity or simplicity of the environment through which travel takes place, people will prefer to retrace routes already taken or known rather than explore new ones; that is, they develop habitual travel behavior. Another assumption is that people will tend to use the same criterion for route selection whether using a map or stored memories (cognitive maps) for guidance.

These assumptions do not always hold. For example, Golledge (1995a) has indicated that as map orientation changes, criteria used to select routes may also change. Golledge used three types of transportation environments: (1) a rectangular network with origin and destination points selected from the grid intersections; (2) a basic rectangular grid modified by diagonal connectors between a subset of intersections; and (3) a modified network in which both rectangular and curvilinear segments were included, as well as some regions designated either as parks or waste dumps (figure 1.4). He found that as the underlying environmental structure and its imbedded reference frame changed from coordinate regularity to curvilinear irregularity, the criterion used in path selection also changed. As the environment changed from simple origin-destination pairings in a regular rectangular array to multiple node and choice points in an irregular mixed network environment, the tendency to perform retraces increased. Presumably this response minimized decision-making efforts. When trip chaining occurred, a complex set of path selection criteria often was chosen.

There can be many reasons for choosing a specific route. These reasons can change over time. Time spans can be as brief as part of a day, daily, weekly, or other longer episodic frequencies. Different routes may be chosen for different trip purposes, even when the final destinations are located near each other (e.g., work and recreation).

The critical problem that remains unsolved at this time concerns which wayfinding processes are used for which purposes. For example, a direct trip home after work may involve path integration or habitualized route retrace. A leisurely drive for recreational purposes might involve search and exploration of an unfamiliar environment, followed by intuitive shortcutting or map-assisted route choice in the home direction. Human wayfinding is very trip purpose-dependent and it is difficult to attribute any specific cognitive psychological process to wayfinding generally. What can be done, of course, is to see if specific processes of path integration, route retrace, landmark recognition, and so on *can* be undertaken with a sufficient degree of accuracy so that one or another of them can be distinguished as being the most probable criterion used. However, many people are completely *unaware* of consciously using specific wayfinding strategies. For example, Golledge (1996)

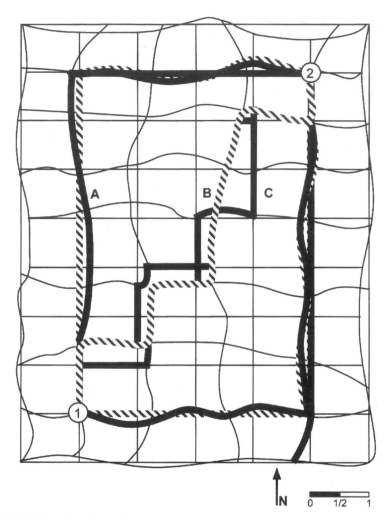

FIGURE 1.4. Regular and irregular networked environments used to determine path selection and to deduce the selection criterion associated with chosen paths. As network complexity increases, decision making is usually simplified by retracing routes on return journeys.

showed a difference between self ratings of what people *thought* they used on a trip, and what the trip revealed to be their most probable criterion; there were many distinct differences between the two (table 1.2).

Thus, in human travel, the purpose of a trip or trip sequence usually plays a major part in selecting the route taken, and consequently the wayfinding process that will be used. For repetitive trips such as journey-

TABLE 1.2 | Comparison of stated criterion versus revealed criterion in route selection

Stated criterion	Ranking of stated criteria most frequently used	Ranking of revealed criteria most frequently used
Shortest path	1	1
Least time	2	6
Fewest turns	3	3
Most scenic or aesthetic	4	9
First noticed	5	2
Longest leg first	6	7
Many curves	7	10
Most turns	8	5
Different from previous route taken (variability)	9	8
Shortest leg first	10	4

to-work, journey-to-school, and journey-for-convenience shopping (e.g., for food), experimentation with different routes is usually minimal and an optimal or satisfactory primary route is usually defined quickly, within five or six trials (Rogers 1970). A significant path selection criterion for these types of activities is usually least time, least cost, or least distance, which are compatible with path integration ideas even if they do not replicate the process. As the purpose behind other activities changes, however, the path selection criterion can change and as a result the route that is followed will change.

Thus, although one may wish to minimize time or distance covered on a trip to work or to school (often a single-stop, single-purpose trip), on multistop multipurpose trips, such criteria may not be followed. For example, one may wish to follow a different route from work to home than from home to work because on the way home multiple activities such as recreation, convenience shopping, comparative shopping, or socializing may become critical components of the total decision process. Each one of these components may require violation of an overarching or global criterion such as minimizing distance or time. Hirtle and Gärling (1992) have suggested that multipurpose trips may consist of a series of locally optimal decisions that may not in total reflect any single dominant path selection criterion (see chapter 3). A trip that involved leaving work and visiting a gym, followed by a visit to a bar or tavern, and then going to a grocery store before returning to home base might involve minimizing left turns on the way to recreation, maximizing aesthetic value on the second leg for socialization purposes, maximizing freeway use on the segment from social to shopping activities, and minimizing time or distance on the final homeward leg. If the return to home journey is considered as a multistop multipurpose trip, then it is unlikely that any single dominant globally optimal criterion will fit this mélange of locally optimal criteria.

In some situations path selection criterion change quite dramatically between the outbound and inbound journeys. For example, household members participating in a Sunday afternoon drive for recreational purposes may wander leisurely through local neighborhoods and suburban streets when exploring new areas and enjoying the aesthetic and the novel experience. At some stage, however, time constraints may emerge and realization dawns that the trip home must commence. In this case a homing vector process might be invoked with the intention of taking a shortcut to the home base. This may not involve taking the optimally defined shortest route, but it would certainly involve taking a more direct route to the home base than a retrace would provide.

Shortcutting is common in both pedestrian and vehicular travel situations. But, because of the nature of the built environments in which humans live, it is rarely if ever possible to take a direct homing vector route as many animals appear to do. This does not mean that a shortcut was not intended. Regardless of its optimality, linearity, and directedness, any form of shortcutting implies knowledge of the current location with respect to the home base or some nearby landmark or anchor point.

In all environments, as the task changes from a simple origin-destination (OD) pairing to trip-chaining (involving a number of intermediate destinations), the minimum distance criterion usually ceases to hold (Gärling & Gärling 1988; Säisä & Gärling 1987). Thus, with human wayfinding, there appear to be a larger number of potential path selection criteria than appear exist in animal movements. Table 1.3 indicates some of the criteria used in human travel. As these change, the path or route selected for travel will often vary. This raises the issues of what path selection criteria are being employed when we notice humans taking different paths between the same origins and destinations, and how we can determine the paths that are likely to be used on any given trip.

Geographers and other transportation scientists are interested in answers to questions such as: What proportions of people actually retrace their routes to and from specific destinations? Which of the criteria listed in table 1.3 are most likely to be chosen for any specific trip purpose? What factors are likely to influence or insure divergence from route retrace? What path selection processes lie behind decisions to take different routes to and

TABLE 1.3 | Types of route-selection criteria

Longest leg first	Ensuring locomotion remains within a given width (corridor) surrounding a straight line connection between origin and destination
Shortest leg first	
Fewest turns	
Fewest lights or stop signs	Maximizing aesthetics
Fewest obstacles or obstructions	Minimizing effort
Variety seeking behavior	Minimizing actual or perceived cost
Minimizing negative externalities (e.g., pollution)	Minimizing the number of intermodal transfers
Avoiding congestion	Minimizing the number of layers of a road, street, or highway system that have to be utilized
Avoiding detours	
Responding to actual or perceived congestion	
	Fastest route
Minimizing the number of segments in a chosen route	Least hazardous in terms of known accidents
Minimizing the number of left turns	Least likely to be patrolled by authorities
Minimizing the number of nonorthogonal intersections	Minimizing exposure to truck or other heavy freight traffic
Minimizing the number of curved segments	

from the same origin and destination pairs? Some of these questions are explored at length in Golledge (1996), where it was shown in one experiment requiring participants to travel between neighboring buildings, that on one route, 60 percent of the subjects retraced their original path on the return trip. Ninety percent of the participants used eight different routes to complete the trip. For a slightly different path between the same origin and destination, only 26 percent of the subjects retraced their original route and 90 percent confined their paths to only four different routes (figure 1.5). The questions listed previously are pursued regularly in transportation science, particularly by those using activity approaches (Ettema and Timmermans 1997).

Having experienced a route and noticed its environs, spatial abilities are used to perform the mental geometry and trigonometry needed to construct a layout of the area. The cognitive equivalent of surveying procedures (e.g., triangulation and route integration) may be implemented to link the various components into a configuration. But if order or sequence of features or route segment is violated, if turn angles are incorrectly remembered, or segments or angles are rotated, reflected, or otherwise transformed in this process, then the cognitively constructed configuration will not match objective reality. The results of these mental integration processes

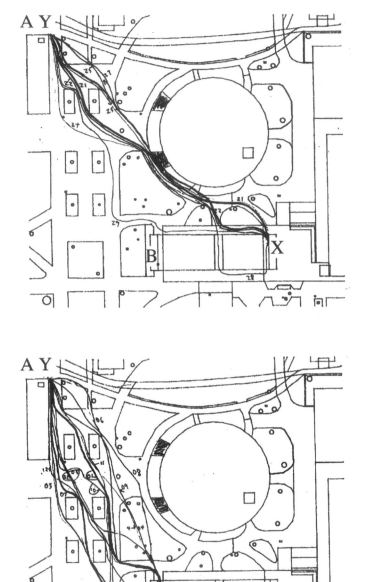

FIGURE 1.5. (**A**) Traces of routes selected from elevator (X) to flagpole (Y). (**B**) Traces of routes chosen between flagpole (A) and entrance to stairs (B).

are the cognitive maps used as anchors for human spatial decision making and behavior.

Sholl (1988) suggests that one important facet of wayfinding is a sense of direction. This is defined as knowledge of the location and orientation of an organism's body with respect to facing direction (heading) and the location of significant nearby or distantly perceived (or memorized) features (landmarks). Such knowledge is critical in choosing which way to travel to an unseen but known destination. Tests of the accuracy with which organisms can point to (or orient their body toward) unseen but experienced locations provide information about both this sense of direction and the accuracy of the individual representation of the distant features.

In a later report, Sholl (1996) suggests that human organisms use a self-referencing system to develop a sense of direction while wayfinding, the basis of which is a sensorimotor level in which body axes provide a reference frame to which all spatially directed motor-activity is referred. Thus, position vectors tie the organism to objects in the visual field providing a body-centered polar coordinate system. As the organism moves along a trajectory, effortless spatial updating occurs with reference to this system. Rieser (1989) indicates that this process also occurs in the absence of vision. Sholl suggests that an organism builds object-to-object sets of relations learned by experiencing an environment into a cognitive map which is stored in long term memory. This preserves the object-to-object relations observed in the real world, and allows the development of self-to-object relations needed to update position while traveling.

The cognitive map preserves the object-to-object relations of the real world independently of any change in perspective resulting from altered location. By superimposing the polar coordinate body centered self-to-object relations over the stored object-to-object relations, continuous positional updating takes place. In air and sea navigation by humans this process is called "piloting" (see Loomis et al., chapter 5), and is referred to by Sholl as "geocentered dead reckoning." Current position and known landmark locations are thus integrated into a representational self-reference system. Travel in a familiar environment is traced onto the cognitive map, thus allowing route retrace or route learning to take place. Dynamic self-tracking systems such as this have been proposed by Gallistel (1993), and O'Keefe and Nadel (1978). Thus, in human travel, kinesthetic cues (including optical flow and motor-joint and vestibular information) are used to provide linear and angular displacements of the body that are recognized via changing relations between self-to-object and object-to-object records. Sholl suggests that an organism can thus update current location at the representational level independently of updating in the self-to-object level.

In a 1987 experiment, Sholl showed that humans can point success-fully to unseen but familiar locations in space. Other landmarks not nec-essarily recorded in the cognitive map (i.e., places of low salience or of local but not global prominence) were used to provide knowledge of current position. Establishing local self-to-object relations and establishing a tie to a superordinate frame (e.g., cardinal directions) could then provide a means for integrating local awareness into the global cognitive map, thus establishing a geocentric position.

Golledge et al. (1993), Loomis et al. (1993), McDonald and Pellegrino (1993), and Rieser (1989) have discussed the ability of persons with and without vision to imagine themselves at a known location and to indicate, by pointing or walking toward, the direction of obscured locations. Accu-racy was seen to vary considerably depending on the participant's famil-iarity with the test area. It can be deduced from this that in unfamiliar environments if directions are not well known or realized, gross errors in wayfinding may occur—in terms either of actual travel or of verbally describing or sketching where to travel. Using Sholl's schema, this means an inability to relate self-to-object representations to object-to-object representations can produce errors in heading (direction) or motion (dis-tance traveled). The at times confusing use of navigation terms from the nonhuman literature (particularly various interpretations of path integra-tion) emphasizes the need to establish a clear and uniform vocabulary for these processes, which Loomis, Klatzky, Golledge, and Philbeck set out to do in chapter 5.

Cognitive Maps as Wayfinding Tools

Although a variety of guidance instruments and materials are available, humans tend to use cognitively stored and recalled information more than anything else for assistance in wayfinding. This is simply because the majority of trips are made in familiar or partly familiar environments. Travel in familiar environments occurs when migrating to a new place or when first exploring a newly settled environment.

When new to an environment, an individual will tend to use informa-tion from other people and from common representations—maps, photos, images, written or verbal descriptions—to begin and facilitate the process of accumulating environmental information. In areas of dense human occupance, this is accompanied by a myriad of signs ranging from street names and house numbers to topographical representations in two or three dimensions (e.g., maps or models). The key processes relate to skills and abilities to read maps, to match real-world features against knowledge schemas of those features (or vice versa), to comprehend scale transforms, and to understand the symbols commonly used to represent real features on maps or models. Information is usually available in iconic form such as scaled representations from specific perspectives as in a photo, analog forms as in the paper and pencil representation of features on map sheets, or symbolic forms as in the mathematical representation of spatial relations such as formulae for showing changes in expected feature occurrence frequency with increasing distance from an origin. Using these types of information, wayfinders plan routes to follow for different purposes using different modes of travel. As pointed out by Levine (1982), if guides such as "You Are Here" maps are misaligned, the wayfinder will find it difficult to match map information with the real world, resulting in an inability to follow an appropriately defined route.

As familiarity with an environment increases there develops a tendency to use cognitively stored information in lieu of external wayfinding aids. As one learns a chosen route to work, the tendency to use a map or to ask directions decreases. More reliance is thus placed on a cognitive representation of an environment. This representation may be incomplete and may include many errors, such as those discussed in prior sections. But human competence often is great in terms of their wayfinding ability.

Using self-referenced evaluation procedures, Lawton (1996) has examined self-ratings of spatial competence and the kinds of anxiety people feel when faced with wayfinding problems in known and unknown places. She found that people were generally confident in their own wayfinding and locational abilities, but that men expressed more confidence than women. Evolutionary psychologists claim that this characteristic is due to the traditional role of men as explorers and hunters of game—activities that often took them to distant unfamiliar places and required large-scale environmental knowledge acquisition. In comparison, the traditional gathering activities of women produced very detailed local area knowledge, but little experience with distant places except when tribal seasonal wanderings occurred. These relations can be speculatively tied to anthropological lore, but at present there are no hard facts based on extensive experimentation with isolated residual hunter-gatherer peoples to support them.

There is evidence, however, that today's women tend to have less confidence in their spatial abilities than do men (Lawton 1996). There is some tentative indication that women often use localized landmark-based wayfinding strategies while men often use more global configurational strategies to follow routes, or to give wayfinding directions to others (Bever 1992;

Couclelis 1996). At this stage, confirmatory evidence of such an hypothesis is scant.

If internal spatial representations are stored in a "map-like" manner, then map reading skills would have to be evident before such cognitive maps could be used. Research in spatial aptitude seems to indicate that, theoretically at least, spatial abilities can be summarized by two dominant dimensions—spatial rotation and spatial orientation (McGee 1979).

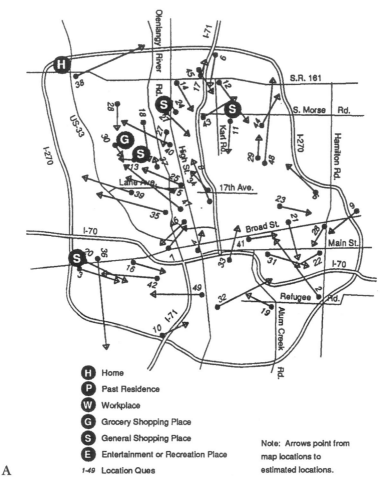

FIGURE 1.6. Activity space elements and locational errors: (**A**) map of activity space elements and actual and estimated sample locations for subject 64; (**B**) map of activity space elements and actual and estimated sample locations for subject 100. (Reproduced from Golledge & Stimson 1997, by permission of Guilford Press Inc.)

Lohman (1979) and Eliot and McFarlane-Smith (1983) offer a third
dimension—spatial relations—to cover critical features of spatial apti-
tudes. Golledge (1995b) has developed a set of primitives that seem to give
substance to this third dimension, particularly in terms of knowing an environment (or a cognitive representation of it). He defines wayfinding as a
higher-order concept derived from a base of understanding location, distance, direction, magnitude, temporal existence, connectivity, and sequence.

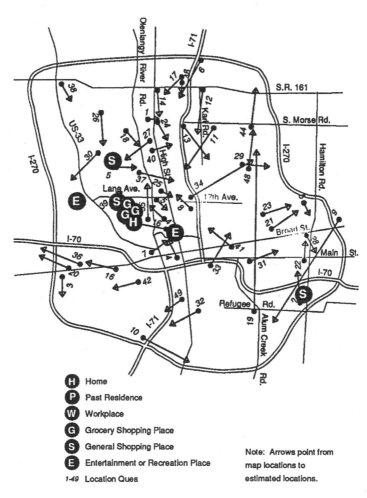

Ⓗ Home
Ⓟ Past Residence
Ⓦ Workplace
Ⓖ Grocery Shopping Place
Ⓢ General Shopping Place
Ⓔ Entertainment or Recreation Place
1-49 Location Ques

Note: Arrows point from
map locations to
estimated locations.

B

FIGURE 1.6 *(continued)*

Other derived concepts include hierarchy, boundary, and region. Understanding these concepts provides a set of procedures that facilitate access to cognized information.

Golledge (1974) illustrated the relative lack of locational distortions of cognized locations within the area of a person's activity space. Beyond this well known and highly familiar space, fragmentation or lack of knowledge dominated in the person's spatial products (figure 1.6). This indicates that travel outside the activity space—or space containing the bulk of regular daily movement patterns—would potentially require guidance supplements, such as maps or given directions, and could be more stressful and prone to error.

HUMAN WAYFINDING ERRORS
Wayfinding errors can take a number of forms, including
1. Movement error resulting from incorrect sensing of velocity, time, or distance with the understanding that such errors can produce over- or underestimates of distances (Böök & Gärling 1981a, 1981b; Cadwallader 1977, 1979; Sadalla, Burroughs, & Staplin 1980);
2. Encoding errors due to distorted frames of reference or poor perceptual recording that can produce orientation or direction errors, turn errors, or result in mismatching choice points and turn angles (Gärling et al. 1986b; Presson et al. 1987, 1989);
3. Decoding errors due to using a distorted representation or cognitive map, incorrectly implementing a correctly encoded behavior such as a turn angle or distance, or incorrectly decoding sets of spatial relations (Hill et al. 1993; Lloyd 1989; Lloyd, Cammack, & Holliday 1996; Passini 1980a, 1980b, 1984a, 1984b); and
4. Wayfinding errors that result from improper internal manipulations such as incorrectly integrating routes to produce layout or configurational information (Golledge et al. 1993), recognition errors such as incorrectly identifying a cue because of a perspective change (Bryant & Tversky 1991; Bryant et al. 1992; Cornell, Heth, & Rowat 1992; Tversky & Schiano 1989) or inadequate familiarity (Gale, Golledge, Halperin, & Couclelis 1990).

Evidence that route knowledge exists can be illustrated if a traveler can successfully reproduce the location or sequence of on- and off-route landmarks, identify choice points, or reproduce the correct order of route segments. Golledge et al. (1993) had participants learn two partially overlapping routes in both forward and reverse directions. Each participant took three trials to complete his or her learning phase. Figure 1.7 illustrates an example

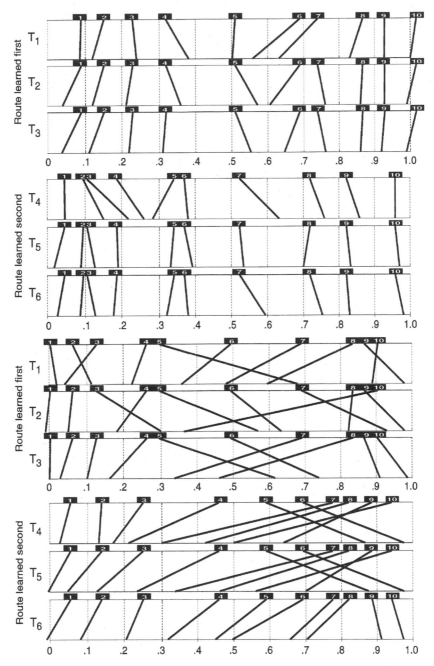

FIGURE 1.7. Example of successful unidirectional learning and not so successful bi-directional learning in this experiment. Angled lines indicate locational displacement; cross-over lines represent sequential order violations as well as location displacement and interpoint distance error. (Reproduced from Golledge et al. 1993, by permission of Academic Press Inc.)

of successful unidirectional learning and not so successful bidirectional learning in this experiment.

The experiment involved identifying the correct sequence of slides of environmental features experienced along a route or from a different perspective on the retrace. Ten slides were viewed, identified, and then had to be arranged in a sequence along a strip of paper representing a linearized version of the learned route. The positioned slides (mounted on small stands) also were to be arranged in proportion to their actual distance apart. The top part of the diagram shows that on both routes in unidirectional cases (i.e., regardless of whether the route was learned unidirectionally from A to B or from B to A), sequencing was perfect. But distancing involved some error. Allen (1982) and Allen and Kirasic (1985) previously suggested that learning the features along a route would be facilitated by chunking the route into segments and determining whether a feature was located within or outside of a segment. In the Golledge et al. (1993) experiment, segment violation occurred infrequently in the unidirectional case. In the bidirectional case, sequencing and distancing of the environmental cues were often violated, as was segment inclusion.

Allen's (1982) chunking hypothesis provides an explanation of accurate sequencing in route and feature learning, although chunking can also introduce errors. Gale, Golledge, Pellegrino, and Doherty (1990) showed that distancing and sequencing can be violated if proximate locations are split between chunks. In their experiment, these authors used triads of cues with one being termed the home base or *referent* and two others being termed *near* and *far* comparisons. Nearness and farness were determined by distance measured over the route traveled. In most cases, even complicated ones where segment inclusion was the dominant principle, the location of cues in their respective segments and perception of nearness and farness was accurate. But where segment inclusion was violated such that a near comparison was located in a segment different from that of the referent, although the far comparison was in the same segment as the referent, considerable confusion and error occurred. Making such errors seems to indicate that route learning was the major process at work, and that general environmental learning or layout type understanding had not been achieved.

An environment may be learned as a result of travel by integrating paths as they are experienced into a configurational whole. Most humans have limited interactions within a given environment, resulting in some routes being well known and others poorly known. Integrating precise and fuzzy sets of information can produce significant configurational distortions.

Human experience is constrained by the paths they have learned to follow in successfully performing specific activities. The best known and most frequently traveled route is usually that between home and work (Axhausen & Gärling 1992). This route provides a primary link between the home and work anchors. Household travel is most frequent over such a route. A single dominant route is usually followed most of the time except when episodic hazards such as accidents, construction, or congestion, force the selection of an alternate path or path segment. The most frequent tendency is to find a temporary way around the obstruction and then rejoin the known path.

When integrating separately learned routes into a configurational structure, certain procedures must be followed:

1. If routes are known to be overlapping they may be positioned with respect to common off-route landmarks, underlying frames of reference, or local or global oriented systems (e.g., the underlying transport network).

2. When routes are partially overlapping they allow configurational representation to take place but such configurations may be rotated with respect to their real formation unless tied to a boundary or overarching frame of orientation.

One useful difference between configurational or layout understanding and route comprehension is that estimates of distance or direction to obscured places on or off a route are possible if the route configuration is learned, whereas if only the ordinal properties of the route itself are learned, two-dimensional concepts may not be understood (Lindberg & Gärling 1981a, 1981b). Simple route-learning procedures cannot provide details of off-route features. If configurational or layout understanding has been achieved, distance and direction can be measured in either crow-fly (straight-line Euclidean) or via the underlying reference frame or network structure (e.g., Manhattan distances or city block distances) (Golledge & Spector 1978; Richardson 1982). If an appropriate cognitive map has been developed of locations and layouts, then it is possible to do internal manipulations by either treating the information as if it were a cartographic map or constructing an equivalent of a cartographic map in working memory so that distance and direction of obscured points can be estimated.

Pointing accuracy is usually a good indicator of whether or not layout understanding has been achieved (Rieser, Guth, & Hill 1982, 1986), particularly if there is a time lag between experiencing the environment and pointing to features within it either in real space or in an imagined or cognitively reconstructed space. For example, Böök and Gärling (1981a, 1981b), Da Silva et al. (1987), and Loomis, Da Silva, Fujita, and Fukusima (1992) all provide evidence that humans can point to obscured, distant, nearby, or passed-by locations with reasonable degrees of accuracy. This can be done both in terms of the immediate act of pointing at an observed or obscured (but known) object, or by walking past it blindfolded while continuing to point to it.

SOME FINAL THOUGHTS

What is it that we appear to know about human wayfinding and the internal spatial representations of environmental features that assist travel? First, neither humans nor animals develop instantaneous, complete, and precise spatial knowledge of a given environment. Experiencing an environment will result in some spatial information being stored in long-term memory. Although there is ongoing debate about how and when such information is stored, accessed, and used, it does appear (by observing human actions and behaviors) that spatial learning corrects errors and inconsistencies in initially encoded spatial information.

It further appears that the degree of knowledge about places, locations, or points along a route varies among people, that is, there are individual differences in the context of environmental knowing. It also seems reasonable to assume that perfect knowledge of a complex environment cannot be achieved, because real world environments are dynamic; they change over time. For some relatively stable environments, however, a remarkable level of spatial accuracy can be achieved, and this can be exhibited by forming spatial products.

Further, it seems that humans do not all behave the same in the same environments, that different people have different levels of familiarity with different environmental settings, and that they give salience to different environmental features. Often, the result is that different spatial products of the same area can incorporate substantially different features and produce different types of errors; for example, one person may have dominantly direction errors, another dominantly interpoint distance errors, and yet another predominantly locational errors. Greater knowledge is achieved by actively rather than passively experiencing an environment. In addition, adults and children do not usually have the same understanding of spatial layout or the configurational structure of different environments (see chapter 2).

There is evidence that humans and other species make errors and develop strategies to compensate for them. As the episodic frequency with which routes are traveled increases, it can be assumed that more accurate knowledge is retained about all features of routes. On successive trials this should imply that travel times are reduced if the appropriate route selection criterion required it. It is also apparent that humans and other species concentrate knowledge at particular places. These represent anchors of the knowledge structure. The most pervasive anchor is the home base. On the surface, because the set of activities they are involved with are more complex, humans appear to develop a wide range of anchor points. Because of the way they have been socialized and developed over time, humans also tend to restrict free wandering by confining movement to constructed ways or channels. This may have resulted in the decay of natural wayfinding ability over time—an ability that was based on using natural aids such as sun angle, celestial compasses, wind direction, and topography. A variety of aids that range in accuracy from highly accurate cartographic maps (which are scale representations of many environmental features), to fuzzy, incomplete, and distorted sketches of the positions of the most easily recalled or most familiar places and features of an environment, have replaced the use of natural features.

When using an environment, certain types of choices (e.g., with respect to travel) must be made. Humans making an incorrect choice can become lost and suffer the psychological and physiological discomforts associated with that state of being. Thus, the greatest amount and highest accuracy of information may be concentrated at points where choices or decisions have to be made, or at locations or places that prime or signal specific activities to take place.

Humans rely on externally recorded representations. These are of particular importance when following complex routes involving many segments and turns. But humans are capable of navigating successfully without learning spatial characteristics such as off-route cues or landmark locations, bounding frames of reference, or general orientation. Evidence of this exists in the fact that vision-impaired or blind people can successfully navigate by following sounds to a destination (beacon following or auditory localization; see chapter 5). They may also use olfactory or tactile cues such as surface texture to help wayfinding or route following. And it has been hypothesized (Loomis et al. 1993; Loomis, Beall, Klatzky, Golledge, & Philbeck 1995a) that even greater navigation success could be achieved by blind pedestrians if more information than is currently available was made accessible via various technologies such as talking signs, verbal landmarks, or personal guidance systems. For example, given that auditory localization is possible, then having environmental features located in either a virtual or real environment identify themselves (e.g., via talking signs) could provide the additional information that brings the blind traveler closer to the sighted traveler in terms of travel competence.

Finally, it is obvious from much of the information in this chapter that successful navigation can take place even if it is not efficient or accurate. Wayfinding in humans has been classified in terms of "efficiency" because many of the mathematical models that have been developed to simulate or explain human movements adopt simplifying assumptions (such as following the shortest path, or expending the least effort, least time, or least cost). These features lend themselves to simple optimization routines, but they may not reflect the true criteria used in individual path following behavior. As is often the case with humans, behaviors can be inexact, and according to Simon (1957) are described more as "satisficing" or "boundedly rational" rather than optimizing. Freundschuh (1991) argues that because of the way we have structured our environment, human navigation and wayfinding does not have to be accurate: it is not essential to make an *exact* retrace, or follow an *exact* route. Thorndyke and Hayes-Roth (1982) concluded that for orientation tasks, those actively experiencing an environment by travel performed better than those learning from maps, and that for distance estimation, map learners were more accurate than route learners. Freundschuh also argued that people do not learn Euclidean metrics, so distances and directions estimated in Euclidean space may be incorrect. If this is the case, then accurate cognitive maps are not needed; this may be a critical point of difference between the cognitive maps or spatial representations of humans and animals. Spatial representations that can provide on demand *sufficient* information to allow a problem to be solved or a spatial task to be completed are all that are required for most human wayfinding purposes.

Although human movement is often constrained to networks of routes such that it is relatively easy to determine mathematically an optimal path according to some specific criteria, the environments through which many nonhuman species navigate and move do not necessarily lend themselves to such comparisons. Even if insects using path integration are capable of flying directly home, they are subject to intermittent prevailing environmental conditions such as the presence of high winds, which may prevent them from doing so. Like early human navigators, they could be blown off course and consequently perish, or come back with grossly exaggerated traces of where they had been and what it takes to successfully negotiate the intervening spaces.

One final important point related to wayfinding is that humans do not necessarily "know" an environment through which a learned route passes. Only some features of the route itself may be learned, stored, and used; the traveler may be ignorant of the environmental surrounds. Such a traveler would not be able to make accurate judgments about the location, distribution, pattern, connectivity, or other spatial relations of features occurring within such an environment.

Once an environment has been experienced either partly or completely, fragments of environmental information (including locations and connections between places) are encoded, stored, decoded, recalled, and used in various tasks. Memories may be fuzzy and imprecise or extremely accurate. Different degrees of precision may be in evidence when different spatial products are developed to externally summarize what is known.

Are cognitive maps necessary for human navigation? The answer appears to be yes. But our cognitive maps do not have to be accurate renderings of the real world.

GARY L. ALLEN

2

Spatial Abilities, Cognitive Maps, and Wayfinding

BASES FOR INDIVIDUAL DIFFERENCES IN SPATIAL COGNITION AND BEHAVIOR

Why are some individuals better than others at wayfinding? Because goal-directed travel is so commonplace, this question may be as interesting to individuals engaged in the routine of everyday life as it is to psychologists, geographers, and other researchers who study it systematically. Scientists and nonscientists probably differ in the type of answer they expect. Non-scientists might well anticipate a definitive answer to such a straightforward query. In contrast, behavioral and cognitive scientists recognize that often the more direct the question, the less clear-cut the answer. The reason is apparent enough; as complex phenomena, cognition and behavior can be influenced by a variety of factors at several different systemic levels. Thus, the question of why some individuals are better wayfinders than are others elicits an amalgam of partial, qualified, and conditional responses rather than a simple, direct answer. Developing a framework for organizing partial, qualified, and conditional answers is an important activity, for ultimately only such a framework can provide a comprehensive scheme for addressing this question satisfactorily.

If this chapter were considered a route, the final destination would be a new approach to the study of individual differences in spatial cognition and behavior. Landmarks along the way would include a functional examination of wayfinding, consideration of the different traditions of conceptualizing spatial abilities, and an overview of evidence relating spatial abilities, cognitive maps, and wayfinding. The candidate conceptual framework that serves as the destination for this journey is an adaptation of the interactive common resources framework developed as a new approach to assessing general intellectual abilities (Kyllonen 1993). There has been no previous link between this approach and the study of spatial cognition and behavior, and, consequently, its merits lie in its potential rather than its empirically demonstrated utility. Hence, this potential is most apparent after considering the *landmarks* along the way.

WAYFINDING

If wayfinding is to be the outcome of focus, then it is reasonable to begin with an explication of what is meant by that term. The word *wayfinding* cannot be found in standard English-language dictionaries, but the increasing frequency of its appearance in the literature of environmental psychology, geography, and even experimental psychology suggests that it is a reasonable addition to the lexicon. For present purposes, it refers to purposeful movement to a specific destination that is distal and, thus, cannot be perceived directly by the traveler (Baker 1981; Blades 1991; Gärling, Böök, & Lindberg 1984; Gluck 1991; Golledge 1992; Heft 1983). This definition includes but goes beyond the act of avoiding obstacles while moving through the environment (e.g., Vishton & Cutting 1995). Obviously, wayfinding as purposive behavior involves interactions between attributes of the traveler and attributes of the environment. Because of short-term variations (e.g., the alertness of a driver, temporary road detours) and long-term changes (e.g., aging-related increases in response time, suburban development) in these attributes, it is reasonable to posit that at some level of consideration no two wayfinding attempts are exactly alike, even those involving repeated travel between familiar destinations. An element of uncertainty is a factor in every effort. Ultimately, successful wayfinding is reflected in the traveler's ability to achieve a specific destination within the confines of pertinent spatial or temporal constraints and despite the uncertainty that exists.

An examination of individual differences in wayfinding will necessarily focus on the attributes of the traveler. This focus is not meant to downplay the role of environmental attributes or to suggest an oversimplification of wayfinding tasks. Indeed, it is the case that the abilities exhibited by humans and other species are quite obviously adaptations to environmental demands and opportunities. Nevertheless, the specific impact of environmental differences on wayfinding is not the focus of the present effort.

WAYFINDING TASKS

As recognition of the importance of wayfinding research grows, so does the need for a taxonomy of wayfinding tasks and means for accomplishing these tasks. Of the many potentially useful ways of categorizing wayfinding tasks, perhaps the following three-category scheme based on functional goals is of the greatest heuristic value: travel with the goal of reaching a familiar destination; exploratory travel with the goal of returning to a familiar point of origin; and travel with the goal of reaching a novel destination.

Travel between known places is very common. Commuting between home and work place is an example drawn from everyday life in contemporary Western societies. Exploratory wayfinding is also a relatively common activity, not only for human beings but for most highly mobile species. People often explore their surroundings after changing their residence or when visiting a new environment. The point of such travel is to reconnoiter and return to the point of origin. Wayfinding to novel destinations is more common among human beings than among other species, especially when it involves arrival at a goal state through reliance on symbolic spatial information communicated to the traveler. Wayfinding guided by maps or verbal directions is the epitome of this type of task. A parallel involving nonhumans would be travel by honey bees to a novel site described in dance by a scout returning to the hive. Because a specific destination is required for wayfinding as defined here, some cases of traveling to new places fit this category only marginally. Examples include travel to novel surroundings when resources in a former habitat become too scarce, when territorial battles make a home range too dangerous, or when mature offspring distance themselves from parental territory.

WAYFINDING MEANS

These three categories of wayfinding tasks can be accomplished by a variety of means, some of which apply primarily to one type and some of which apply across tasks. Oriented search is a simple and sometimes problematic means of reaching a destination. In oriented search, the traveler first orients according to a source of information and then searches, either idiosyncratically or systematically, until the destination is achieved. Although various species are sensitive to distal visual (including solar, lunar, and stellar), tactile (including wind and water currents), geomagnetic, and olfactory information, humans tend to rely most heavily on visual or, in cases of visual impairment, on auditory, vestibular, and proprioceptive information. Oriented search may have its greatest utility in instances of exploratory travel, in which the traveler orients with respect to perceptual information on the outward journey and then reorients to return to a familiar point of origin. Nevertheless, it can also be used in travel to familiar and novel destinations. The chief problem is that it is efficient only when the distance traveled is relatively small, the environment is stable, and systematic rather than idiosyncratic search heuristics are employed.

Following a continuously marked trail is a reliable means of wayfinding that can be applied to all three types of wayfinding tasks. Examples range from the mundane activity of commuting to work along a freeway to the unusual hobby of underwater cave exploration, in which divers typically rely on a tether in order to make their way back to a point of origin. Trails have also been used to mark routes to common destinations in expansive, poorly differentiated built environments that host short-term visitors. Color-coded trails painted on corridor floors in large medical complexes serve as a case in point. By design, trail following reduces cognitive demands on the traveler while reducing uncertainty to a bare minimum. Constructing continuously marked trails is an expensive enterprise, however, and the cognitive demands of trail following escalate if multiple trails are indicated, as in the case of complex highway interchanges.

Wayfinding by means of *piloting* between landmarks, a common means of wayfinding for many species, is a third method that is equally applicable across the three types of wayfinding tasks. In landmark-based piloting, the traveler relies exclusively on sequentially organized knowledge; one landmark is associated with specific direction (and perhaps distance) information that leads to another. This description closely resembles the standard definition of route knowledge. Landmark-based piloting is an efficient means of traveling to familiar and novel destinations in a well known environment, and may be the standard method of locating a novel destination in unfamiliar territory, in view of the fact that verbal directions consist primarily of condition-action lists. Success in landmark-based piloting depends upon recognition of landmarks, either from previous experience or from descriptions, and remembering the spatial relations associated with them. Piloting of this sort is also common in exploratory wayfinding. In such cases, however, the traveler must select landmarks rather than rely on familiar or prescribed features. Heavy reliance on learning in such circumstances suggests the advantage of using memory strategies during travel (Cornell, Heth, & Alberts 1994; Cornell, Heth, & Broda 1989).

Habitual locomotion is a means of wayfinding that pertains only to familiar destinations. With increasing experience, a traveler relies more and more on procedural knowledge. Under conditions of consistent environmental structure, extensive repetition of a locomotor pattern can lead to its automatization. A fully automatic pattern would require no attention to environmental circumstances, and movement would be stereotypic. Rarely would environments and contextual circumstances be sufficiently consistent to permit completely automatic wayfinding, but habitual locomotion is involved to some extent in all activity that involves familiar routes. Many commuters arrive home with a minimum of attention paid to the trip that brought them there.

Path integration by reliance on information from visual, auditory, proprioceptive, or vestibular sources can be an effective means of "homing" after exploratory travel, but it can also be employed to reach a familiar destination by a more direct route. Path integration relies on monitoring self-movement, either dead reckoning on the basis of velocity information or inertial navigation on the basis of acceleration information. Such information is used to compute direction and distance to a specified location. Humans appear capable of crude path integration even without vision (Loomis et al. 1993); small rodents are quite competent (Etienne 1992; Potegal 1982, 1987); and ants and bees are extremely adept (Wehner & Menzel 1990).

The most sophisticated means of wayfinding involves reliance on an integrated internal representation of relationships among places, frequently referred to as a cognitive map. This means of oriented travel could conceivably be used for all three types of wayfinding tasks, although the learning required for locating a novel destination in a novel environment and perhaps for "homing" after exploratory travel could make these activities a considerable challenge. The internal representation involved in this case could be either a vector map derived from vector information or a topographic map derived from landmark information. Either type of knowledge structure could provide the basis of travel to a destination by means of novel routes. Evidence indicates that humans can engage in navigation of this type, but distinguishing wayfinding by these means from wayfinding by other means is a significant challenge for researchers (see McDonald & Pellegrino 1993).

IMPLICATIONS FOR UNDERSTANDING
INDIVIDUAL DIFFERENCES

Table 2.1 shows the possible utility of proposed wayfinding means for various wayfinding tasks. Although the match of means to tasks is largely speculative and based on cursory descriptions, it advances three general

TABLE 2.1 | Possible utility of proposed wayfinding means for various wayfinding tasks

Wayfinding means	Wayfinding tasks		
	Travel to familiar destinations	Exploratory travel	Travel to novel destinations
Oriented search	X	X	X
Following a marked trail	X	X	X
Piloting between landmarks	X	X	X
Path integration	X	X	
Habitual locomotion	X		
Referring to a cognitive map	X	X	X

propositions. First, multiple means can be used for the same wayfinding task. Second, most means can be used for multiple tasks. Third, more means can be applied to travel to familiar destinations than to exploratory travel, which in turn involves more means than does travel to novel destinations. In other words, there is flexibility and redundancy in solving each of the wayfinding tasks, but there is greater flexibility and redundancy in travel to familiar destinations than in exploratory travel, and more in exploratory travel than in travel to novel destinations.

Even if the specifics of the preceding taxonomy were found to have their shortcomings, acknowledgment of a variety of wayfinding tasks and a number of means that can be employed in those tasks has important implications for framing questions about individual differences. In addressing the question of why some individuals are better than others at wayfinding, it is useful to ask about the nature of the wayfinding task. Individuals may differ in their success on a wayfinding task because they use different means, for example, path integration versus landmark-based navigation to return home after exploratory travel. But they may also differ in their ability to apply a particular means, for example, good versus poor landmark-based navigation. If both of these types of differences are to be considered, then a suitably expansive conceptual framework must be developed.

SPATIAL ABILITIES

The term *spatial abilities* carries with it a variety of connotations because this rubric has been used in different empirical traditions that are tangentially related, if indeed they are related at all. Seeking common conceptual or empirical threads among these traditions is a formidable challenge. One approach to this challenge is to ask some questions concerned with function. Is it possible to group spatial abilities according to their function, that is, according to the tasks or situations in which they come into play or according to the purpose they may serve as adaptations? With a functional perspective, it may be possible to group abilities according to a "family resemblance" (Rosch & Mervis 1975) heuristic. In other words, it may be possible to identify groups of spatial abilities on the basis of their relation to a common function (for example, abilities involved in object identification) rather than on the basis of criterial properties or processes that determine group membership (e.g., visualization).

THE PSYCHOMETRIC TRADITION

The study of spatial abilities has long been part of the psychometric tradition in psychology. This tradition, which is the conceptual home of the term "individual differences," has provided a delineation of abilities using factor analysis and various schemes derived through multidimensional scaling or cluster analyses to explicate how abilities are related to each other. Psychometric research has clearly identified a domain of spatial ability consisting of a number of different factors (Guilford & Zimmerman 1941; Lohman 1988; McGee 1979; Thurstone 1938), all reflecting facility in perceiving, remembering, or mentally transforming figure stimuli.

Lohman (1979, 1988) provided an informative listing of the factors most frequently described as spatial. The most widely recognized among them are visualization, speeded rotation, and spatial orientation. Visualization is concerned with the ability to imagine or anticipate the appearance of complex figures or objects after a prescribed transformation; examples of marker tests include form board, paper folding, and surface development. The speeded rotation factor, formerly referred to as spatial relations, involves the ability to determine whether one stimulus is a rotated version of another, particularly when simple rotations (such as simple figures in two-dimensional space) are involved; marker tests include flags and cards. Spatial orientation may be considered the ability of an observer to anticipate the appearance of an object or object array from a prescribed perspective; marker tests include aerial orientation and chair-window. Sample items from marker tests can be found in Ekstrom, French, and Harman (1976) and Eliot and McFarlane-Smith (1983).

Conceptually differentiating these abilities is straightforward; empirically deriving them as distinct factors can be rather challenging. Test items requiring complex rotation or perspective-taking resemble visualization problems, and some simple orientation problems are amenable to solution by means of rotation. An effort to differentiate among spatial factors must include multiple marker tests of each factor.

Other spatial factors include abilities known as flexibility of closure, which involves the detection of figural stimuli embedded in a "noisy" visual context (e.g., embedded figures); speed of closure, which involves the perception of integrated familiar figures when only parts of the figures are visible (e.g., gestalt completion); visual memory, which refers to remembering the arrangement, location, and orientation of figural stimuli (e.g., building memory); and spatial scanning, which pertains to rapid visual exploration of a large or complicated array (e.g., maze tracing). Less familiar are factors Lohman (1988) referred to as serial integration, the ability to identify a figure on the basis of temporally separated constituent views (e.g., successive perception), and kinesthetic, the ability to make rapid left-right discriminations (e.g., hands).

How these abilities are related to each other has been suggested by factor analytic studies (e.g., Eysenck 1967; Guilford 1967), multidimensional scaling analyses (e.g., Snow, Kyllonen, & Marshalek 1984), and experiments (e.g., Lohman 1988). The evidence points to a hierarchical organization based on complexity. At the lowest level in the hierarchy are abilities requiring only visual perception, attention, or temporary memory, such as speed of closure, flexibility of closure, spatial scanning, and visual memory. At an intermediate level of complexity are abilities requiring accommodation, or translation, of the spatial relationship between observer and stimulus or between two stimuli. Speeded rotation and spatial orientation fit this general description. At the highest level is visualization, which requires transformations involving surfaces or other components of stimuli. Logic and evidence lead to the conclusion that the basis for individual differences would not be consistent at different levels within the proposed hierarchy, a point that was clarified by research on visualization and mental rotation in the information-processing tradition.

It is important to note that for decades the psychometric view of human abilities was influenced by the limitations of assessment technology, with paper-and-pencil testing being the norm. Computerized assessment has ushered in a new era in this regard (Embretson 1987). In the domain of spatial abilities, computerized assessment has provided the means for identifying dynamic spatial ability (Hunt, Pellegrino, Frick, Farr, & Alderton 1988; Schiff & Oldak 1990), described as the ability to judge the velocity and trajectory of objects moving in the visual field. Ability of this type appears to be factorially distinct from the assorted abilities described previously, and it may well be an important link between abilities comprising the cognitive domain (Guilford 1967) and abilities comprising the psychomotor domain (Fleishman 1975).

THE INFORMATION-PROCESSING TRADITION

The information-processing tradition incorporates a normative approach to the study of cognition that is generally driven by task analysis. In other words, an attempt is made to characterize cognitive phenomena in terms of a set of generalizable constituent processes. The spatial phenomena that have most frequently been subjected to this type of analysis include visualization and mental rotation as reflected in psychometric tests of spatial ability, visual-spatial memory, mental imagery, and spatial perspective-taking and orientation.

Visualization and Mental Rotation

The information-processing approach to the study of visualization and mental rotation abilities was part of a general effort to describe cognitive abilities in terms of combinations of basic processes. Tasks resembling items on psychometric tests were used in experiments designed to differentiate fundamental processes on the basis of the time necessary to respond accurately and, in some instances, error analysis. Overall, this approach was a qualified success. Among its liabilities were problems in characterizing fundamental processes in an informative way and determining if or when processes were employed serially or in parallel. Furthermore, critical differences among individuals, particularly in the performance of visualization or complex rotation tasks, were found to be in the application of task-specific strategies. Describing, classifying, operationally defining, and empirically validating these strategies remain a difficult part of task analysis.

These considerations notwithstanding, the effort to analyze performance on spatial ability tests in terms of basic processes did yield useful insights. Task components on which individuals did not differ were readily identified. Also, the complexity of psychometrically determined spatial abilities was shown to be a direct reflection of the number and type of processing steps demanded by test items. For example, visual memory, as a simpler ability, requires only encoding and comparing stimuli, either two-dimensional figures or three-dimensional objects. At the other end of the hierarchy visualization involves encoding the stimulus, performing a series of transformations of the encoded stimulus, and comparing stimuli to determine the consequences of the transformations. The higher the number of stimulus features or surfaces, the greater the burden on temporary memory to apperceive the consequences of transformations (Lohman 1988; Mumaw & Pellegrino 1984; Pellegrino, Mumaw, & Shute 1985).

Three major sources of individual variation indicated by information-processing research were speed of processing (a nonstrategic factor) and application of frames of reference and execution of transformations (two strategic factors). The speed with which processes were performed was a significant source of individual differences. Beyond its obvious impact on speeded rotation performance, speed of processing has implications for time-sensitive processes, such as maintenance of a representation in temporary memory (Mumaw, Pellegrino, Kail, & Carter 1984). Success in maintaining such a representation is a second source of individual differences in spatial task performance. In complex tasks, imposing a frame of reference—either stimulus-based or viewer-based—may be considered part of the process of representing the problem (Just & Carpenter 1985). The ability to execute prescribed transformations on the represented stimulus (e.g., displacement of intact stimuli, rearrangement of stimulus components) is particularly important with more complex rotation and visualization tasks (Pellegrino, Alderton, & Shute 1984; Pellegrino & Kail 1982). The execution of transformations introduces strategy use as a source of individual differences. Early studies involving verbal accounts of performance (e.g., Barratt 1953; Myers 1957) and contemporary efforts using analyses of eye movement data (Just & Carpenter 1985) indicate clearly that strategy differences can influence differential success on tests of complex spatial abilities. In summary, it appears that individuals who are very successful on rotation and visualization tasks encode relevant task information into temporary memory rapidly, retain it successfully, and execute necessary transformations efficiently with the aid of relatively demanding strategic procedures (Carpenter & Just 1982; Just & Carpenter 1985; Pellegrino et al. 1984).

Visual-Spatial Memory
Hasher and Zacks's (1979) contention that spatial attributes of stimuli are encoded automatically caused something of a stir—due perhaps in part to the sizable number of academicians who habitually lose track of their respective office keys. Nevertheless, it appears that spatial location is committed to memory with very little effort, if not altogether automatically (Naveh-Benjamin 1987). Recent evidence has indicated that accuracy is achieved by encoding fine-grained locational information as well as categorical information based on regions or subdivisions of the relevant space (Huttenlocher, Hedges, & Duncan 1991). Although this conclusion was

reached on the basis of research involving locations in small two-dimensional arrays, it has been extended to include performance in locomotor spaces (Franklin, Henkel, & Zangas 1995).

Despite the relative ease with which location is encoded in memory, the substantial differences found on psychometric tests of visual memory suggest that there is more to the story. In fact, visual memory tests often involve multistimulus arrays that require memory for the specific stimuli as well as memory for the locations of those stimuli. Pezdek and Evans (1979) demonstrated nicely that remembering that a certain location was occupied by an object is different from remembering which particular object was in that location. The latter requires a degree of specificity of association that is absent in the former. Some researchers have referred to this specificity of association as psychological binding (see Chalfonte & Johnson 1996). Once attributes such as object and location are bound, they can serve as cues or primes for each other, for example, location serving as a cue for the object that was observed there.

The work of Thorndyke and his colleagues (Thorndyke 1981; Thorndyke & Hayes-Roth 1982; Thorndyke & Stasz 1980) on individual differences in memory for information from maps further illustrated the role of strategic factors in remembering objects in locations. These investigations revealed that individuals who learned map information quickly tended to subdivide the map into regions during learning, focus effort on locations yet to be remembered, and use a variety of means to encode spatial information. It is noteworthy that good and poor map learners were differentiated by their memory for spatial relations among map features rather than by memory for verbal labels attached to those features. Not surprisingly, it was found that performance on the map-learning task was positively correlated with scores on a psychometric test of visual memory.

It is worthwhile to point out that visual-spatial memory as it pertains to wayfinding is much more likely to involve remembering scenes (i.e., objects in complex settings) than remembering arbitrary arrangements. Ample evidence documents the role of syntactic and semantic factors in processing the information comprising a scene (e.g., Biederman 1972; Biederman, Mezzanotte, & Rabinowitz 1982). Individual differences in scene recognition show the influence both of specific object-setting relations and knowledge-based expectations. A viewer's prior knowledge of a setting can play a critical role in memory for objects in positions, as illustrated by Chase and Simon's (1973) work on expert versus novice chess player's memory for game board situations (see also Gobet & Simon 1996).

The implications of information-processing research on visual-spatial memory for understanding individual differences in spatial abilities pertain chiefly to various strategies brought to bear in remembering what goes where in multiobject arrays. No doubt, individuals vary in their application of general strategies (such as "chunking" objects according to region), but such differences may be even greater with specific knowledge-based strategies (such as using an analogy between a set of locations to be learned and a set of locations within a well-known configuration). Scenes involving objects in complex settings may provide additional opportunities to employ effective situationally specific, knowledge-based strategies.

Mental Imagery

With *Image and Mind,* Kosslyn (1980) brought the study of imagery back into the mainstream of scientific psychology. His theory explicitly named and described the processes required to form and manipulate surface-level images from deep-structured representations. Processes included production (e.g., picture, image, lookfor), alteration (e.g., scan, pan, zoom, rotate), and comparison (e.g., find) transformations. The specification of these processes suggested potential sources of individual variation, a suggestion that was followed up with some success (Kosslyn, Brunn, Cave, & Wallach 1984). A subsequent effort attempting to validate more "neurologically plausible" processes was perhaps even more successful (Kosslyn, Van Kleek, & Kirby 1990).

It would seem quite natural to expect sizable correlation between imagery-related processes and performance on psychometric tests of visualization and speeded rotation. Such correlation, however, ranges from low to moderate (Poltrock & Brown 1984). This finding is consistent with the decidedly modest correlation between reports of visual imagery and scores on psychometric tests of spatial abilities frequently reported in the literature (Drose & Allen 1994; Poltrock & Agnoli 1986; Snow, Lohman, Marshalek, Yalow, & Webb 1977). In some instances, this situation may reflect difficulty in operationalizing the various imagery processes. Another factor that must be considered is that visualization and complex rotation problems are amenable to solution by nonvisual propositionally based strategies (Carpenter & Just 1982; Lohman 1988).

The study of imagery is very much alive and well in contemporary cognitive psychology (De Vega, Intons-Peterson, Johnson-Laird, Denis, & Marschark 1996), although individual differences have not taken center-stage in this enterprise. Perhaps the characterization of imagery as "those memories that have salient sensory-perceptual and spatial features"

(Intons-Peterson 1996:20) will help to build a bridge between the image in Kosslyn's (1980) *Image and Mind* and the image in Lynch's (1960) *The Image of the City.*

Spatial Perspective-Taking and Orientation

This eclectic category includes lines of experimentation that are not obligatorily cross-referenced. As in the case of many experimental procedures, the study of spatial perspective-taking has Piagetian origins, and its developmental implications will be addressed subsequently. The ability to identify the appearance of a stimulus array from a designated perspective has been examined using table-top displays (e.g., Huttenlocher & Presson 1973, 1979), room-size environments (e.g., Hardwick, McIntyre, & Pick 1976), and urban neighborhoods (i.e., Kirasic, Allen, & Siegel 1984). If one were pressed to make sweeping generalizations relevant to individual differences on the basis of these dozens of studies, emphasis would be placed on two processes, namely, encoding the array in memory and executing necessary transformations. Encoding the array is not necessarily just a storage problem. As mentioned in the context of information-processing accounts of mental rotation and visualization, encoding includes employing a frame of reference such as a viewer-based, object-based, or abstract scheme. The number of objects in the array and the scale of space involved have important implications for the ease with which different frames of references are applied. In turn, frames of reference have implications for the ease with which prescribed transformations are executed. Kozlowski and Bryant's (1977) findings that self-rated sense of direction correlated positively with valid distance and direction knowledge in a campus environment suggest that individuals have direct awareness of their abilities to employ a frame of reference successfully.

Information-processing studies of visualization, mental rotation, and perspective-taking thus appear to converge on common conclusions about the basis for individual differences. This convergence is exactly what would be expected on the basis of theories such as the one advanced by Olson and Bialystok (1983), who posited a common core of representation and transformation processes for visualization, mental rotation, perspective-taking, and spatial orientation abilities. Research by Hintzman, O'Dell, and Arndt (1981) suggests, however, that generalizations should be made with caution. After detailed analysis of performance in a number of analogous tasks involving different scales of space and even different sensory modalities for acquiring spatial information, Hintzman et al. (1981) could posit no general information-processing model to account for solving problems requiring the preservation of spatial relations among locations. Clearly, fixing a frame of reference was a critical component of task solution in each case, and the data indicated that distance and direction information were processed separately. Yet, perhaps because different scales of space tended to elicit different frames of reference, a unifying scheme for explaining results across experiments was elusive.

Orientation ability has also been studied in the context of locomotor tasks. In an extensive series of studies, Gärling and his colleagues (Böök & Gärling 1980, 1981b; Gärling, Böök, & Lindberg 1984, 1985; Gärling, Lindberg, & Mäntylä 1983; Lindberg & Gärling 1981a, 1983) examined variables critical to the process of maintaining orientation in familiar and unfamiliar environments. Their findings indicated that movement monitoring, accomplished through an analysis of visual, vestibular, or proprioceptive information with some involvement of "central information processing" to compute direct paths to designated locations, was essential to the maintenance of orientation in unfamiliar surroundings. With regard to orientation in familiar environments, it was assumed that recognition memory plays an important role, especially in relating a traveler's specific perspective within an environment to a larger frame of reference.

Overall, Gärling et al. (1985) pointed to three ways of updating location after movement based on perceptual information: recurrent "on-line" computation, computation based on reconstructive memory for movements, and transformations of visual perspectives. Subsequent research by Loomis et al. (1993) showed that even on-line updating of location relative to a starting point is memory-based in the sense that it refers to a record of movement, a temporary representation that is not automatically discarded after updating. With regard to the third way of updating location, Cutting and his colleagues (Cutting, Springer, Braren, & Johnson 1992; Cutting & Vishton 1993; Vishton & Cutting 1995) have demonstrated that differential parallactic displacements and inward displacements can serve this purpose.

Looking for commonalties within this diverse aggregation of perspective-taking and orientation studies requires some conceptual flexibility, and seeking general implications for individual differences from these sources requires not only conceptual flexibility but perhaps a vivid imagination as well. This point taken, it may be imagined that perspective-taking and orientation skills are influenced by sensitivity to and memory for information obtained through sensory systems, facility with different spatial frames of reference, and the successful computation of directional vectors and distance judgments.

Map Interpretation

Establishing a correspondence between a cartographic representation of space and the space that it represents is quite different from simply memorizing the locations of symbols on a map. Map interpretation involves, first, apperceiving a "stand for" relationship between environmental features and cartographic conventions used to symbolize them, and second, establishing a "you are here" relationship between the map user and the environment. Sensibly, most of the research on learning cartographic conventions has been done by cartographers (see Lloyd 1993). The spatial conventions that most frequently come into play in cartography (e.g., spatial autocorrelation, spatial heterogeneity, hierarchical patterning) rarely, if ever, serve as the focus for research on spatial abilities (Gilmartin & Patton 1984; Self, Gopal, Golledge, & Fenstermaker 1992), despite the fact that such concepts are essential for establishing the "stand for" relationship between environment and map.

In comparison, the "you are here" relationship between map user and environment has received more attention. A variety of studies have shown that, in general, correspondence between the map user's view of the environment and a point on a map is greatly facilitated by aligning the map with the environment so that, for example, looking to the left on a map reveals map features corresponding to environmental objects actually seen to the individual's left (Levine 1982; Levine, Marchon, & Hanley 1984). Success in using misaligned maps requires facility at perspective transformation, much the same operation as is involved in perspective-taking tasks.

Generalizations

Across these diverse studies, the information-processing approach to perspective-taking and orientation affords useful but loosely connected inferences about individual differences. In short, it is plausible that individuals differ in their sensitivity to information from various perceptual systems, their ability to store information obtained from these systems, their inclinations regarding different frames of reference, and their facility in computing distance and direction estimates to target locations. It is not much more of a stretch to suggest that individuals differ in some aspect of "central information processing" that supports simultaneous storage and processing of task-relevant information, that is, working memory.

THE DEVELOPMENTAL TRADITION

The contributions of developmental theory and research to the study of spatial abilities are extensive and formidable. The only way to broach this topic without being absurdly superficial is to use the type of tasks discussed previously as a brief checklist. The research literature—much of it from an information-processing perspective—shows age-related improvement on tasks requiring mental rotation and visualization, spatial perspective-taking and orientation, and visual-spatial memory from early to late childhood. An age-related decline in performance on such tasks has been documented for late adulthood.

Mental Rotation and Visualization

The standard mental rotation task, with its requirement for the rapid comparison of abstract figure stimuli, is based on the ability, apparent in infancy, to detect invariant features comprising rigid objects. Beyond this basic requirement that shape be recognized (or perhaps "conserved" in the Piagetian sense) despite rotation, the chief factor responsible for age-related improvement is the speed with which the figures are compared. Rate of comparison increases over the course of childhood (Kail 1991), and declines over late adulthood (Salthouse 1992). The standard visualization task also begins with feature recognition and, again, the importance of the concept of conservation of mass or area seems self-evident. Given that visualization tasks place a load on temporary memory and require strategy implementation, it is not surprising to see improvement over childhood (Lohman 1988). Speed and, to some extent, temporary memory limitations are implicated in performance decrements with increasing age in late adulthood (Salthouse 1992).

Visual-Spatial Memory

Visual-spatial memory, especially in tasks involving a stationary observer, is relatively robust from an early age. Memory for where an event occurred is functional in early infancy, and by early childhood learning object-location relations is not particularly problematic. Indeed, researchers have warned parents that children tend to be formidable opponents in memory games with significant visual-spatial components, such as *Concentration* (Mandler, Seegmiller, & Day 1977). But performance is influenced strongly by task demands. For example, remembering a match between a specific object and a well-defined location is less demanding than remembering where an object was located in an undifferentiated field. As Huttenlocher, Newcombe, and Sandberg (1994) demonstrated, accuracy in the latter task is enhanced by parsing the field into segments. There is an age-related increase in parsing of this type over childhood, but it is strongly influenced by the size and shape of the field in which locations are to be recalled (Huttenlocher et al. 1994). Also, in tasks requiring the reconstruction of an array of objects from memory, it is clear that performance is enhanced by the use of strategies such as grouping or the use of semantic information (Herman 1980; von Wright, Gebhard, & Kartunnen 1975). Curiously, the literature on cognitive aging shows age-related decline on visual-spatial memory tasks that require matching a set of objects to a set of locations (Light & Zelinski 1983; Pezdek 1983; Schear & Nebes 1980). Presumably, this decline is based on decreased speed with which strategies are implemented—rather than ignorance of such strategies—and increased susceptibility to interference during retrieval.

Perspective-Taking and Orientation

A great deal of research has been done using spatial tasks developed by Piaget to demonstrate support for his theory positing Euclidean spatial concepts as the product of a lengthy developmental process that begins with infantile egocentrism (Piaget & Inheldler 1967; Piaget, Inhelder, & Szeminska 1960). Studies on the coordination of perspectives and the representation of horizontality and verticality are two very relevant examples. As mentioned previously in the context of sex-related differences, the water level test has Piagetian origins as one of a series of tasks designed to show the development of the concept of horizontality. Age-related change toward correct demarcation of horizontality is a persistent finding, although most contemporary interpretations of this trend emphasize the conflict between frames of reference that is inherent in the task. Thus, performance is taken to reflect which of several possible frames is relied upon rather than the emergence of a concept.

The situation is not that different in the case of the coordination of perspectives. Most of the research on children's perspective-taking ability was inspired by Piaget's interpretation of performance on the three-mountain task in which children attempted to select photographs that portrayed the view of a display from a prescribed perspective. The failure of children to perform accurately in this task prior to late childhood led to the conclusion that young children lacked perspective-taking ability (Piaget & Inhelder 1967). Although dozens of studies followed, the most insightful critique of this conclusion was based on research by Huttenlocher and Presson (1979), who showed that difficulty in solving problems of this type stems from conflicting frames of reference. Posing perspective problems in terms of the anticipated location of a specific object in an array rather than in terms of the anticipated appearance of the entire array greatly enhances accuracy. Newcombe and Huttenlocher (1992) demonstrated that posing questions in this manner reveals perspective-taking skill in children in their preschool years. It may be useful to note that solving problems on the basis of near-far relations was an easier matter than was solving them on the basis of left-right relations and that, as Piaget and Inhelder had demonstrated decades before, selecting a photograph depicting the entire array from a prescribed perspective was extremely difficult. Consistent with what was mentioned in the context of other tasks, performance on perspective-taking tasks has generally been shown to be slower and more errorful in older adults than in younger or middle-aged adults, even when the tasks involve questions about specific observer-item relations (Herman & Coyne 1980; Ohta, Walsh, & Krauss 1981). Holding spatial relations in memory while simultaneously calculating the effect of observer movement is a likely source of this difficulty.

The Piagetian claim of spatial egocentrism early in life, that is, inability to differentiate between current and potential perspectives, also stimulated numerous studies of children's ability to track spatial locations after movement. Across a host of studies using various methods, the evidence indicates that infants and young children can use a variety of frames of reference (e.g., body-based or object-based) to orient themselves with respect to an object's location (Acredolo 1981; Bremner 1982; McKenzie, Day, & Ihsen 1984; Presson & Somerville 1985). Staying oriented with respect to an object, then, is a matter of employing a reliable and functional frame of reference in the context of a specific task. Nevertheless, it is very important to note that what is reliable and functional has much to do with response capabilities and everyday experience. A body-based scheme will do nicely if one is incapable of locomotion and the scenery changes little, but it is inefficient when one is mobile and exploring new vistas from a familiar base.

A strong case has been made for the importance of self-produced movement in the normal development of perspective-taking and orientation skills (e.g., Bremner & Bryant 1985; Campos, Svejda, Campos, & Bertenthal 1982), but empirical studies involving individuals whose physical disabilities severely impair mobility have been few. In such studies, it is often the case that the participants have been children and that their motor disabilities stem from a variety of causes, including cerebral palsy, muscular dystrophy, malformation of limbs, and severe arthritis. Predictably, comparisons involving a group of this heterogeneity are difficult. This reservation noted, it can be reported that children with severe motor impairment have been found to have considerable difficulties on a variety of spatial tasks when compared to groups of peers matched on age, sex, and other abilities.

Examples included Kershner's (1974) study of perspective-taking skill using table-top displays and Foreman, Orencas, Nicholas, Morton, and Gell's (1989) examination of children's spatial knowledge of their classroom using map drawing, landmark placement, and direction estimation tasks. Although many qualifications must be made regarding the interpretation of these data, they do point to the quite credible conclusion that experience with active locomotion is a common facilitator of perspective-taking and orientation skills.

Of obvious relevance to wayfinding is the theoretical stance that the ability to conceive of the environment as objects in Euclidean space emerges over the course of childhood (Piaget & Inhelder 1967). Siegel and White (1975) made the explicit claim that children's cognitive representational ability follows a developmental sequence that begins with the recognition of landmarks, proceeds through the linear (or one-dimensional) representation of sequences of landmarks as routes, and concludes with the multidimensional representation of Euclidean relations among landmarks as configurations. Some supportive findings have been reported (e.g., Cohen & Schuepfer 1980; Cousins, Siegel, & Maxwell 1983), but unequivocal evidence has been difficult to obtain. Also, contrary to the Siegel and White position, some researchers report that Euclidean knowledge of environmental configurations can be found in very young children when appropriate methodologies are used (Conning & Byrne 1984; Spencer & Darvizeh 1981). Resolving this issue is not an easy matter. There has been some inconsistency in defining and operationalizing the concept of configurational knowledge, and it is clear that many spatial tasks can be performed by a variety of means (see Pick 1993).

The selection of landmarks is one aspect of spatial cognition in large-scale environments that shows unambiguous signs of age-related improvement during childhood. Laboratory studies (Allen, Kirasic, Siegel, & Herman 1979; Allen & Ondracek 1995) and field experiments (Cornell et al. 1989) have shown that older children select more distinctive and informative landmarks along real-world routes than do younger children and that these selections are related to better performance on tasks requiring distance and direction knowledge. Allen and Ondracek's (1995) finding that age-related improvement in this selection process during childhood was not related to increments in memory performance or information-processing speed suggest that it is a knowledge-based skill acquired through wayfinding experience. This interpretation is difficult to reconcile with the finding that older adults show a decrement on this same type of task when compared to young adults (Kirasic, Allen, & Haggerty 1992). One plausible possibility is that older adults take longer to scan a scene for a target stimulus than do younger adults, which has been labeled the spatial localization hypothesis (Plude & Hoyer 1985). If scanning for landmarks is more time consuming for older adults, then the process is perhaps more susceptible to problems stemming from incomplete visual search and from interference.

Map Interpretation

As mentioned earlier, map interpretation involves applying cartographic conventions and establishing a correspondence between a map user's view of the environment and a place on a map. Knowledge of cartographic conventions, such as map symbology, is acquired through experience with maps; thus, it is not surprising to find age-related increases in facility with these conventions (Downs & Liben 1985; Liben & Downs 1986). The matter of establishing correspondence between map user and environment has generated a modicum of debate. Some researchers claimed that this ability is quite simply innate (see Blaut & Stea 1971; Landau 1986), but a considerable body of evidence from clever and informative studies indicates that an understanding of the basic "stand for" relationship between environments and symbolic representations of those environments emerges in early childhood (DeLoache 1987). Thus making use of this understanding in map interpretation tasks is a matter of relevant experience, practice, and task demands (Blades & Spencer 1986, 1987; Downs & Liben 1985; Uttal 1994; Uttal & Wellman 1989).

THE NEUROPSYCHOLOGY TRADITION

In contrast to other traditions, abilities in the neuropsychological tradition are inferred from cognitive or behavioral dysfunction, or differential failure to perform a particular task (DeRenzi 1982; Zola-Morgan & Kritchevsky 1988). The overall motivation for this approach is to establish reliable correlation between cognitive or behavioral deficits and specific foci of neurological damage. Such correlation has been sought for over a hundred years, but recent technological advances allowing the in vivo localization of subtle brain damage has provided great impetus for contemporary work.

Efforts to define specific spatial abilities on the basis of human neuropsychological evidence combine available results with informed speculation. A succinct illustration of the outcome of this combination is Kritchevsky's (1988) framework for "elementary spatial functions" (see also DeRenzi 1982; Ratcliff 1982). This framework includes five categories of elementary functions reflecting a range of complexity. The less complex categories are spatial perception, spatial attention, and spatial memory. Spatial perception includes object localization, line orientation detection, and spatial synthesis (i.e., the ability to identify a complex object by examining its component features) as elementary spatial functions.

Object localization is described as involving both left and right occipital and parietal areas (but see Logothetis & Sheinberg 1996), line orientation the right parietal, and spatial synthesis the right temporal and parietal areas. The spatial attention category involves attention to the left hemispace, mediated by the right parietal area, and attention to the right hemispace, mediated by the left parietal area. Short-term and long-term spatial memory are included as elementary spatial functions under the category of spatial memory. Short-term memory function is attributed to the right hippocampus and the medial thalamus; the brain areas responsible for long-term memory have been difficult to determine because tests that might assess such memory also involve other elementary spatial functions (e.g., short-term memory, spatial attention to left and right hemispace).

The more complex categories of elementary spatial functions as proposed by Kritchevsky (1988) are spatial mental operations and spatial construction. Mental rotation is the only example of the spatial mental operation category for which substantial data are available. Mental rotation, as the name implies, refers to the ability to resolve the angular discrepancy between two perspectives on the same stimulus or object. Evidence suggests this function involves right parietal areas. The final category, spatial construction, refers to constructional praxis, the ability to assemble an object manually from its constituent parts. Substantial evidence indicates that both right and left parietal areas are involved.

Kritchevsky's taxonomy underscores the point made by Morrow and Ratcliff (1988) that neuropsychologists have focused on abilities in small-scale space (i.e., tasks performed by a stationary observer involving objects within reach) and know little about abilities in large-scale space (i.e., tasks performed by a mobile observer surrounded by stable environmental features). With regard to the latter, evidence indicates that a useful distinction can be made between the inability to find one's way through novel surroundings, which is associated with unilateral lesions of the right hemisphere, and the inability to remain oriented in familiar environments, which is associated with posterior bilateral lesions (Benton 1969; Morrow & Ratcliff 1988). Noteworthy dissociation has been reported for learning a manual versus locomotor maze (Ratcliff & Newcombe 1973) and for spatially mediated versus verbally mediated wayfinding (Hécaen, Tzortzis, & Rondot 1980). In summary, it appears that the right hemisphere in general and the posterior cortical regions in particular are important for remaining oriented in new environments.

SPATIAL ABILITIES: SOME FUNCTIONAL DISTINCTIONS

It seems justifiable to divide the current collection of spatial abilities into three families. One of these families concerns a stationary individual and manipulable objects; another involves either a stationary or mobile individual and moving objects; and the third has to do with a mobile individual and large, stationary objects.

The abilities most often studied by psychometricians and neuro-psychologists are concerned with interactions between a stationary observer and an object that is small relative to the size of the observer. Quite simply, these abilities involve establishing the identity of an object based on its constituent features under various conditions, such as when the intact object is rotated, when it is only partially visible or manipulable, when it is embedded in a complex background, or when the object is disassembled. All of the evidence suggests that these abilities may be considered to be related in a hierarchy reflecting complexity. Visually based matching of intact objects (or parts of objects) is at the base of the hierarchy and anticipation of the appearance of transformed objects is toward the top. It seems reasonable to posit that such abilities evolved as features of a visual-tactile system for detecting the affordances of small objects in proximity to the observer. Evolutionarily speaking, such a system would be very useful in the service of foraging efforts and extremely efficacious in the development of tool use. The ability to envision a tool in the materials from which it may be fashioned is a notable innovation.

As the data show, individuals differ considerably in their performance of tasks requiring these abilities. Sensitivity to visual or tactile information about objects is assumed as the foundation for task performance. One source of variation is the speed with which task-relevant features of objects are detected; another source is success in retaining these features in memory temporarily, as necessary. Temporary memory is especially important in more complex tasks in which prescribed changes in the appearance of objects must be anticipated; in such tasks, information is retained temporarily while prescribed transformations are performed. This dual function meets Baddeley's (1986) definition of working memory. Also, individuals differ in the strategies they employ when complex transformations are required (Lohman 1988). Regardless of task complexity, prior experience results in knowledge that dramatically influences performance. Repeated experience with specific objects, which is much more common in everyday life than in psychological research, leads to faster recognition, rotation, and transformation of those objects. Similarly, repeated experience in recognizing, rotating, or transforming a variety of objects results in generalizable skills so that novel objects can be recognized, rotated, or transformed in a specific way.

A second family of spatial abilities is concerned with observers and mobile objects. These abilities are reflected in tests of dynamic spatial skills requiring the anticipation of a target's velocity or trajectory. Actually, for some purposes it may be useful to distinguish between two varieties of dynamic skills, one that involves tracking targets in the visual field and one that pertains to monitoring relations between a mobile individual and a mobile object. The distances involved in these abilities are typically defined by perceptual range (e.g., line of sight in the modal case of vision), and the size of the objects is constrained somewhat by the fact that they are mobile. Given that a majority of mobile objects in the evolutionary history of our species were animate, it requires little imagination to propose that this family of spatial abilities emerged in the context of human beings' roles as predator and prey. Computing an intercept course for an object moving through the visual field is an effective adaptation for hunting, particularly when projectile tools are involved. Monitoring spatial relations between one's own location and the location of another moving creature seems essential for acquiring prey and avoiding predators.

Logically, the most important source of individual variation in these skills is perceptual-motor coordination, central to which is timing. Anticipating an intercept course depends on extrapolating from perceived speed and direction of movement. But there is a bit more to the story. In the simple case of an object moving through the visual field, anticipating a trajectory intersecting that of a moving object is necessary but insufficient to accomplish the desired results. The more challenging component of the task is to accommodate the speed of the intercepting projectile so that its arrival coincides with the arrival of the moving object at some point in space, hence the critically important role of perceived temporal duration. The problem is complicated by observer motion (see Cutting, Vishton, & Braren 1995). Judging an intercept course—regardless of whether the goal of the mobile individual is to maintain or avoid such a trajectory—while simultaneously avoiding other objects and perhaps engaging in other activities fits the description of working memory described earlier. Individuals with prior experience with task-specific objects, either the same moving object or the same projectile, should demonstrate greater skill in such tasks because of greater accuracy in velocity calculations. Similarly, when observer movement is an additional variable, familiarity with task setting and concomitant activities can have the effect of easing the burden on working memory, thus making possible more rapid and accurate computation of an intercept course.

The third family of spatial abilities suggested by the review involves mobile individuals within an encompassing environment containing large, immobile objects. These are the abilities that come to mind when wayfinding is mentioned. As can be seen from the discussion of various research traditions, sensitivity to visual, auditory, vestibular, and proprioceptive sources of information serve as the foundation for orientation skills. As pointed out earlier, as a particularly rich base of information about spatial relations among immobile objects, vision may be considered the modal perceptual source of observer-independent knowledge of environmental layout (Gibson 1979; Vishton & Cutting 1995). A method for acquiring observer-independent knowledge of this type is a useful adaptation for species who return to a home site after traveling far beyond objects that can be perceived immediately from that site. Another useful adaptation for mobile species is the ability to compute the direction and distance to the home site by referring to a record of movement comprising the outward journey, that is, path integration.

Sensitivity to available perceptual information—visual, vestibular, tactile, or proprioceptive—provides the foundation for this family of abilities. Individual differences in learning environmental layout may be based on differential speed in identifying distinctive environmental objects for use as landmarks and success in retaining these objects in temporary memory long enough to establish their spatial relations to previous encountered reference points. In the case of path integration, differences in working memory capability are particularly salient to success because calculation of the homing vector depends on a record of previous turns and distances. Knowledge stored in long-term memory is also a salient source of variation among individuals. Overt strategies based on prior experience, such as the selection of a frame of reference, can greatly facilitate orientation efforts. Repeated strategy application in specific situations can give rise to orienting schemas (see Neisser 1976), expectations that enhance the salience of certain types of information. The benefits of extensive experience can also be conveyed symbolically, through cartographic or linguistic means, to assist travelers who are less experienced or who need specific information. Obviously, knowledge of cartographic and linguistic conventions regarding the communication of spatial information is another source of individual differences.

It is not suggested that the three families of spatial abilities proposed here are independent of each other or of other abilities. It is, however, plausible that they evolved in response to conceptually separable environmental pressures and that facility with one of these groups has little implication for facility with the others. Nevertheless, they may draw on some of the same general constituent capabilities or resources. This possibility of general interactive resources will be expanded upon presently. First, however, it is useful to consider one of the most enduring theoretical constructs in cognitive psychology.

COGNITIVE MAPS

Are some individuals better than others at wayfinding tasks because they have more accurate cognitive maps or are more facile at cognitive mapping? If *cognitive map* means memory for spatial layout, then the answer has to be yes. People with accurate memory for layout should indeed be successful wayfinders, just as individuals with a large vocabulary are bound to be pretty good readers. But individuals vary tremendously with regard to reading ability. By analogy, it is reasonable to posit that variation in cognitive mapping skill is also significant.

THE THEORETICAL CONSTRUCT

The term *cognitive map* has a long history in psychology, readily traced to Tolman's (1948) contention that rats in a maze-learning task acquired knowledge of the spatial relation between point of origin and destination rather than—or in addition to—a series of stimulus response associations. Decades of research have shown that rats are indeed capable of place learning, but the story is a little more complicated than was evident 50 years ago (Restle 1957; Rudy, Stadler-Morris, & Albert 1987; Schenk 1985). As McDonald and Pellegrino (1993) concluded after reviewing the relevant literature, rats can acquire stimulus-response associations, stimulus-place associations, and place-place expectations. What is learned is determined by the developmental status of the organism, the affordances of the environment, specific demands of the task to be performed, and the extent of the organism's prior experience with the task. The original connotation of the expression *cognitive map* pertains primarily to place-place expectations acquired in differentiated surroundings after ample experience.

The definitional boundaries of the term *cognitive map* became quite variable as the concept was applied widely to human spatial cognition (Allen 1985; Downs & Stea 1973b; Gärling, Lindberg, Carreiras, & Böök 1986b;

Kitchin 1994; Kuipers 1982; Moore & Golledge 1976b; O'Keefe & Nadel 1978; Siegel & Cousins 1985; Siegel, Kirasic, & Kail 1978; Spencer, Blades, & Morsley 1989). Thus, although there is consensus that a cognitive map is a knowledge structure, it may be advantageous to reassert some limits on its connotation. A good place to start is O'Keefe and Nadel's (1978:86) definition as "a representation of a set of connected places which are systematically related to each other by a group of spatial transformation rules." It is frequently indicated that the set of environmental places that is represented in memory cannot be seen in its entirety from a single point of observation within the environment, and certainly the cognitive map construct is more interesting with this stipulation. The spatial transformation rules referred to in the definition permit generalizations and inferences beyond the specific spatial information gained through direct experience.

Cognitive maps are synonymous with survey (Shemyakin 1962), configurational (Siegel & White 1975), and vector knowledge (Byrne 1982). Each of these proposed types of knowledge involves multidimensional information about the spatial relationships among environmental features. Knowledge of this type can be distinguished from unidimensional information about the temporal spatial sequence of environmental features, referred to as route knowledge (Shemyakin 1962; Siegel & White 1975) or network knowledge (Byrne 1982). Some theorists have proposed that route knowledge is a primitive form of cognitive map that precedes configurational knowledge (Siegel & White 1975). Others have proposed that route knowledge and cognitive maps involve different types of learning mediated by different neural structures (O'Keefe & Nadel 1978). Clearly, travel through a large-scale environment is a sequential experience, and recognition of a landmark's temporal-order relation to other landmarks may precede knowledge of its Euclidean relation to those landmarks (Kuipers 1982). This seems particularly likely when temporal order is discovered through direct perceptual or perceptual-motor experience (Magliano, Cohen, Allen, & Rodrigue 1995). In such instances, Euclidean relations between places must be determined through vector addition based on path integration or through topological mapping of locations. But spatial relationships among places can also be learned through symbolic means such as cartographic maps and verbal descriptions.

Rescarch on spatial information acquired from cartographic maps has shown that experience of this type does result in an internal spatial representation that can support generalization and inferences, key criteria for the existence of cognitive maps (Presson, DeLange, & Hazelrigg

1989; Presson & Hazelrigg 1984; Sholl 1987; Thorndyke & Hayes-Roth 1982). However, knowledge gained from maps frequently differs from similar knowledge gained through locomotor experience. Map study often leads to orientation specific knowledge (Evans & Pezdek 1980; Sholl 1987; Thorndyke & Hayes-Roth 1982), whereas locomotor experience often leads to orientation-free knowledge. Also, distance and direction estimates made on the basis of actual locomotor experience are likely to be influenced by the amount and variety of information encountered by the traveler (Gärling, Säisä, Böök, & Lindberg 1986c; McNamara, Ratcliff, & McKoon 1984; Thorndyke, 1981).

Studies of spatial representations acquired from verbal descriptions have shown that, as in the case of map study, such descriptions led to the type of generalization and inference making that characterizes cognitive maps (Perrig & Kintch 1985; Taylor & Tversky 1992a, 1992b). The most noteworthy finding arising from this research was evidence that readers were able to acquire orientation-free environmental representations from two different types of descriptions, one of which described a traveler's-eye view of a sequence of locations (i.e., a route description) and the other of which provided a hierarchically organized overview (i.e., a survey description). The chief means of assessing spatial knowledge has been the verification of inferences and the accuracy of sketch maps. Direct comparisons of performance based on locomotor experience and performance based on verbal descriptions have been rare.

The point of a cognitive map is to represent a great deal of information in a flexible format with an economy of effort. This principle is reflected dramatically in computer simulations of cognitive maps. Although many examples are available (Gopal, Klatzky, & Smith 1989; Kuipers & Levitt 1988; Leiser & Zilbershatz 1989; Mataric 1990), one of the most interesting is Chown, Kaplan, and Kortenkamp's (1995) PLAN model. As with other models of this type, PLAN includes the basic elements of landmarks linked by routes and includes local areas embedded in regions that are, in turn, embedded in a global overview. Its innovativeness stems from the fact that it incorporates a scene-based scheme for establishing and maintaining orientation. In this sense, scenes correspond to what Gibson referred to as vistas, that is, the world as visible from a path of observation. Travel along such a path involves "the opening up of the vista ahead and the closing in of the vista behind" (Gibson 1979:198). Each new vista has implications for the traveler's orientation with respect to proximal and distal places. It is interesting to note that although cognitive maps—and indeed the entire "muddle of memory"—had no place in Gibson's theory, theories are obliged to establish a link to perception if they are to be regarded as plausible accounts of everyday experience. In this regard, the effort to integrate specific perceptual events into simulations of cognitive maps and cognitive mapping represents a qualitative step forward (see also Kuipers 1978, 1983; Kuipers & Levitt 1988).

INDIVIDUAL DIFFERENCES

For the most part, research on cognitive mapping has been aimed at understanding general processes rather than individual differences in those processes. Defining cognitive maps as sets of places and transformation rules for relating these places suggests two obvious sources of individual variation, one involving memory and the other rules. If a cognitive map is basically a registry of known places, then individuals differ greatly in the content of their cognitive maps and, according to studies of environmental learning, they also differ in the process of cognitive mapping. This observation raises an obvious but interesting question. How are cognitive maps and cognitive mapping different from other knowledge and other learning processes? There is an ample number of theorists who posit that spatial cognition is specialized, modularized, or otherwise distinguished (e.g., Bever 1992; Neisser 1985; O'Keefe & Nadel 1978). Yet, experimental studies suggest that learning environmental layout is influenced by many of the same factors that influence other types of declarative learning, such as learning facts or events. For example, perceptual speed, working memory capability, and strategy application are sources of individual differences in declarative learning (Kyllonen 1993), and as summarized previously, various studies indicate that these abilities also influence success in experimental tasks requiring perspective-taking and orientation performance.

Ultimately, the data rather than the process differentiate spatial learning. The data are the distance and direction information that reveal the geometry of the environment (Kuipers 1982). Individuals may differ in their comprehension of this geometry because of limitations in speed of processing and working memory capability, limitations that can be overcome to some extent by the use of strategies such as experience-based orienting schemas or experimentally induced frames of reference. The analogy to reading comprehension is one that more than a few researchers have pondered. Language is governed by syntax and semantics, and language is read (or heard) in a sequential stream. As a passage is read, important concepts and relations are processed through working memory so that the reader can interpret later developments in terms of previous ones. The goal is comprehension of the passage as a whole. Frequently, comprehension can lead to inferences based on the passage but not expressed directly. Environmental layout has structure and meaning, and during locomotion it is experienced in a sequential stream. During travel, important places and relations among places are processed through working memory so that the traveler can relate past locations to present location. The goal is comprehension of spatial relations among observed places, which frequently allows inferences about spatial relations not known through actual travel. This analogy invites further discussion of common interactive resources.

SPATIAL ABILITIES, SPATIAL KNOWLEDGE, AND WAYFINDING

Few studies have been concerned directly with relationships among spatial abilities, spatial knowledge, and wayfinding behavior. The typical finding is that spatial abilities as measured with psychometric tests have little in common with spatial performance on tasks in large-scale space. This outcome is illustrated by Lorenz and Neisser (1986), who administered a number of spatial ability tests along with a variety of wayfinding and orientation tasks to a large number of research participants. Their results indicated a single factor for spatial ability tests and three different facets of environmental knowledge, including landmark knowledge, route knowledge, and awareness of cardinal directions. Three conclusions seemed warranted. First, the types of abilities assessed by psychometric tests were not closely related to those involved in acquiring environmental knowledge. Second, the distinction between landmark and route knowledge, which is fundamental to many theoretical descriptions of environmental learning, was empirically validated. Third, information acquired about a specific environment was distinguished empirically from general spatial knowledge. Results from a recent study by Lovelace and Montello (1997) appear to substantiate a basic distinction between abilities as assessed by psychometric tests and abilities as reflected on tasks in geographic-scale environments.

There are rare exceptions that involve visual memory. Significant relations between scores on tests of visual memory and performance on various wayfinding or orienting tasks have been inferred or reported in Kirasic's (1991) field experiment involving young and elderly adults' spatial learning in a supermarket setting and Thorndyke and Hayes-Roth's (1982) examination comparing learning from maps and learning from environmental experience. The importance of visual memory received additional emphasis from Allen and Ondracek's (1995) study showing that age-related change in this ability was related to children's success in recognizing scenes from a pictorialized route. But studies showing correlations involving psychometrically identified spatial abilities other than visual memory are virtually nonexistent.

The weakness of direct links between spatial abilities and spatial learning, however, does not rule out the possibility of indirect or mediated relationships between the two. This possibility was the focal point of Allen, Kirasic, Dobson, Long, and Beck (1996), in which participants were administered a series of psychometric tests of spatial abilities, two experimental tasks in the laboratory, and a number of tasks involving the acquisition of spatial knowledge in an actual novel environment. Results showed that performance on one of the experimental tasks (maze learning) was a significant mediator of the relationship between psychometric spatial abilities, which were identified as a single factor, and environmental learning, which also was identified as a single factor. A second experimental task (perspective-taking) was a significant mediator of a similar relationship in one of two studies. Overall, these findings, portrayed in figure 2.1, suggest that basic

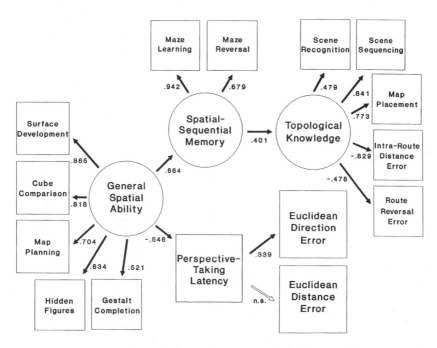

FIGURE 2.1. Path analysis showing two constructs (spatial-sequential memory and perspective-taking latency) mediating the relationship between spatial abilities as assessed by psychometric tests and environmental knowledge as assessed by a variety of tasks. (Reproduced from Allen et al. 1996, by permission of Ablex Publishing Co.)

spatial abilities are related to environmental cognition but that the relationship is an indirect route. In other words, basic spatial abilities are related
to information-processing skills, such as spatial-sequential memory or
perspective-taking, which are, in turn, related to the acquisition and use
of large-scale spatial knowledge.

The idea of mediated relationships appeared again in Kirasic's (1997)
study of relationships among basic spatial abilities, spatial learning in a supermarket environment, and behavioral efficiency in that setting. Her findings indicated that spatial learning mediated the relationship between
basic spatial abilities as assessed by psychometric tests and behavioral efficiency as reflected by a simulated shopping task. Despite the apparent similarity between these findings and those of Allen et al. (1996), important
differences remain to be explained. For example, in Kirasic's study, experimental measures of hypothesized information processing skills did not stand
alone as mediated constructs between basic abilities and environmental learning, but instead loaded on the spatial ability factor as if they were additional
scores from ability tests. Nevertheless, these studies reflect the type of path-
analytic approach that is extremely informative in investigating relationships involving basic abilities and more complex cognitive and behavioral
outcomes.

Although these studies showing mediated relations constitute an important contribution, they did not provide a strong conceptual or theoretical
basis for the links between fundamental spatial abilities and performance
on other categories of spatial tasks. A firm basis for hypothesizing such relations is possible, as shown in an elegant series of studies by Sholl (1989),
who examined the relation between the ability to judge horizontality and
the ability to update changes in location after passive transport. In addition
to substantiating a common basis for performance on the Piagetian water-
level task and the rod and frame test, she found that women who had difficulty judging horizontality in these tasks were very inaccurate on a path
integration task involving passive conveyance. Sholl (1989) interpreted these
results as suggesting a common mechanism involved in judgments of
horizontality and vestibularly guided path integration. In both instances,
poor performance could reflect either relative insensitivity to vestibular input
or visual modulation of vestibular information when visual and vestibular
input must both be considered.

Overall, studies of relations among spatial abilities, spatial knowledge,
and wayfinding help to underscore the previous speculative analysis regarding three different families of spatial abilities. Common denominators for
the family of abilities pertaining to object identification and the family of

abilities concerning orientation in large-scale environments are temporary visual-spatial memory, enduring memory for objects, and strategies for anticipating the effect of changes in viewing angle on objects' appearance. It is not surprising, then, that the only significant correlation typically found between psychometric spatial abilities and environmental learning involves visual-spatial memory. Similarly, it makes empirical and conceptual sense that the link between performance on abstract tests of spatial ability and learning the layout of a particular environment is mediated by skills involving the temporal-spatial representation of locations or perspective transformations.

COMMON INTERACTIVE RESOURCES: A CANDIDATE FRAMEWORK FOR FUTURE RESEARCH

A hybrid of psychometric and information processing research has recently brought about a new conceptual approach to the study of individual differences in cognition. A common resource model (Kyllonen 1993) posits that individual differences in cognitive performance across a wide range of tasks can be attributed to distinctive but potentially interactive resources applicable to three content areas. The resources are working memory capability, information processing speed, declarative knowledge, and procedural knowledge; the content areas include verbal, spatial, and quantitative domains. Adapted to address specific issues, this model could be used to explore the bases for individual differences in spatial cognition and behavior.

The preceding overview of spatial abilities suggests a potentially informative adaptation of an interactive sources framework that would include perceptual capabilities, fundamental information-processing capabilities, previously acquired knowledge, and motor capabilities. Perceptual capabilities refer to sensitivity to visual, auditory, vestibular, tactile, and proprioceptive sources of information about self-produced movement or environmental structure. Fundamental information-processing capabilities include speed, internal timing, and working memory, both spatial and verbal. Consistent with Baddeley's (1986) model, spatial working memory is characterized by a visual-spatial "sketch pad" for the temporary retention of images. Previously acquired knowledge can serve as a resource in several ways. Obviously, memory of specific environments (declarative knowledge) and memory for specific routes (procedural knowledge) can be considered resources. In addition, there are generalizable concepts (cardinal directions), strategies (orienting schemas), and skills (knowledge of cartographic conventions) that can be pressed into service during various wayfinding tasks.

It is useful to think of these resources as supporting a number of wayfinding means as adaptive processes. Together, basic resources and adaptive processes could represent the primary sources of individual variation in spatial cognition and behavior. The claim to be verified by future investigation is that these resources, individually, in combination, and in interaction with each other, go quite a distance in answering the question of why some people are better than others at wayfinding tasks.

There is some research to suggest that this approach has merit. Ondracek (1995) and Kirasic (1996) have examined the relationships between basic resources and the ability to acquire spatial knowledge maps from verbal descriptions. In both studies, working memory capability has been found to be a significant predictor of the ability to infer spatial relationships among described locations. The missing link in these cases is between this spatial knowledge and actual wayfinding performance. Equally important in its implications for the suggested framework are findings showing that individual differences in environmental knowledge and wayfinding ability as assessed by multidimensional scaling techniques are related directly to basic cognitive skills and extent of environmental experience (Golledge, Rayner, & Rivizzigno 1982; Golledge, Richardson, Rayner, & Parnicky 1983; Richardson 1982; Rivizzigno 1976).

INTERACTIVE RESOURCES AND WAYFINDING MEANS:
SOME SPECULATION
Figure 2.2 provides an overview of suggested relations between interactive resources and wayfinding means. Perception, knowledge, and information-processing capabilities are portrayed as fully interactive resources, as are perception and motor capabilities. An account of how basic resources and adaptive processes may be related to the various wayfinding means described previously reflects some of the potential utility of this approach to the study of individual differences. Oriented search, for example, would rely primarily on perception of the information source used to orient and either procedural knowledge involving stereotypic search or strategic knowledge to direct planful search. Following a continuous trail entails only perception of the trail and locomotion. Habitual locomotion relies exclusively on procedural knowledge and locomotion.

Piloting between landmarks, path integration, and navigation based on a cognitive map all involve reference to specific prior knowledge, either declarative or procedural, especially when the environment is familiar. Furthermore, path integration and navigation involve computations, thus implicating roles for working memory capability and speed of processing

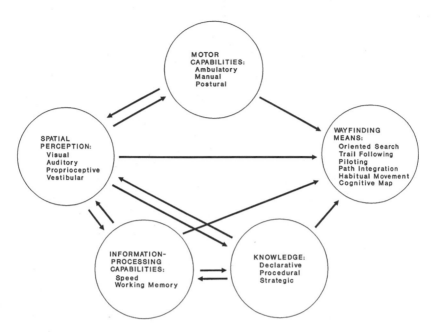

FIGURE 2.2. Speculative model of relations among interactive resources (spatial perception, previously acquired knowledge, information-processing capabilities, and motor capabilities) and wayfinding means.

in support of the adaptive process of inference making. When unfamiliar surroundings are involved, these same basic resources support declarative and procedural learning in the context of wayfinding efforts.

Wayfinding on the basis of information conveyed through verbal directions or maps places additional demands on cognitive resources. Verbal comprehension is supported by declarative knowledge of spatial terms and linguistic pragmatics, as well as by verbal working memory. Forming an image on the basis of verbal instructions may also involve spatial working memory. Certainly, map interpretation should be supported by declarative knowledge of map conventions and spatial working memory. In these enterprises, basic resources can support the adaptive process of declarative learning, which in this instance refers to a set of expectations related to environmental structure.

Considering individual differences in terms of fundamental resources and adaptive processes further provides a framework for examining the very diverse wayfinding challenges faced by a variety of populations. Wayfinding by visually impaired travelers, for example, might be characterized in terms of reliance on alternative perceptual resources, additional procedural knowledge (such as cane use), and specific declarative knowledge (such as the location of specific crosswalks and sources of information available aurally or written in Braille). Travelers with motor impairments are assisted by declarative knowledge of environmental features such as curb cuts, ramps, functional doors, and elevators, as well as procedural knowledge required for maneuvering in wheelchairs. Mentally retarded individuals face wayfinding with limitations on previously acquired knowledge, working memory capability, and speed of processing. Persons showing early signs of dementia may have difficulty with declarative learning or with accessing recently acquired declarative knowledge, although they show no difficulties retaining older declarative and procedural knowledge. Elderly travelers may have their wayfinding efforts affected by diminished working memory capability, slower rate of information processing, and locomotor limitations. These reduced resources would be manifested in slower declarative learning in new environments but unimpaired performance in familiar surroundings (see Kirasic 1991). It should be noted that the resource limitations mentioned in each of these examples have implications for the success of intervention efforts to improve wayfinding in these diverse groups.

SUMMARY

Little research on the relationship of basic spatial abilities, spatial knowledge, and wayfinding has been conducted to date, and the studies that have been done have not sampled spatial abilities or wayfinding tasks in an ordered or systematic way. Filling this research void represents an informative future research agenda. Such research will be most useful if it involves a comprehensive framework featuring a taxonomy of abilities and tasks. Wayfinding is a logical focus for research of this type: It is an everyday activity for a variety of species, thus affording the opportunity for a truly comparative psychology. It also includes a range of cognitive and behavioral phenomena diverse in their complexity, thus inviting the collaborative efforts of psychologists, geographers, neuroscientists, cognitive scientists, and linguists.

TOMMY GÄRLING

3

Human Information
Processing in
Sequential Spatial Choice

Spatial extent is a conspicuous aspect of all environments. Its important role in human adaptation processes is recognized by research in environmental psychology on environmental cognition (Evans & Gärling 1991; Gärling 1995a; Golledge 1987). This research is distinct from, although complementary to, psychological research on how people perceive the spatial layout of environments (Gibson 1979). Rather than focusing on the perception of the relative locations of events and objects in space, it attempts to understand how successive perceptual inputs are integrated into a cognitive representation or cognitive map of the environment. Also addressed is the related question of which properties such cognitive maps have.

When people locomote they use spatial information to maintain orientation, navigate, and find their way in familiar as well as in unfamiliar environments. Moreover, research has aimed at revealing means that are used by

The author's research reported in this chapter was financially supported by grants from the Swedish Council for Research in the Humanities and Social Sciences, the Swedish Council for Building Research, and the Swedish Transportation and Communications Research Board. The author acknowledges the important contributions made in particular by Anders Böök and Erik Lindberg, but also by several other collaborators and students, including Anita Gärling, Stephen Hirtle, and Jouko Säisä. Thanks are also due to Reginald Golledge and Herbert Pick, Jr., for valuable comments on previous versions of the chapter.

human beings, as well as by nonhuman species, for these activities (see other chapters in this volume).

Locomotion is usually part of executing a travel plan. As discussed by Gärling, Böök, and Lindberg (1984, 1985), spatial orientation, navigation, and wayfinding are, therefore, primarily means of monitoring this execution. A travel plan entails a spatial choice of one or several destinations of travel. The primary focus of this chapter is how such choices are made when there are alternative options. In general, information about the options are retrieved from a cognitive map of the particular environment. Planning travel also entails several other things, which are discussed in other chapters of this volume.

In the next section it is noted that spatial choices entailed by travel plans are multiattribute, sequential, and stated rather than actual. Implications for methodology are also stressed. Travel plans are distinguished from other plans in that the spatial dimension is fundamental. Spatial attributes influencing spatial choices entailed by travel plans are discussed later in this chapter. In this connection some research is presented on how people solve the traveling salesman problem of finding the shortest distance between destinations. Nonspatial attributes also influence spatial choices. The final section presents research on how the nonspatial attributes of time and priority are traded off against spatial attributes.

FEATURES OF SPATIAL CHOICE

MULTIATTRIBUTE SPATIAL CHOICE

It is common that choice refers to overt behavior, that is, actual execution of behavior. A spatial example is that an individual moves from one place to another optional place. It may be concluded that such a choice reveals the individual's preference for one *spatial* location over another. But an expressed spatial preference is also correlated with nonspatial attributes, implying that there are differences in attractiveness of locations that would be experienced the same irrespective of their physical distances. If limited to spatial factors, physical distance is only one of several attributes that may determine the choice. Travel time, ease of wayfinding, safety, or scenic beauty of routes are examples of other attributes.

In current research on judgment and decision making (for review, see Payne, Bettman, & Johnson 1992), it is common to conceive of choice alternatives as being multiattribute. According to one such conceptualization, termed *multiattribute utility theory* (e.g., Keeney & Raifa 1976), a decision maker faces a choice between options that are described on several attributes. A question addressed in research is how these attributes are combined in an overall evaluation. The answer is often confined to the determination of how much weight a decision maker places on the different attributes when choosing or judging an option. In empirical studies, observed judgments or choices are related to the attribute values by means of statistical regression analysis.

In applied research on spatial choice pursued in geography and related disciplines (for review, see Timmermans & Golledge 1990), several attributes are used frequently to describe optional locations, including physical distance or ease of access. Residential locations, recreational facilities, and shopping centers are examples of sets of options that have been investigated with the aim of finding out which role ease of access plays for spatial choices relative to nonspatial attributes. In this research, regression analysis has been employed for the mathematical-statistical modeling of the relationship between judgments or choices and attributes.

A more thorough understanding of multiattribute spatial choice requires a description of how information is processed when the choices are made. Shortcomings of regression analysis in yielding detailed descriptions were noted early on (Payne 1976; Svenson 1979), and new process-tracing methods were proposed. Today the new methods constitute valuable means for studying cognitive processes in problem solving (Ericsson & Simon 1993) and in judgment and decision making (Payne, Bettman, & Johnson 1993). As noted by Einhorn, Kleinmuntz, and Kleinmuntz (1979), these methods are complementary to regression analysis.

Questions concerning decision strategy are addressed by means of process-tracing methods. A decision strategy is defined as a sequence of mental operations that bring a decision maker from the initial state of indecisiveness concerning possible courses of action to a final state, in which one course of action is chosen. A mental operation refers to acquisition of pieces of information about an alternative (usually conceived of as attractiveness or utility of attribute values), comparison of utilities of attribute values, adding or calculating differences between utilities of attribute values, or simplifying operations like eliminating attributes. Decision strategies (Payne et al. 1993; Svenson 1979) include the additive utility rule or its derivatives, according to which choices are made of the alternative that has the largest sum of attractiveness or utilities of attribute values. This rule implies that one attribute value compensates for another. An example is that the attractiveness of an activity at a location compensates for a longer distance. There is also a class of decision strategies that do not require compensation. According to one such noncompensatory decision strategy, termed *lexicographic*, choices are made of the alternative with the most attractive value on the most important attribute that discriminates between the alternatives. In other words, all attributes except one are eliminated and a value on one attribute is, therefore, never traded off against a value on another attribute. In our example distance is perhaps never taken into account, only attractiveness of the activity at the location. Perhaps more realistic is that an attribute value is ignored if it is larger or smaller than some cutoff level. This so-called satisficing heuristic (Simon 1956, 1990) implies that a location would be chosen if it was attractive enough and not too distant. Again, no tradeoff is made between attribute values.

By means of computer simulations of sets of compensatory and noncompensatory decision strategies, Payne et al. (1993) found that a reciprocal relationship holds between accuracy and effort (assumed to be roughly equal to the number of required mental operations). Empirical research has then shown that human decision makers trade off accuracy against effort in choosing decision strategy (Ford, Schmitt, Schechtman, Hults, & Doherty 1989; Payne et al. 1993). For instance, for a required accuracy, the decision strategy that takes the least effort is chosen which achieves this accuracy.

SEQUENTIAL SPATIAL CHOICE

Current research on decision making predominantly investigates isolated choices. However, travel plans entail several related spatial choices which are made sequentially. A question which research therefore needs to address (see Gärling, Karlsson, Romanus, & Selart 1997a) is to what degree, how, and why such choices are dependent. An obvious form of dependence is that a preceding choice delimits or changes the alternatives of subsequent choices (upper diagram in figure 3.1). This occurs, for instance, when a location in a sequence of locations is chosen since the distance to other locations is thereby changed.

Another dependence arises if a decision maker evaluates the consequences of subsequent choices on the particular choice he or she is facing (lower diagram in figure 3.1). In this example the attractiveness of all possible combinations of the alternatives of each choice may be determined before the first choice is made. A spatial example is that the total distance of all possible sequences of locations are evaluated. Thus, as implied by the definition of planning provided by Hayes-Roth and Hayes-Roth (1979: 275–76) as "the predetermination of a sequence of actions leading to some goal," the choices are made *simultaneously* although the process of making them may extend in time.

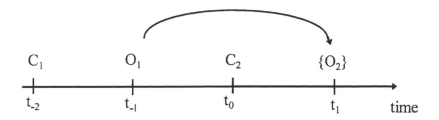

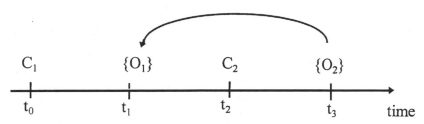

FIGURE 3.1. Influences of the outcome (O) of a current choice (C) on the set of outcomes ({O}) of a subsequent choice (*upper diagram*), and influences of the set of outcomes of a subsequent choice on the evaluations of the set of outcomes of a current choice (*lower diagram*).

Because of a limited capacity to process information, problems arise when the number of choices or the number of alternatives of each choice faced by a decision maker are large. In such situations a primary question is, which principle or principles are used by the decision maker in reducing the combinations of alternatives?

ACTUAL VERSUS STATED SPATIAL CHOICE

It may be argued that actual observed choices are more valid indicators of preferences than *stated* choices, that is, if subjects are asked to state which of several locations they would choose. A problem in studying actual choices that are sequential is that no information is obtained concerning when the decision was made and whether it consisted of several decisions. Thus, if detailed information is sought about decisions preceding sequential spatial choices entailed by planning, it is preferable to investigate stated choices. Although much research has suggested that stated choices are often externally valid (Levin, Louviere, & Schepanski 1983; Payne et al. 1992), such results may not necessarily generalize to sequential spatial choices.

Two studies of the external validity of stated sequential spatial choices were reported in Säisä and Gärling (1987) and Gärling and Gärling (1988). In the former study undergraduates serving as subjects in field tests first stated which choices they would make when located at a starting point; then they executed these choices. Choices were made between two locations in each of two choice sets. Thus, the sequence location 1 in choice set 1 followed by location 1 in choice set 2 may be chosen; or the same locations in the reverse order may be chosen; or any other combination of the two locations in each choice set in one of the two possible orders (1 and 2, 2 and 1, or 2 and 2) may be chosen. The results indicated that subjects chose the shortest route even though it required that they did not start with the closest location. Of more interest here, their choices were similar to another group of undergraduates who made the same choices in the laboratory rather than in the actual environment. Furthermore, in the field tests subjects seldom revised their choices when subsequently carrying them out during a walk. In the second study (Gärling & Gärling 1988) downtown shoppers were asked to indicate in interviews which locations they had visited and in which order they had visited them. The results were again found to be similar to those from laboratory studies of sequential spatial choices.

Although the results concerning external validity are compelling, they may of course not generalize to other conditions. Caution must, therefore, be exerted in drawing conclusions about decisions or plans subjects make in the laboratory if not committed to actually carrying them out.

SPATIAL ATTRIBUTES

DISTANCE

If distance is the basis for spatial choice, several questions need to be answered concerning how distance is cognitively represented. A number of studies have investigated distance cognition, that is, judgments of distances that cannot be perceived directly but must be retrieved from memory. Montello (1991b) recently reviewed this research focusing on methodological problems.

What relationship cognitive distance has to objective distance is one obviously important question that has been examined (Baird, Wagner, & Noma 1982). In general it has been concluded that the relationship is nonlinear. Another related question is whether cognitive distances are Euclidean, that is, if they have the property of remaining invariant over different changes. For instance, Burroughs and Sadalla (1979), Byrne (1979), and

Lee (1970) showed that distance judgments are asymmetrical, that is, that the distance from i to j (d_{ij}) is judged to be different from the distance from j to i (d_{ij}). In a similar vein, judgments of distance sometimes violate the triangle inequality $d_{ik} \leq d_{ij} + d_{jk}$ (Cadwallader 1979).

Another important question to ask is what type of distance is being judged. This question was raised by Gärling, Säisä, Böök, and Lindberg (1986c); Péruch, Giraudo, and Gärling (1989); and Säisä, Svensson-Gärling, Gärling, and Lindberg (1986). Specifically, Säisä et al. (1986) assumed a progression from one type of distance to another as people learn about an environment. In particular if information has been gained from a map (Gärling, Lindberg, & Mäntylä 1983; Levine, Jankovic, & Palij 1982; Thorndyke & Hayes-Roth 1982), a person in an unfamiliar environment may only have a vague feeling of where in space different objects are located. Thus, only approximate straight-line distances may be known. After repeated actual encounters with the environment, first paths and later a system of paths may be learned (Gärling, Böök, & Ergezen 1982). At this stage a person is much better at judging travel distances. Regularly occurring events such as traffic congestion may also be learned, leading to more accurate judgments of travel times. But travel distance is perhaps still the most important determinant of judgments of travel times.

Partial empirical evidence for these hypotheses were obtained in Säisä et al. (1986). Subjects judged straight-line distances that were proportional to actual distances, travel distances that were longer than judged straight-line distances, and travel times that were proportional to (although not highly correlated with) judged travel distances. Distances were judged between locations in downtown with which subjects should be familiar. But similar results were found for other parts of the city, which were assumed to be less familiar. Thus, little evidence was found for the hypothesis that judged travel distances and travel times are identical to judged straight-line distances in unfamiliar surroundings. In another study Péruch et al. (1989) found that taxi drivers' judgments of travel distances differed from lay people's, although they were shorter rather than longer. It was suggested that taxi drivers knew short-cuts lay people did not. Although not in the expected direction, the results demonstrated a relationship between the type of distance judged and the cognitive representation of the spatial layout.

What role do different types of distances play in sequential spatial choice? Gärling et al. (1986c) assumed that straight-line distance is first minimized. On this basis the order between locations is determined. Route choices that minimize travel distance are then made in the chosen order between locations. In experiment 1 of Gärling et al. (1986c) subjects in one condition were asked to first order locations on the basis of straight-line distances before reporting the shortest route between the locations in the indicated order. Small differences were found compared with a control group that did not receive these instructions but was simply asked to report the shortest route. Latencies did not contradict the hypothesis that order choices preceded route choices.

A related question is whether subjects minimize total distance or distance at each stage of a sequential choice process. As already noted, Säisä and Gärling (1987) found that subjects often did not choose the closest location among optional locations if this would lead to a longer total distance. In the traveling salesperson problem (Hoffman & Wolfe 1985), the order between a number of locations should be chosen so that total distance is minimized. Choosing the closest location each time (called the nearest-neighbor heuristic or NNH for short, see Barr & Feigenbaum 1981) will not necessarily lead to the shortest total distance. For several reasons human subjects may nevertheless do this. One reason is that they actually believe that the NNH minimizes total distance, and in fact it often either does or leads to a close approximation. Another reason is that the NNH imposes little demand on information processing. For instance, if there are n locations to be visited once, the number of possible orders increases exponentially with n. The number of locations need not be large until the task of calculating and comparing the total distances for all permutations becomes intractable. Naturally, the task is made even more difficult if all the information must be retrieved from long-term memory and the calculations and comparisons carried out in working memory.

In experiment 2 in Gärling et al. (1986c), subjects were presented with problems consisting of a varying number of locations and asked to choose the order that minimized (straight-line) distance. The results suggested that subjects used the NNH most but not all of the time. In addition, it was found that information processing was affected both by retrieval from long-term memory and load on working memory. Table 3.1 shows an interaction between instructions to minimize total distance or to use the NNH, memory condition (locations that were perceptually available, committed to working memory, or retrieved from long-term memory), and the number of locations. As memory load increases, a reduced instructions effect leads to less minimization of total distance. But as memory load and the number of locations increase, subjects also become less able to use the NNH due to difficulty in discriminating distances. An important difference between the conditions should be noted. In the long-term-memory condition, the locations were actual ones the subjects had encountered in the city where

TABLE 3.1	Percentage of choices of order conforming to the nearest-neighbor heuristics (NNH)							
	Set size for instructions to use NNH				Set size for instructions to minimize total distance			
Condition	3	4	5	6	3	4	5	6
Perceptual	94	88 (90[a])	85 (77)	70 (67)	94	83 (99)	72 (87)	56 (83)
Short-term memory	91	87 (86)	71 (64)	39 (44)	88	69 (81)	49 (56)	33 (32)
Long-term memory	97	62 (65)	47 (51)	33 (34)	92	65 (77)	39 (51)	24 (35)

Source: Data from Gärling et al. 1986.

[a]The percentages within parentheses indicate agreement with orders that minimized total distance. If the NNH had been applied consistently, these percentages would in the perceptual and short-term memory conditions equal 100, 83, 83, and 67 for set sizes 3, 4, 5, and 6, respectively, and in the long-term memory condition, they would equal 100, 67, 50, and 33.

they were living. In the other conditions the locations were dots on a vertical computer screen, either directly perceived or committed to short-term memory. Substantiated by latencies and the results of an additional experiment, perhaps the determination of the closest location was accomplished by spiral scanning of the screen on which the locations were displayed or recalled to be displayed (Wilton 1979), whereas long-term memory retrieval of actual locations entailed serial comparisons of pairs of locations. The important implication is that subjects in the latter condition were unable or less able to view the locations simultaneously.

It may be proposed that the NNH is used if locations are represented either in a maplike format, making simultaneous perception of and direct comprehension of spatial relationships possible (Downs 1981b; Kosslyn 1980), or in some other format that makes this possible. Performance may improve only in the former case, however, because it is conceivable that subjects discover spatial configurations of locations making it obvious to them that another order is shorter. An example is given in figure 3.2. Here the sequence ABC represents successive choices of the closest locations from the starting point. If the figure is available for inspection, it is apparent that BAC represents a shorter total distance than ABC because in the latter case AB is traversed twice. Thus, subjects may change their choices. That they did when a picture like the one in the figure was available as shown in Gärling (1989). In fact this order was chosen in 81 percent and 94 percent of all cases. On the other hand, if subjects only had access to numerical information,

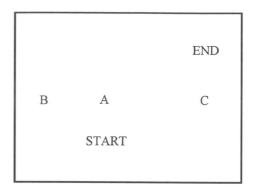

FIGURE 3.2. An example of co-
linear locations (A, B, and C) for
which the use of the NNH does
not lead to the shortest distance.

they appeared to use the NNH (so that only from 15 to 54 percent of all chosen orders minimized total distance) unless requested to draw a picture, in which case performance slightly improved (to 58 percent and 65 percent). Collinear locations are one of possibly several features that, if discovered, under certain conditions suggests to subjects that the NNH does not minimize total distance. Changes from the NNH thus appear to capitalize opportunistically on recognition of patterns that can be made to bear on the goal of minimizing total distance. It is also interesting to note that a pattern making such recognition possible may be constructed on the basis of fragmented information. Thus, even though long-term memory of actual locations in large-scale space may not provide the required simultaneous view, such a view is perhaps possible to construct for at least limited parts of the space.

SPATIAL CONFIGURATION

Not only distances but also how the spatial configuration or relative locations of places are represented in cognitive maps may affect spatial choices. A method for revealing the spatial configuration of locations represented in cognitive maps is multidimensional scaling (MDS) (Golledge 1976; for a more recent methodological review, see Gärling, Selart, & Böök 1997b). In studies employing this method (for examples, see, e.g., Baird, Merril, & Tannenbaum 1979; Gärling, Böök, Lindberg, & Arce 1991; Golledge, Briggs, & Demko 1969), ratings or rank orders of distances or directions between pairs of places (conceived of as points or vectors) are input to a MDS algorithm. A virtue with MDS is that it makes testable the assumption that the cognitive map has geometrical properties. Thus, the MDS algorithm assesses the goodness of fit of a system of points organized in k (usually 2) dimensions. This is in contrast to other methods, such as sketch-map drawing (Canter 1977; Lynch 1960) where Euclidean-geometrical two-dimensional space is a constraint.

With MDS it has been possible to examine in great detail distortion of cognitive maps, for example, size transformations, directional compression, and elongation (e.g., Golledge & Hubert 1982). Regionalization as a consequence of clustering around landmarks as anchor points has also been proposed (Couclelis, Golledge, Gale, & Tobler 1987). Additional distortions were demonstrated by Tversky (1981, 1992) and Tversky and Schiano (1989) who suggested that they reflect the operation of simplifying storage heuristics. Inaccurate perceptions or inferences are other possible causes.

It has also been suggested that a cognitive map is hierarchically organized, consisting of nested levels of spatial (and nonspatial) information (Hirtle & Jonides 1985; McNamara 1986; McNamara, Hardy, & Hirtle 1989; Sadalla, Burroughs, & Staplin 1980). Methods such as measuring priming reaction times and clustering in free recall have been used to infer the existence of a hierarchical structure. Furthermore, it has been demonstrated that a hierarchical structure may distort judgments of directions (Stevens & Coupe 1978) and distance (Hirtle & Jonides 1985).

A hierarchical organization may be conceived of as a simplifying storage heuristic (Tversky 1981, 1992). In a similar vein, it may be asked if a hierarchical structure plays a role in sequential spatial choices if the number of locations increases. For instance, if (spatial) clusters are formed, they may be treated as single locations in a stage of the spatial-choice process in which the order between clusters is determined. In a subsequent stage, choices may be made of the order between single locations within clusters. Evidence for such clustering effects in spatial choices were reported by Hayes-Roth and Hayes-Roth (1979).

In a test of the cluster hypothesis, Hirtle and Gärling (1992) presented subjects with sets of 6, 10, and 18 locations. The locations were presented like maps on pages in a booklet. For each set several maps were constructed by rotating and reflecting two key maps that differed with respect to how much the locations were spatially clustered. Subjects were asked to repeatedly report the order between the locations that minimized total distance. The results indicated that whereas the NNH was used when there were six locations, it was almost never used when the number of locations increased. If, however, spatial clusters were inferred by means of cluster analysis (Jain & Dubes 1988), the results were consistent with the interpretation that the NNH was first used to choose the order between clusters, then the order between single locations within clusters.

The results were less clear-cut for the locations that were not clustered. Here subjects sometimes appeared to make choices based on a zigzag pattern, passing back and forth across the entire space.

In the study by Hirtle and Gärling (1992), the locations were presented simultaneously in a map. If the locations instead had been retrieved from long-term memory of encounters of actual locations, it is likely that subjects would not had discovered any zigzag patterns. An intriguing question not yet addressed is whether an effect of spatial clustering on sequential spatial choices would prevail and whether there would be any relationship to spatial clustering in recall.

TRADEOFFS AGAINST NONSPATIAL ATTRIBUTES

If distance is a single attribute, the predetermination of a sequence of actions implies that choices are made of locations or the sequence of locations that minimize distance (assuming that distance is inversely related to utility). When other attributes are also relevant, according to multiattribute utility theory (Keeney & Raifa 1976), maximizing utility requires that these attributes are traded off against distance. Some nonspatial attributes apply to locations or the activities performed in these locations, others to routes between locations. The attribute levels may or may not depend on either time or order or both. In any case, if choice is multiattribute, such attributes will affect which locations are chosen and how the chosen locations are sequenced. Reported in this section are several series of experiments with the aim of investigating how subjects make order choices in the traveling salesperson problem when locations also differ in other respects.

TIME

Combining attributes requires tradeoffs. Such tradeoffs should, however, be less difficult between attributes like time and distance because they are closely related to each other. In fact, travel time is frequently proportional to travel distance. Also, as described previously, Säisä et al. (1986) observed that judged travel time was proportional to judged travel distance.

In real-life planning problems, there are also time constraints such as the requirement to be in time at a particular location (e.g., because a store closes). Actually, choosing a fast route in part derives its significance from this requirement. Although usually not framed as a constraint, it is probably also important for people to avoid wait time. Thus, being neither too late nor too early is perhaps an important goal.

In a series of laboratory experiments, Gärling (1994) asked subjects to imagine that they were in the pedestrian part of a fictitious central business district. They were asked to choose the order in which to visit a number of stores where they had planned to run different errands. In the simplest case, subjects were shown the schematic map illustrated in figure 3.3. Their task was to indicate if they wanted to visit store No. 1 before store No. 2 or the reverse. From the map it was obvious that choosing to walk to store No. 2 before store No. 1 was a longer distance. By manipulating opening hours, the results of one experiment showed, as expected, that trading off distance against time did not pose any problem to subjects. More surprising results were obtained when subjects knew that they would have to wait longer in store No. 1 than in store No. 2. Although there was no time pressure, the subjects' choices of walking the shortest distance (store No. 1 before store No. 2) were affected by the difference in wait time. To some extent subjects preferred to perform the errand in store No. 2 instead of waiting in store No. 1, even though it meant that they had to walk longer.

START

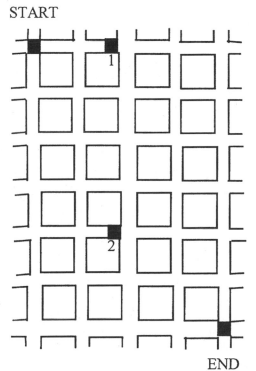

FIGURE 3.3. Schematic map indicating locations of stores (Nos. 1 and 2) in relation to start and end points.

END

TABLE 3.2 | **Percentage of minimization of walking distance**
           | **related to various factors**

	Probability of wait time and differences in walking distance								
	.10			.50			.90		
	200 m	400 m	600 m	200 m	400 m	600 m	200 m	400 m	600 m
Prioritized errand farthest	81	98	90	42	77	77	15	29	37
Prioritized errand closest	83	96	92	48	98	96	35	75	85

Source: Data from Gärling 1995b.

This result was replicated in Gärling (1995b), in one experiment (see table 3.2) where the probability of having to wait in the closest location rather than wait time was varied.

It should once again be noted that avoiding wait time in the closest location did not imply any time gain. Thus, the results did not simply reflect a tradeoff between travel distance and travel time. As also suggested by the results of other experiments, wait time was more aversive early than late in the choice sequence. This raises the question of what is negative about wait time. In many cases wait time increases the total time and may therefore jeopardize a plan. Even if this is not the case, waiting rather than performing a required action increases the likelihood of obstacles arising that will prevent performing the action. For this reason, waiting is also a threat to the execution of the plan, the more serious the earlier it occurs in plan execution. On this basis, it may be suggested that wait time is avoided in the beginning of an action sequence because it threatens completion of the execution. Waiting may clearly also be inconvenient for psychological reasons such as boredom, uncertainty, sense of time urgency, and impulsiveness. However, none of these reasons seems to account for the fact that wait time is more aversive in the beginning.

Another issue concerns the tradeoff between distance and wait time that subjects apparently make. Because walking longer requires more effort, it is plausible that there is a limit on how much effort subjects were prepared to invest in order to not jeopardize the action plan. A more important factor was probably that walking too long may have been seen as a threat against reaching the end location in time. Although this threat was hardly real because of the time parameters chosen in the experiment, subjects might have experienced it as such. It may be speculated that trading off distance against wait time also reflected a common objective of creating a feasible plan.

PRIORITY

Even if attractiveness of a location or an activity performed in this location is independent of when or in which order it is chosen, people may still have order preferences that are reflected in their choices. A case in point is that priority of an errand perhaps influences the choice of sequence. Whereas Hayes-Roth and Hayes-Roth (1979) found that high-priority errands were planned first, Gärling (1995b) suggested that such errands may also be *executed* first in a plan even though it leads to longer walking distances and wait times. As table 3.2 shows, subjects participating in experiment 2 in Gärling (1995b) more often chose to minimize distance when the prioritized errand (buying medicine) was in the closest location and the unprioritized errand (buying a newspaper) in the farthest location, than when the reverse was true. It may also be seen that subjects traded off probability of waiting, distance, and priority.

There is an obvious similarity between choosing high-priority before low-priority errands and minimizing wait time early or avoiding long distances. In the latter case the choices lead to an increased likelihood that the plan will be executed; in the former case the choices increase the likelihood that the most important parts of the plan are executed. Thus, feasibility of the plan is perhaps a ubiquitous common objective that underlies the observed tradeoffs.

It should be realized that trading off different attributes whose utilities may depend on the sequence of chosen alternatives is not easy. Even if it were possible, on the basis of each attribute, to choose one of $n!$ possible sequences, it is not clear how the attributes should be combined in the final choice of one sequence. The problem of choosing a sequence of actions is perhaps often solved if the actions are possible to rank order on some attribute. For instance, in choosing the sequence of locations that minimizes total distance, the locations may be rank-ordered with respect to distance to the starting point. Similarly, the locations may be rank-ordered with respect to wait time and with respect to priority of the errands to be performed in each.

If each location is first rank-ordered on each of three attributes, in a following stage a summed rank order may be determined for each location by adding these ranks. This problem is not different from that which a decision maker encounters when choosing a preferred action from several alternatives. A recurrent research finding is then that contingent on the conditions, different decision strategies are used to integrate attribute information (Payne 1982; Payne, Bettman, Coupey, & Johnson 1992; Payne et al. 1993; Svenson 1979). Compensatory decision strategies that take into account all information make possible tradeoffs between attributes. Such decision strategies are predominantly used when task demands are low (i.e., few alternatives or attributes), whereas noncompensatory decision strategies are used when they are high (Ford et al. 1989).

Another series of experiments reported in Gärling (1996) were aimed at investigating how tradeoffs are actually made between distance, wait time, and priority in sequential choices. The experiments used the information-search technique (Ford et al. 1989; Payne et al. 1993; Svenson 1979) in which subjects request every piece of information they need to make their choices. Which pieces of information, how many, and the sequence in which they are requested are registered. In Gärling (1996), the choices subjects made and when they made them were also registered. Figure 3.4 illustrates the information board that was presented to subjects on a computer screen.

The use of a compensatory decision strategy is indicated by a pattern of search that is predominantly characterized as intra-alternative (or inter-attribute), that is, subjects search all information available about one alternative (i.e., priority, wait time, and location) before they start to search information about another alternative. If the number of alternatives and attributes increases, the search tends to become intra-attribute (or inter-alternative). In this case information about alternatives on one attribute is searched before information is searched on another attribute. Patterns of intra-attribute search where some information is ignored suggest the use of a noncompensatory decision strategy.

ERRAND	PRIORITY	WAIT TIME	LOCATION	ORDER
Chemist's	Buy medicine	10 minutes	X, Y	1
News stand	Buy newspaper	20 minutes	X, Y	2

opened before

FIGURE 3.4. Illustration of the information board.

If subjects first rank-order locations on each attribute, one should perhaps expect intra-attribute search to dominate. The results very clearly did not support this expectation. As has been found when the number of alternatives and attributes are few, the search patterns were most frequently intra-alternative. In other words, subjects first searched all information for one alternative before they proceeded to the next. Both the fact that the information search was complete and the statistical modeling of the choices suggested that subjects traded off the different attributes against each other. Thus, a compensatory decision strategy appeared to be used.

The observed search patterns were inconsistent with the hypothesis that subjects first rank-ordered the alternatives on each of the attributes, then summed the ranks before making order choices. Of course, since the search was complete, subjects might have remembered a previously searched value on an attribute when searching on this attribute again. Speaking to this is that choices were not made before all information had been searched. A more straightforward interpretation is still that the attribute values were combined for each alternative before alternatives were compared. Unfortunately, it is not clear how subjects accomplished this. The procedure made it impossible for subjects to infer with certainty a value on an attribute from another value on the same attribute. Thus, rank-order information should not have been available until all values on an attribute were known. It is possible that the absolute distances from the starting point, the absolute wait times, and some assigned absolute priorities were instead summed for each alternative, employing the same set of weights in each case.

A final question concerns the tradeoffs that subjects make. Conceivably, depending on the particular planning task, there are many possible attributes that may be used to infer a preferred sequence of actions. The preferred order on a given attribute has presumably been learned. Sometimes, however, the preferred order is not obvious. This was demonstrated in a study by Loewenstein and Prelec (1993) where subjects preferred to perform a less attractive before a more attractive action. At a more fundamental level, however, scripts or causal models of events may be employed to infer logistic constraints. The ability to make such inferences of *enabling* relations between actions is apparently developed in preschool children (Hudson, Shapiro, & Sosa 1994). Feasibility of the plan is possibly a criterion with a similar status. Further research should focus on such fundamental attributes attended to when the logistics of actions is at issue. Because of the importance of these attributes, tradeoffs or compromises are perhaps not likely to be frequently observed.

CONCLUSIONS

This chapter has reviewed the results of a series of experiments with bearings on several important issues concerning how human beings process information when they make sequential spatial choice. The picture emerging from this review is somewhat fragmented. This suggests that further theoretical and empirical work is needed.

One theoretical issue concerns the relationship between forming travel plans, orientation, navigation, and wayfinding in an environment. This issue was treated in some detail in Gärling et al. (1984, 1985). A promising line of theoretical work building on this treatment is computer simulation of navigation (e.g., Chown, Kaplan, & Kortenkamp 1995; Gopal, Klatzky, & Smith 1989). Such computer simulations explicitly model not only the decision-making process but also the different knowledge representations it draws on and how they are acquired. In other words, computer simulation holds the promise in this area to provide an integrative framework as it has done in other areas (e.g., Newell & Simon 1972). But with the exception of Hayes-Roth and Hayes-Roth (1979), spatial choice has so far not been explicitly modeled in computer simulations.

Several suggestions were made for further empirical research. A most important one is raised by the issue of what attributes make possible sequencing of actions, in particular fundamental spatiotemporal attributes. Furthermore, there is a need for additional investigation in regard to when, how, and why tradeoffs are made between such attributes. More study is also needed of the role of recognition of spatial patterns in decisions about the shortest route in the traveling salesperson problem.

ELIAHU STERN
JUVAL PORTUGALI

4

Environmental Cognition and Decision Making in Urban Navigation

Urban navigation is concerned with wayfinding in citywide transportation networks (e.g., Bovy and Stern 1990). It involves search and decision making and its two essential components are environmental cognition and route choice. The navigator is continuously busy in a sequential process of decision making whose essence is to match internal with external information as it comes. This chapter presents two complementary conceptual frameworks that explain the dynamics and high variability of persons navigating in the urban environment. Each framework deals with a different component. Environmental cognition is interpreted through the concept of an inter-representational network (IRN) as developed by Portugali (1996) and Haken and Portugali (1996); the choice mechanism is conceptualized through decision field theory (DFT) as developed by Busemeyer and Townsend (1993) and elaborated by Stern (1998).

According to the notion of the IRN, the relevant cognitive network to interpret the navigation process is composed of elements in the mind, representing external environments, and elements in the environment, representing the mind. Urban navigation can thus be seen as a synergistic interplay between external and internal inputs and outputs, ordered by one or a few order parameters, which evolve in the process and enslave the interacting representational subsystems.

DFT integrates information coming from the external environment and from the individual's associative memory as determinants of possible actions to be evaluated in a deliberation process. The theory is primarily aimed at understanding the motivational and cognitive mechanisms that guide this deliberation process, especially when decisions under uncertainty are involved.

Both theoretical frameworks are based on specific spatial knowledge, which includes information and experience as inputs in the generation of a cognitive base on which navigational decisions are made. This chapter first defines these components of spatial knowledge and then presents their use within the IRN and DFT conceptual frameworks. Later, the application of each framework is examined through conceptual simulations based on similar scenarios, and the possibility of their integration is discussed. An integrated framework should provide more details to explain the decision processes involved in urban navigation.

URBAN NAVIGATION AND SPATIAL KNOWLEDGE

Urban navigation is a sequential process of decision making concerning route choice, whose essence is to match internal with external information as it becomes available, while moving between a predefined origin and destination. Decision-making and choice behavior in urban navigation is affected by four components:

1. The trip purpose that actually determines the frequency of the individual's navigation. Commuting to work or daily travel to school, for example, would be the most frequently practiced navigation, while visiting a city in a foreign country for the first time would be the least practiced form of navigation.
2. The navigator, including his or her personality, sociodemographics, and especially spatial knowledge and experience. These personal characteristics would govern, in various ways, the retrieval of information, the spatial ability of the navigator, and the speed of data processing.
3. The means of navigation, which entail various constraints on the physical options of choice. Owing to legal or physical restrictions, truck drivers, for example, would have fewer routes available in the city than drivers of small vehicles, but more routes than cyclists.
4. The specific situation in which navigation is practiced, which determines the choice set and the choice setting. The situation may refer, for example, to location (i.e., choice setting) in space and time, thus

determining the number of available and feasible alternate routes (i.e., the choice set).

The urban navigator is reacting to dynamic situations and, depending on the purpose of navigation, decisions are made under various levels of time restrictions. The navigator makes decisions in some cases under time pressure (usually while commuting under congestion) and in others under stress (usually developed as a result of the accumulation of wrong decisions). When the commuter has to deviate from a habitual route due to non-recurrent events, or the tourist has to choose for the first time a route to a given destination, both are also faced with some degree of uncertainty. When decisions are made under uncertainty or time pressure, the probabilities of success can be inferred mainly from knowledge and accumulated experience (e.g., Stern 1998).

It has been documented that the travel-related decision-making process is strongly based on the individual's level of spatial knowledge (e.g., Bovy & Stern 1990), which, in turn, partially depends on familiarity with a given environment. Familiarity is highly related to experience (Chalmers & Knight 1985; Evans, Marrero, & Butler 1981; Stern 1983), although they are not mutually exclusive (Hanson 1977). A person can drive every day in a given environment and still be familiar with only certain components of it. In general, however, a positive relationship between familiarity and navigation frequency can be expected (see also Hanson 1984).

Familiarity includes two components: specific experience of given locality; and global experience of city structures, hierarchy of roads, traffic and directing signs, and the like. These types of direct experiences are important but are only part of the individual's total spatial knowledge (e.g., Stern & Leiser 1988). The other part is information indirectly acquired from maps, media, friends, and so on.

Because decision making is strongly based on the individual's level of spatial knowledge, a change in the relative use of the knowledge components (i.e., the amount and type of information retrieved from one's associative memory) can be expected as a function of navigation frequency. The daily commuter, for example, will mainly retrieve specific experience from associative memory and occasionally will complement it either with global experience or current information (figure 4.1). On the other hand, a visitor to a strange city would base navigational decisions mainly on information, complemented by global experience (Stern & Azrieli 1995). As the frequency of navigation increases, the relative use of information and global experience decreases and the relative use of specific experience increases.

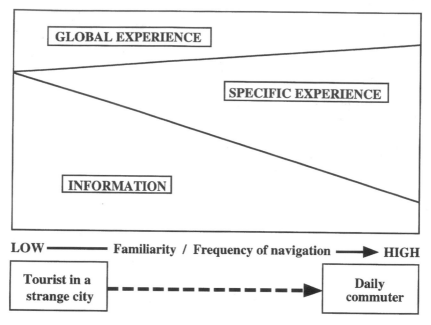

FIGURE 4.1. Components of spatial knowledge and frequency of urban navigation.

The integration of direct information—from the external environment—and indirect information—from the individual's associative memory or experience—into a navigational decision-making process is examined by two conceptual frameworks, each of which employs the same two external scenarios. The following section defines these scenarios.

NAVIGATION SCENARIOS

The most frequently practiced type of urban navigation is the daily commute to work or school. Drivers are likely to take the same route unless they are informed about existing or expected nonrecurrent delays. If this information were available prior to the trip, the driver would have several options of response, including change of departure time, change of travel mode, change of route, and even a change of the entire preplanned daily activity schedule in both time and location. If the information were, however, received enroute, the only possible response would be to deviate and change the route. When this response is chosen, the driver starts a sequential process of decision making while navigating to the desired destination. The driver is navigating in a relatively familiar environment and decisions are made at each intersection that are considered to be part of the individual's recognized set of possible alternative routes. This type of choice behavior is also known as the myopic approach (Sheffi, Mahmassani, & Powell 1982). We shall consider this commuter's choice situation as the first external (most frequent) navigation scenario.

The least frequently practiced type of urban navigation is the occasional first-time, self-driving tourists in a strange city. In such cases information is accumulated prior to the trip, and the route is often preplanned (Bovy & Stern 1990). Nevertheless, deviations from the preplanned route are very common owing to either nonrecurrent obstacles or a gap between the cognitive network and the objective situation. Such a gap encourages the driver to look for strengthening clues, usually by matching elements in the mind with elements in the environment (e.g., landmarks). The ability to "confirm" the route raises the driver's self-confidence, and consequently fewer navigational decisions are made. Otherwise, the driver deviates from the preplanned route and navigational decisions must be made more often. The choice situations of the self-driving tourist are considered as the second external (least frequent) navigation scenario.

The two navigation scenarios just discussed can, in fact, be combined into a single integrated scenario: a stranger is coming to a new city (second scenario), learns some routes, drives them, and after some time becomes an ordinary daily commuter in the city (first scenario). In the discussion that follows we shall use the two separate scenarios to elaborate the DFT approach and the integrated scenario to elaborate the notion of an IRN.

INTER-REPRESENTATIONAL NETWORKS

The notion of IRNs suggests that the cognitive system in general, and the one associated with cognitive maps in particular, extends beyond the individual's mind-brain into the external environment (Portugali 1996). The idea is that the cognitive system is composed of internal elements representing the external environment and external elements representing the mind-brain. External elements refer to the human ability to represent ideas, emotions, thoughts, and so on externally, by means of mimetic, linguistic, and artifact-making capabilities. Although this idea has never dominated the mainstream of cognitive sciences, it was nevertheless always present (implicitly or explicitly), for example in the studies of Vygotsky (1978), Gibson (1979), Rumelhart, Smolensky, McClelland, and Hinton (1986), Lakoff (1987), Donald (1991), and Edelman (1992). Some experimental examples of the operation of IRN come from the serial reproduction scenarios devised by Bartlett (1961) in his book *Remembering*. The general structure of the Bartlett scenarios serve here as a paradigm case to convey the notion of IRN.

A typical Bartlett scenario starts when a test person is shown a figure or a text, asked to memorize it, and then asked to reproduce it externally from memory, that is, to produce an external representation of it. This external representation is given to another test person who is asked to repeat the operation, and so on. It is typical of such scenarios that at the beginning there are strong fluctuations, or major differences, from reproduction to reproduction. Then, at the later stages, the reproduction of the text or the figure is stabilized and does not change much from reproduction to reproduction. The interpretation offered by Portugali (1996) is that the Bartlett scenarios provide an example, first of a cognitive network composed of internal representations (the memorized figures or text) and external representations (the figurative or textual reproduction) and second of a cognitive process that develops as a sequential between the internal and external representations of the network. Third, this process of sequential internal-external interplay develops as a typical process of self-organization: it starts with strong fluctuations and then settles into a steady state. According to the synergetic interpretation of self-organization, the strong fluctuations at the start indicate a competition between emerging configurations of texts or figures, which then give rise to a certain order parameter that enslaves the external elements and representations of the system and brings the system to a steady state. This is explained in some detail in the next section.

SYNERGETICS INTER-REPRESENTATIONAL NETWORKS

Synergetics is a theory developed by Haken (1983) to deal with open, complex, self-organizing systems. The theory was applied to several research domains, including cognition, with the implied suggestion that the brain-mind is a self-organizing system. The typical case here is pattern recognition. It has been demonstrated that a partial section of a pattern offered to a brain or a computer triggers a competition among several stored configurations of patterns. This competition is eventually solved when, by means of associative memory, one or a few configurations become order parameters that enslave the system so that a recognition is established.

A similar process takes place in the construction of cognitive maps (Portugali 1990; Portugali & Haken 1992). Because of its size, a person can never directly see the whole pattern (i.e., a city) that must be cognitively mapped. Consequently, as in Haken's experiment on pattern recognition, the cognitive system must construct a whole pattern or map out of a partial set of features available to it. This is achieved when a certain mapping principle or mapping order parameter enslaves the various features and brings the map into a steady state. Compared with ordinary pattern recognition, cognitive map formation concerns large patterns (e.g., cities). This quantitative difference has several qualitative implications, among them the central role assigned to external representations and storages in the process of cognitive map formation. The notion of IRN aims to capture this property.

The reformulation of IRN within the framework of synergetics (Haken & Portugali 1996) starts with the realization of Haken's (1991) synergetics as a three-layer network (figure 4.2). In the context of the present study, this network can symbolize a decision maker, where the input layer represents input information, the inner layer of order parameters represents alternative decisions, and the output layer represents action or behavior along a route, for example. If one looks at the network of figure 4.2 from the side (as indicated by the arrow) one arrives at figure 4.3. Figure 4.4, which is a graphic representation of a synergetics inter-representational network (SIRN), is an elaboration of figure 4.3. Here the decision maker is subject to two kinds of inputs, internal (a cognitive map) and external (the environmental information as it comes in), and is producing two kinds of outputs, again internal and external. The external output is action and behavior in the environment, whereas the internal output is feedback information to the person's cognitive map about the route and its properties. The middle node symbolizes the several order parameters that have been established in the process, each of which refers to a certain decision and its resultant action and behavior. It is important to note that the same order parameters may govern quite different external outputs (such as a drawing, as in the Bartlett scenarios) or some other action (such as movements, navigation, and so on) that may lead to an external storage in, say, maps, geographic information systems (GISs), and the like.

In their 1996 paper, Haken and Portugali have demonstrated how the model of figure 4.4 can be applied to various aspects of cognitive mapping, including navigation, environmental learning, the formation of collective cognitive maps, and their connection to city dynamics. All these applications were based on three prototype submodels of serial reproduction derived from the general model of figure 4.4. They are intrapersonal, interpersonal, and interpersonal with a common reservoir. The model relevant to the present discussion is the intrapersonal prototype.

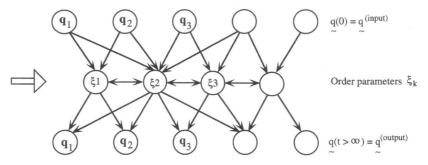

FIGURE 4.2. A three-level network of the synergetic computer. The first (upper) layer consists of model neurons that receive the input. This first layer projects on the second layer, which represents the order parameters. The third layer represents the output from the order parameter layer. Though formally similar to a neural computer arrangement, the algorithm of the synergetic computer is quite different; for example, the model neurons are interacting by means of soft nonlinearities. Note that learned patterns are encoded in the connections between the first and second layers and those between the second and third layers. In the case of static patterns, the connections between the order parameters are of the same universal form, whereas in the case of dynamical patterns the order parameter connections may depend on the movement patterns to be generated. These remarks hold for all of the following IRN-related figures.

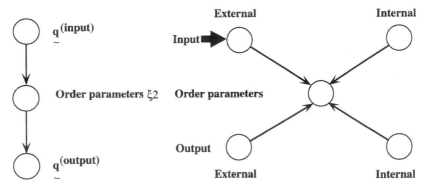

FIGURE 4.3. Figure 4.2 seen from the side, as indicated by the arrow.

FIGURE 4.4. The simplest cast of an IRN model with its external in- and outputs and its internal in- and outputs. The middle area represents the order parameters. Note that in analogy to figure 4.3, a network corresponding to figure 4.2 is seen from the side, so that each circle represents a whole set of model neurons.

ENVIRONMENTAL LEARNING, COGNITIVE MAPS, AND NAVIGATION

The question of environmental and route learning—that is, the way in which people learn a route and in the process construct a cognitive map of it—has been discussed in several research domains, including the development of spatial abilities, way finding, navigation, and the like (see McDonald & Pellegrino 1993 for a discussion and bibliography). A basic assumption in many of these studies is that the environment is out there, and that the task of people is to learn the environment by encoding and internally representing what is already out there. The notion of IRN and its formulation in terms of synergetics in the model discussed previously imply a different view. Environmental and route learning is seen as a process by which the individual constructs patterns in the external environment and corresponding patterns in the mind. The result is an IRN cognitive map, part of which are external representations in the environment, and part of which are internal representations in the mind-brain. The process as a whole evolves as a synergistic interplay between external and internal inputs and outputs, ordered by the order parameters, which evolve in the process and enslave the interacting representational subsystems.

Consider the above-noted integrated scenario of an individual who comes to a new city. Figure 4.5 illustrates graphically how the process develops in terms of an intrapersonal SIRN model. The individual starts with some previous, internally represented knowledge about cities in general and about that city in particular (given that previous learning about it was derived from maps, stories, tourist guides, and the like). Starting from the left side of figure 4.5, it can be seen that at the beginning the individual's cognitive system is subject to two flows of incoming information: a flow of external input that comes in as the individual advances in space, and a parallel flow of internal input stemming from some initial previous environmental knowledge stored in the individual's memory. The interaction between the external and internal flows of input, ordered and enslaved by the order parameters that emerge in the process, also entails an interactive interplay between internal and external flows of output. In this sequential interplay between external and internal representations, objects and patterns in the external environment are being determined, marked, and internally represented as the IRN of the emerging cognitive map. The output of the first excursion along the route is the first cognitive map of the area, and it provides part of the input for the second excursion, and so on iteratively. As can be seen in figure 4.5, this map is composed of an external output $q(e)$

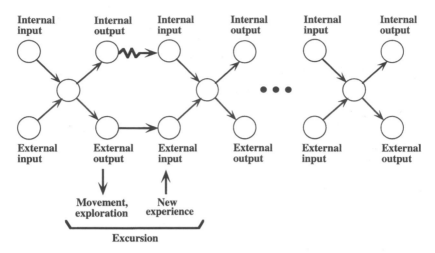

FIGURE 4.5. Graphic representation of a person exploring a route. Each individual diagram represents the network within the person and illustrates the different stages according to which the states of the network change.

in the form of external landmarks, and an internal output $q(i)$ in the form of internally remembered and represented landmarks. The important feature of this process is that the cognitive system is not just photographing what is out there in the environment, but it is actively constructing the external and internal networks that make the cognitive map.

It is important to note again that at the start of the process the initial internal input need not be related directly to the new environment being learned. It is enough if the individual stores in memory some general environmental concepts and categories, such as "houses," "streets," "pavements," "traffic lights," and "forest." Furthermore, nonenvironmental internal information can also play an important role in the process. For example, an Israeli making a first-learning excursion in a European city is likely to be attracted by a Hebrew sign, which will then become an externally and internally represented landmark of that individual's cognitive map.

Each trip along the route is constrained by the time and energy available to the individual. The outcome of the first trip is thus a partial and incomplete cognitive map that then becomes the starting point, or input, for the next excursion, and so on in a sequential process. Each new trip adds more details to the previously constructed cognitive map. As conceptualized by our SIRN model, in the first excursions there are likely to be marked differences among the cognitive maps reproduced from iteration to iteration. Such fluctuations are interpreted as a competition between the order parameters of the emerging cognitive map. From a certain point onward, an order parameter of the area is established, enslaves the various externally and internally represented patterns, and brings the cognitive map to a steady state. Once this state is reached, the marginal informational contribution of every new trip diminishes, and the system (i.e., the cognitive map) enters a steady state. This stable state is reached not necessarily when the cognitive map corresponds to the accurate cartographic map, but instead when it enables the person to survive and function in that specific environment. Thus, as was found in various studies, different people (children, taxidrivers, pilots, and so on) tend to construct their own personal and specific cognitive maps.

The previous description might lead to the impression that cognitive maps constructed as above are highly personal and subjective. This is indeed so, but only up to a point. As noted in previous studies and the preceding discussion (Portugali 1990, 1996; Portugali & Haken 1992), many of the concepts, categories, schemata, and patterns that one uses in constructing personal cognitive maps come already self-organized and enslaved by a complexity of intersubjective collective order parameters.

DECISION FIELD THEORY

CONCEPTUAL FRAMEWORK

The decision field theory (DFT), developed by Busemeyer and Townsend (1993), is a synthesis of two prior and independent psychological theories: approach-avoidance theories of motivation and information-processing theories of choice response time. The theory is aimed at understanding the motivational and cognitive mechanisms that guide a deliberation process involved in making decisions under uncertainty. It integrates information coming from the external environment and information coming from the individual's associative memory as determinants of possible actions to be evaluated in a deliberation process. Deliberation is a time-consuming cognitive process that involves information seeking about the consequences (i.e., events, E_i) of recognizable actions A_i, weighting of the payoffs (Y_i) of the events, and comparing the weighted consequences to establish a preference (choice) state. The deliberation process is characterized by vacillation until a choice is made and an action is taken (Svenson 1992).

The basic ideas of the theory are summarized in a sequence of seven incremental stages that were mathematically derived from the synthesis of the two previously mentioned theories. The mathematical formulation will not be presented here. Rather, only the conceptual structure and the underlying assumptions of the decision field theory are discussed.

Beginning with the stages of deterministic and random subjective utility theories (e.g., Busemeyer 1985; Edwards 1962; Savage 1954), in each subsequent stage a new and more general theory is formed by incorporating a new processing assumption into the previous theory. Each new processing assumption is needed to represent a fundamental property of the deliberation process involved in decision making. The last stage completes the entire

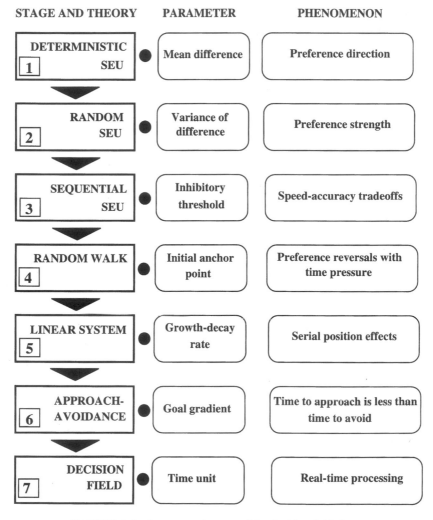

STAGE AND THEORY   PARAMETER   PHENOMENON

DETERMINISTIC SEU [1]	Mean difference	Preference direction
RANDOM SEU [2]	Variance of difference	Preference strength
SEQUENTIAL SEU [3]	Inhibitory threshold	Speed-accuracy tradeoffs
RANDOM WALK [4]	Initial anchor point	Preference reversals with time pressure
LINEAR SYSTEM [5]	Growth-decay rate	Serial position effects
APPROACH-AVOIDANCE [6]	Goal gradient	Time to approach is less than time to avoid
DECISION FIELD [7]	Time unit	Real-time processing

FIGURE 4.6. Incremental stages of the decision field theory.

description and introduces the processing time of deliberation. Lower-stage theories are thus special cases of higher-stage theories in which the decision field stage is the highest. This chapter is not intended to provide a comprehensive description of the decision field theory, so only the sequence of the theory's steps is presented (figure 4.6). The last stage is described in more detail.

Decision field theory is thus an attempt to formalize the deliberation process. The theory introduces a time element, representing the amount of time that it takes to retrieve and process one pair of anticipated consequences of an action before shifting attention to another pair of consequences (of another action). The anticipated consequences are retrieved from associative memory. The total decision time, namely deliberation time, equals the total time required to retrieve, compare, and integrate all the comparisons of the anticipated consequences. The calculations involved in such a procedure are assumed to be realized by an underlying neural system. Decision field theory is an abstract representation of its essential dynamic properties.

The integration of all the stages summarized in figure 4.6 into one conceptual scheme is presented in figure 4.7. Potential gains and losses are given by corresponding payoffs in the approach-avoidance subsystems. These payoffs are inputs to the valence system, connected to the current considered actions by attention weights W. The weights change over time as a result

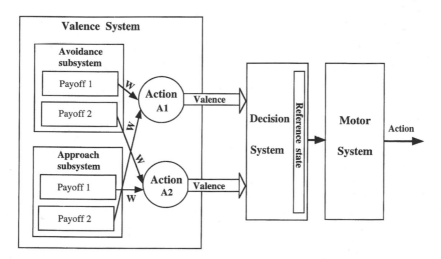

FIGURE 4.7. A schematic outline of decision field theory (based on Busemeyer & Townsend 1993).

of shifts in the decision maker's attention to the various consequences during deliberation. The valence of an action at any moment represents the anticipated value of an action at that moment, and it is produced by the weighted sum of the input values connected to that action. The valences produced within this system are fed into the decision system that compares the valences and integrates the comparisons over time to produce a momentary preference state. Finally, the preference state drives a motor system that inhibits responding until a threshold is exceeded and an action is taken.

This theory is examined as a conceptual framework for modeling choice behavior in the context of urban navigation, commonly characterized by decision making under changing degrees of uncertainty and time pressure or accumulating stress.

All the travel-related choice models of the black-box type lack a choice process mechanism and use all the variables expected by the researchers to affect the output without determining their effects on the choice process itself. Unlike those models, figure 4.8 presents an expanded framework of a choice mechanism based on the seven stages of the decision field theory with proposed exogenous variables. These variables are assumed to affect the choice process itself and, therefore, not all black-box variables are shown at this stage. For example, sociodemographics are excluded because it is assumed, at this stage, that personality attributes directly affect the cognitive mechanism involved in choice situations facing the urban navigator. But these are not substitutes for sociodemographics.

The theoretical framework presents a system in which exogenous variables affect either the deliberation process or the threshold. In psychological theory, the threshold is expressed by an inhibitory criterion presenting a given value of a preference state. In a travel-related situation the threshold could be defined, or expressed, by actual or expected delay time. According to decision field theory, the deliberation is a sequential process of calculating and comparing valences of N subjectively identified actions. The time spent on the calculation of a given action (hA in figure 4.8) depends on the attention given to the events of this action. The attention given to the consequences of the events of alternative action B is not necessarily equal to that of action A, and therefore hA is not necessarily equal to hB. The result of one deliberation cycle is a momentary preference state that is chosen only when it reaches a threshold. Until then, vacillation is observed, thus activating another deliberation cycle. The smaller the distance between the preference state and the threshold, the smaller the number of deliberation cycles. Without discussing the theoretically anticipated effects of the exogenous variables on either the deliberation or the

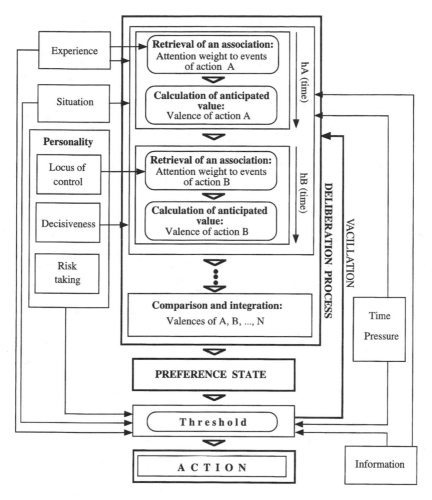

FIGURE 4.8. The decision-based framework of choice behavior.

threshold, the following section uses the decision framework to describe conceptually the route choice process involved with each of the urban navigation scenarios described earlier.

CONCEPTUAL SCENARIO-RELATED CHOICE PROCESS
Basic Conceptual Components
As mentioned earlier, decision field theory combines the approach-avoidance theory of motivation and the information-processing theory of choice response time. Approach-avoidance theory reflects the relationship between the distance from one's goal and the weight given to the payoff

in the deliberation process. This relationship changes depending on whether the consequence of an action is a reward (*approach*) or a punishment (*avoidance*). The basic idea is that the attractiveness of a reward or the aversiveness of a punishment is a decreasing function of the distance from the point of commitment to an action. Thus, the consequences of an action become more salient as the preference state for that action approaches its threshold. As the possibility of performing an action increases, the attention (e.g., the weight) assigned to its consequences increases. The theory goes further in proposing that avoidance-avoidance (av-av) decisions produce longer mean deliberation times than do approach-approach (ap-ap) decisions. When av-av decisions are made, the preference state vacillates and slows down the decision process. When ap-ap decisions are taken, the preference state is racing toward the threshold and thus shortens the decision time.

The information-processing theory considers knowledge, familiarity, experience, and information as resources from which the individual retrieves and implies consequences of subjectively identified possible actions. In the theoretical framework (figure 4.8), there is a distinction between experience and information. It is only experience from which the consequences of actions are weighted. According to the random walk subjective expected utility theory, the initial preference state starts at some anchor point, biased by prior knowledge and experience. More experience may thus shorten the deliberation time. Information only defines the possible actions and sometimes adds new objective input to the subjective calculation of the valences, although both affect the threshold through expectations. Expected delay time, for example, is affected by experience and real-time information. For instance, availability of real-time traffic information on congestion levels of alternate routes would lower the threshold and trigger actions. These two basic components of the decision field theory exert the major effects in the following navigation choice scenarios.

Choice Process of the Daily Commuter
Figure 4.9 presents six alternative routes of travel between origin O and destination D, one of which is the daily habitual route. These routes are generated by four decision points and two joining points. Commuters leaving origin O receive enroute information about anticipated delays owing to nonrecurrent congestion. While approaching decision point (junction) 1, both real-time information and specific experience (familiarity) are used to define the possible actions (the choice set) at point 1. Retrieval of both long-term and specific experiences is then used to evaluate weights of each

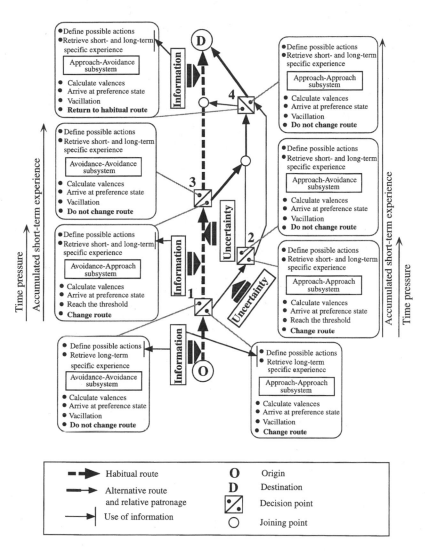

FIGURE 4.9. A conceptual route choice process of a daily commuter according to DFT.

of the events of the subjectively recognizable actions. Based on the exogenous variables presented in figure 4.8, some of the commuters will adopt an av-av approach and some will adopt ap-ap behavior (figure 4.9). Consequently, the ap-ap commuters will reach a threshold and change their route while the av-av commuters will vacillate and continue to drive along their habitual route.

Beyond decision point 1 each of the commuters starts to accumulate short-term experience. Provided information is relevant only for commuters on the habitual route, those heading to decision point 2 are faced with a higher level of uncertainty. Possible actions are then defined on the basis of long-term specific experience and weights are inferred at point 2 from both long- and short-term specific experience. The choice at point 2, to continue along the habitual way, is made under risk, so at point 3 the possible actions are still defined on the basis of real-time information (see also Maule & Svenson 1993). All commuters heading to points 2 and 3 are also under growing time pressure. According to the DFT, time pressure increases the selectivity of retrieved information and salient attributes are given higher weights. Consequently, vacillation decreases and actions are taken more often. We should thus expect that relatively more commuters will change their route at point 3 than at point 1 (see arrows in figure 4.9). No such distinction can be made at point 2 because those commuters have already adopted an ap-ap behavior. It should also be mentioned that analog empirical evidence has already been found to support partially the relatively higher deviation at point 3. Commuters driving from point 2 to point 3 are assumed to face an increasing level of congestion. In a longitudinal study (1989–1992) conducted in the Tel Aviv metropolitan area, Stern (1996) found a 20 percent increase in the number of drivers deviating enroute, in contrast to a 28 percent increase in the total number of vehicles, with almost no parallel growth in the number of roads.

At decision point 4 there is a possibility of returning to the original, habitual route. A study by Khattak, Schofer, and Koppelman (1992) showed that commuters who made longer trips were significantly more likely to return to habitual routes after diversion. Moreover, commuters using radio information and those with higher stated preference for diversion were more likely, but not significantly so, to return to their habitual routes. Commuters oriented to "adventure and discovery" were less likely than others to return to their original routes. Based on the DFT and these empirical findings, those who will return to the habitual route at point 4 (figure 4.9) will most likely base their decision on specific information and long-term knowledge of the network system. Commuters who deviated at point 1, and moreover commuters who diverted at point 2, are more likely to be risk prone and consequently less likely to return at point 4 to their habitual routes. Being risk prone, those commuters will also be more likely to adopt an ap-ap strategy.

The network patronage that results from the assumed choice process, represented by the relative width of the arrows in figure 4.9, closely resembles the actual traffic load pattern around central metropolitan cities: the closer to the metropolitan core, the faster congestion is diffused to all possible routes leading to that core. In other words, as the core is approached (e.g., as a general destination) fewer free alternatives are available (e.g., Bovy & Stern 1990).

Choice Process of the First-Time Visitor
The same routes used to simulate the route choice process of the daily commuter are used for the first-time visitor (figure 4.10). There are, however, several differences between the two scenarios. The visitor, coming to the city for the first time, accumulates information prior to arrival and plans the particular trip between O and D in advance (preplanned route in figure 4.10). Similar to the daily commuter, after leaving O the visitor also receives en-route information about anticipated delays owing to non-recurrent congestion (situational information in figure 4.10). Unlike the commuter, upon approaching decision point 1, the visitor has not only to define possible actions in response to congestion (or any other unexpected obstacle) but also to confirm that the right route is being followed. Route confirmation is based on previously acquired route information and the very short accumulated specific experience. Possible actions are based on situational information but are also affected by route confirmation. Action weights are then based mainly on global experience.

DISCUSSION

Decision field theory and inter-representational network theory, along with their respective scenario-related conceptual simulations, seem to emerge from different scientific streams. In decision field theory the emphasis is on an information-processing approach. It thus follows the mainstream of cognitive science by regarding cognition as essentially any internal process and by seeing what happens in the external world as raw data that must be analyzed by the brain as part of the decision-making process. It is not a rule-based model, but rather an explanatory conception for the variety of human preferences. Navigation in a dynamic urban environment is, accordingly, a sequence of decisions made and executed.

The approach of inter-representational network theory is somewhat different. It suggests that the external world, like the internal one, is full of representations, that is to say, meaningful entities that transmit information (and not just raw data). It further suggests that the internal and external representations, and thus the individual's cognitive maps, are constructed

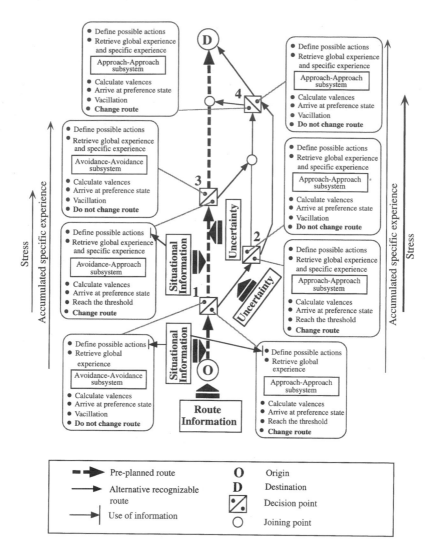

FIGURE 4.10. A conceptual route choice process of a first-time visitor according to DFT.

when the individual interacts with the environment. For example, when learning and navigating a route, a person landmarks objects in the environment and in the process transforms them into external representations. Simultaneously, the person remembers them (marks them in the brain) and transforms them into internal representations. The resulting cognitive map of the route is thus constructed from this inter-representational interaction.

As can be seen, the main emphasis of DFT is on the process of decision making, whereas the emphasis of IRN is on the construction of cognitive maps. From this perspective the two theories complement each other. IRN suggests how cognitive maps, which are the basis of decision making, are created, and DFT explains the process of the decision making itself.

Both theories have a quantitative base that was not discussed in this chapter. To enable empirical testing and predictions, there is a need to integrate the theories and to structure the combined conceptual framework mathematically. IRN can thus be a submodel in the DFT deliberation module. We believe that such an integrated framework, which provides a testable theoretical explanation for the navigational decision-making process, will be a further step in revealing the contents of the black box.

II

Perceptual and Cognitive Processing of Environmental Information

In this part, three chapters explore cognitive processes and human navigation in a variety of contexts, including an extensive investigation of path integration by humans covering wayfinding without vision, updating an object's orientation and location during nonvisual navigation, exploring the geometrical constraints and calibration of action-representation couplings, and relating perceptual processes to various navigation requirements. Focusing primarily on aspects of human perception and cognition with respect to wayfinding, these authors explore non-traditional domains to show the versatility of relevant theories and concepts. Although vision is accepted to be the spatial sense par excellence, there is no doubt that blind or vision-impaired humans can become competent independent

travelers using simple cognitive processes and aids such as the white cane, guide dog, or a variety of recently developed auditory navigational aids.

In the first of these chapters (chapter 5) Loomis, Klatzky, Golledge, and Philbeck discuss the process of human navigation by path integration, a process that until recently was recognized more in the nonhuman domain. They begin by clearly defining two types of processes influential in wayfinding—piloting and path integration. The recent literature is replete with misconceptions of the nature of these processes, but little is left in doubt following their clear and comprehensive discussion.

Navigation by humans, animals, and machines is accomplished using two distinct methods. Piloting is the determination of current position and orientation using landmark information in conjunction with a map, either external or internal. Path integration is the updating of position and orientation on the basis of velocity and acceleration information about self-movement. The chapter begins with a consideration of a number of models of path integration. Following is a review of the empirical research on human path integration, with a focus on controlled experimental investigations. Such investigations have been carried out using two distinct tasks: return-to-origin after the passively guided traverse of an outbound path, and perceptually directed action, whereby the person sees or hears a target and then, with the target extinguished, attempts to indicate its position by actively locomoting toward it or by pointing in its direction during locomotion that passes by the target.

These authors explore a number of experiments in which people without vision or with restricted vision perform a variety of wayfinding and shortcutting tasks. The ability to take a shortcut is generally believed to be an indication of the existence of survey, configurational, or layout understanding of the environment. Loomis and colleagues suggest that dead reckoning or path integration might be a natural way for humans to go directly to an origin after constrained or random walking, without necessarily learning landmark or feature locations either along their course or in the surrounding area. This approach obviates the need for remembering the outward path and the cumulative errors associated with the search memorization.

In the next chapter (chapter 6) Amorim discusses a neurocognitive approach to human navigation. He suggests that human navigation performance is viewed as a result of the interplay of neurocognitive functions. Spatial updating and frames of reference constitute the two concepts of maximum interest in this work. He provides experimental evidence on the role of reference frames in computing locations in space, as well as on the effect of two processing modes for the updating of an object's location and orientation. Amorim uses an information-processing approach (commonly used in cognitive psychology) in an effort to understand human processes of updating an object's location and orienting it with respect to a bounding frame of reference.

To localize a person in the environment as well as localize an object the environment contains, Amorim suggests that the acquisition, coding, and integration of sensory information (both perceptual and representational) are necessary. Building on the model of visuo-spatial cognition proposed by Kosslyn (1991), Amorim offers two studies; one investigates the role of reference frames in computing locations in space, whereas the other compares two processing modes for the updating of an object's location and orientation in space. In interpreting the results of these experiments he evaluates the neurocognitive approach to the study of the pathological causes of topographical disorientation. His system is based on the viewer's perspective of the world on the one hand and in the object-centered frame on the other. In both cases, the objects or places are represented with respect to their intrinsic axes (gravity and prominent landmarks are used in the case of places). Thus, he suggests that localizing and reaching are operations that require a viewer-centered processing mode, whereas identifying, grasping, and manipulating require an object-centered processing mode. Amorim pursues the idea of *spatiotopic mapping*, tying it to either viewer-centered or object-centered reference frames. As with the Loomis et al. chapter, mentally updating the location and orientation of previously seen people or objects while walking in complete darkness represents the experimental mode.

He concludes that distortions and fluctuations in topographic memory constitute one major reason for disorientation when navigating. But disorientation might also be linked to perception while encoding environmental features during one's travel. Considering problems of topographical disorders subsequent to brain lesions, Amorim suggests using the term *topographical amnesia* to represent disorders arising from either an inability to access the memory for places or a loss of the memory content. In concluding, he suggests that neurological patients may display syndromes suggestive of a disassociation between two spatial information-processing modes—object-landmark versus walked trajectories. For example, persons suffering from constructive topographic agnoisia orient themselves well with respect to landmarks but are unable to combine the different structures located in an external space in the right order. Such persons would have a deficit in implementing a trajectory-centered mode of processing. They would, however, benefit from using an object-centered mode of processing. In conclusion, Amorim's chapter invites researchers to consider human navigation performance with respect to neurocognitive functions, using approaches similar to that suggested by Kosslyn and his colleagues. By doing so, and by further investigating the effect of brain dysfunctions related to navigation, Amorim suggests that we may find the path to solving the problem of how human wayfinding takes place and thus improve our understanding of the spatial behavior of humans.

In chapter 7, Rieser examines action-representation couplings, focusing in particular on the geometrical constraints on such calibrations. He argues that perception and action are coupled, so that motoric actions result in dependable changes in the actor's perspective. For example, during locomotion the structure of an actor's perspective visibly rotates and translates in directions and at rates that fit with the geometry and rate of locomotion. This coupling provides a chance for perceptual-motor learning. While walking with vision, people learn the covariation of optical (and possibly nonoptical) flow and afferent-efferent input associated with the biomechanical activities of walking. This learning, in turn, provides the basis for the coupling of representation and action. Representation is coupled with action in working memory in analogous ways. When acting without vision, people are knowledgeable about the resulting changes in their perspective. So for example, after viewing their surroundings and then walking without vision, people are able to keep up to date on the changing self-to-object distances and directions relative to their remembered surroundings.

Chapter 7 focuses on two issues concerning the coupling of representation and action. One is the degree to which representation-action couplings show the same constraints as perception-action couplings. Perception-action couplings are hypothesized to provide the basis for the representation-action coupling. An implication is that they should show many of the same constraints. The other issue is the calibration of locomotion and its organization. Given that people can view objects in their surroundings and then close their eyes and walk to remembered object locations, how is it that they know how far to walk and to turn to reach the remembered targets? Rieser and his colleagues hypothesized that the calibration of walking without vision to remembered targets is based on visual-motor learning that occurs while walking with vision. Experimental studies, conducted with simple rotations as well as simple translations, are consistent with this hypothesis. Continuing studies are focused on the generality and organization of these calibrations. The evidence indicates that they are functionally organized.

Many investigators of spatial cognition implicitly accept the naïve view that perception produces a replica of real-world objects and environments. In this view,

cognitive processes, such as memory, are not generally considered germane to perceptual processes. A common assumption is that memory access occurs only after visual perception has produced a three-dimensional representation of the external world (except perhaps in the face of ambiguity or other breakdowns, in which case cognitive processes are thought to supplement perceptual processes).

This part thus examines the cognitive and perceptual processes involved in human navigation. Not surprisingly, the principal authors are all psychologists, and they have a profound interest in the role of perceptual processing of data while moving via optical flow, the efferent feedback obtained from moving with guides or independently, the linkage of perception-action pairs, the role of perspective in human-object relations, and the comprehension of surrounding frames of reference. The link between environmental perception and spatial cognition—as experienced by moving through different environmental settings—is pursued at length. And the possible use of extant skills such as path integration and homing—actions that are frequently reported in animal studies, but rarely reported in studies of human movement—is vigorously explored. So, do humans, like other animals, still use path integration to solve some types of wayfinding problems? Or must we search elsewhere for explanations of the acquisition of spatial knowledge and its use for travel? Your reading of this section will either provide answers or open a Pandora's box of questions.

JACK M. LOOMIS
ROBERTA L. KLATZKY
REGINALD G. GOLLEDGE
JOHN W. PHILBECK

5

Human Navigation
by Path Integration

Navigation involves the planning of travel through the environment, updating position and orientation during travel, and, in the event of becoming lost, reorienting and reestablishing travel toward the destination. This chapter is concerned with human navigation performance using a method of navigation known as path integration. We focus on the functional properties of path integration as manifest in navigation performance.

NAVIGATIONAL TERMS

We begin by defining a few important terms in connection with figure 5.1. The instantaneous velocity of a traveler, depicted by the velocity vector in the figure, has two components: the direction of motion over the ground, referred to as *course*, and the velocity magnitude, referred to as *speed*. *Course* is defined with respect to some reference direction, such as magnetic north or true north. The facing direction of the traveler's body or vehicle

The authors are grateful to John Rieser and Tommy Gärling for their helpful comments on an earlier draft of the chapter.

is referred to as *heading,* also defined with respect to the reference direction. For aircraft, watercraft, flying organisms, and swimming organisms, course and heading frequently differ because of crosswinds and currents; for terrestrial organisms and vehicles, course and heading are often decoupled by skidding, sidestepping, and backstepping.

Bearing refers to the direction from one point to another, measured with respect to the reference direction; in figure 5.1, the bearing of a landmark from the traveler's position is shown. If one knows the bearing of an unseen desired destination, one steers a course equal to its bearing. *Bearing difference* refers to the difference between two bearings from a common origin; thus, for two visible landmarks, the bearing difference can be measured without knowledge of the reference direction. *Heading-relative bearing* refers to the direction from a traveler to some other location, measured with respect to the traveler's heading (Beall & Loomis 1996). Thus, the heading-relative bearing of a visible landmark seen from an aircraft is its direction relative to the longitudinal axis of the aircraft. *Course-relative* bearing refers to the direction from a traveler to some other location, measured with respect to the traveler's course (Beall & Loomis 1996). Thus, the traveler proceeding to a visible landmark would need only to null the course-relative bearing (i.e., align his or her motion with the landmark). Finally, *relative course* is the inverse of course-relative bearing and refers to the direction of motion with respect to identifiable locations in the environment; as such, its determination does not require access to

FIGURE 5.1. Depiction of navigational terms. *Course* is the direction of the traveler's velocity vector, measured with respect to a reference direction, such as north. *Heading* is the traveler's facing direction with respect to the reference direction. *Bearing* refers to the direction from the traveler to an object, such as a visible landmark. *Course-relative bearing* is the bearing of an object with respect to the traveler's course. *Heading-relative bearing* is the bearing of an object with respect to the traveler's heading. (Adapted from figure 1 of Beall & Loomis 1996, by permission of Pion Ltd., London.)

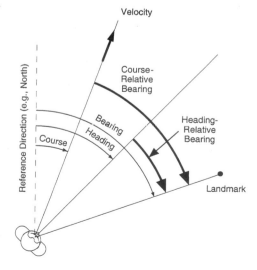

a fixed reference direction. In recent years, there has been a great deal of research on the visual perception of relative course based on optic flow (e.g., Crowell & Banks 1993; Cutting, Springer, Braren, & Johnson 1992; Royden in press; Warren & Hannon 1990).

METHODS OF UPDATING POSITION AND HEADING

The updating of heading and course can be accomplished in either of two ways—direct sensing and integrating turn rate. If a reference direction or azimuthal reference is afforded by an extended terrestrial feature (e.g., a mountain range) or some celestial feature (e.g., the sun), heading can be determined on the basis of sensing that feature (Mittelstaedt 1985). Heading might also be directly sensed from the earth's magnetic field, but the evidence that humans have magnetoreception is equivocal (see the collection of articles in the volume edited by Kirschvink, Jones, & McFadden 1985). Like heading, course is determined by sensing relative course with respect to the environment and then relating this direction to the azimuthal reference. If an azimuthal reference is not available, the traveler needs to integrate turn rate over time to determine the turn since the last known heading. Present heading is then the sum of initial heading and turn. Consequently, present course is given by present heading plus the angular difference between course and heading.

Methods of updating position can be classified on the basis of kinematic order: position, velocity, and acceleration (see also Loomis in press; Loomis et al. 1993). Position-based navigation (called *pilotage* in aviation and *piloting* in boating and in the animal literature) relies on external signals indicating the traveler's position; such signals would include those from visible, audible, and odorous landmarks. Piloting to locations that are beyond the current sensory field requires that the traveler have an external or internal map of the environment to be traversed. Methods of self-localization ("position fixing") include being coincident with a single landmark, computing position using distances to multiple landmarks (trilateration), computing position using bearings or bearing differences to multiple landmarks (triangulation), and computing position using bearing and distance to a single landmark; Gallistel (1990) discusses several other position fixing techniques.

Velocity-based navigation (referred to as *dead reckoning* or *path integration*) relies on external or internal signals indicating the traveler's course and speed. Determining displacement from the origin is done by integrating velocity, and, as such, requires no map. The measurement of velocity and turn rate can be based on either external (allothetic) signals or internal (idiothetic) signals (Mittelstaedt 1985). Allothetic information for locomoting organisms includes optic flow and acoustic flow; idiothetic information includes a copy of the efferent commands issued to the musculature and afferent proprioception from the muscles and joints. For many species, idiothetic information about turn rate is provided by the semicircular canals.

Acceleration-based navigation (referred to as *inertial navigation* or *path integration*) involves the sensing of linear and rotary accelerations and doubly integrating these values to obtain translational and rotational displacements relative to the starting position and orientation (Barlow 1964; Potegal 1982). When conducted without an azimuthal reference, inertial navigation has the virtue of not requiring allothetic information; as such, it is independent of weather and other environmental conditions. Inertial information is available to humans in the form of vestibular and somatosensory signals (e.g., mechanoreceptor responses to pressure against the skin).

Electronic navigation systems used for aircraft, ship, automobile, and now pedestrian travel use a combination of position-updating methods. Radio navigation, including the use of global positioning systems (GPS), is the dominant method, but when reception of transmitter signals fails, dead reckoning and inertial navigation serve as important backups (e.g., Loomis, Golledge, & Klatzky, 1998; Loomis, Golledge, Klatzky, Speigle, & Tietz 1994). The latter methods are not sufficient by themselves, for they accumulate error that can be corrected only by position fixing.

Like electronic navigation systems, human and nonhuman species make use of a combination of these methods (Able 1980; Baker 1981; Barlow 1964; Gallistel 1990), subject to similar limitations. Fully understanding navigation by humans and other species will come about only by recognizing these constituent methods and the functional properties they confer on navigation performance. This is the premise underlying much of the current research on nonhuman species.

PATH INTEGRATION:
DEFINITION AND FUNCTIONS

Path integration is the inclusive term referring to the updating of position on the basis of velocity and acceleration information (Etienne 1992; Mittelstaedt & Mittelstaedt 1982; Wehner & Wehner 1986). Within the classificatory scheme presented path integration is navigation based on means other than position fixing. In the spirit of the latter meaning, we wish to broaden the meaning of path integration to encompass navigation that is not based strictly on the sensing of velocity and acceleration.

Consider the situation in which a traveler is walking through an unfamiliar town of narrow streets and the traveler's view is limited to the buildings on the street along which he or she is walking. If the traveled path traverses a complex route without repeating route segments, the traveler can know his or her location relative to the origin only by way of path integration. Suppose that, instead of walking the route, the traveler is shown a succession of photographic slides taken at "viewpoints" along the route separated by 5-m intervals. Although the person no longer receives velocity and acceleration information, he or she might be able to determine the displacements and heading changes between successive viewpoints by computations on the perspective views within each pair. On the scale of locations within a street segment (i.e., those that can be seen from one slide to the next), we would say that the person is using piloting to determine changes in location and heading. On the scale of the entire town, however, the person uses these local displacements and heading changes to determine current position and orientation relative to the travel origin. Thus, we maintain that the larger-scale process of updating position is one of path integration even though velocity and acceleration signals are not explicitly involved. To generalize, path integration is the process of navigation by which the traveler's local translations and rotations, whether continuous or discrete, are integrated to provide a current estimate of position and orientation within a larger spatial framework.

In his treatment of navigation concepts and animal navigation, Gallistel (1990) noted several important functions of path integration. First, path integration allows one to venture into unfamiliar territory for the purposes of seeking a destination. Second, as one explores an unfamiliar region of space, it is path integration that provides the traveler with an ongoing estimate of current position and, on this basis, allows the traveler to gradually integrate the isolated perspective views encountered into an internal representation (cognitive map) suitable for subsequent piloting.

PATH INTEGRATION
IN NONHUMAN SPECIES

Far more research on path integration has been done on nonhuman species than on humans. Most of this research has examined return-to-origin behavior observed in connection with natural foraging or in experimental tasks devised by researchers. With respect to the former, the animal travels outbound from the nest, actively searching for food; upon finding it, the animal is often observed to travel along a direct path back to the nest. With respect to the latter, the animal is induced to travel outbound along the path of the experimenter's choosing; then, at some drop-off point along the path, the animal attempts to return on its own to the origin. Some of the species in which path integration has been implicated using return-to-origin behavior are the desert ant (Müller & Wehner 1988; Ronacher & Wehner 1995; Wehner & Wehner 1986, 1990), the honeybee (Esch & Burns 1996; Srinivasan, Zhang, Lehrer, & Collett 1996), the funnel-web spider (Mittelstaedt 1985), the golden hamster (Etienne, Maurer, & Saucy 1988; Séguinot, Maurer, & Etienne 1993), the gerbil (Mittelstaedt & Glasauer 1991; Mittelstaedt & Mittelstaedt 1982), and several species of geese (Mittelstaedt & Mittelstaedt 1982; von Saint Paul 1982). The key to implicating path integration is the exclusion of piloting as a possible basis for the animal's behavior. This is typically accomplished either by eliminating any positional cues for piloting or by some manipulation that makes the return response by piloting discrepant from the return response by path integration and then showing that the animal makes the return predicted by path integration. In the following, we briefly discuss three examples of return-to-origin behavior to illustrate the diversity of conditions under which path integration operates.

One of the most impressive examples of animal path integration is the foraging behavior of the desert ant (genus *Cataglyphus*) as studied by Wehner and his colleagues (Müller & Wehner 1988, 1994; Ronacher & Wehner 1995; Wehner & Wehner 1986, 1990). Under the unremitting sun of the Tunisian desert, the ant forages for dead insects tens of meters from its nest in traverses extending hundreds of meters. In the absence of landmarks suitable for piloting over such a large foraging range, evolution has provided the desert ant with an exquisite path integration mechanism. Wehner and his colleagues have provided definitive proof that the ant does use path integration, for after the ant has found the carcass of a dead insect, the ant is passively transported to a grid marked on the desert, whereupon it traverses a path to a location that would be coincident with the nest had it not been displaced. The ant indicates its arrival at the estimated nest location by beginning a systematic search pattern; further research has shown that the ant continues doing path integration throughout this search (Müller & Wehner 1994). That the ant can perform reliable path integration over thousands of strides indicates that the path integration process is virtually noise-free. Several points should be noted about this demonstration of path integration: (1) the ant actively determines both its foraging for food and its search for the nest; (2) the ant can determine its course and heading using the sun as an azimuthal reference; and (3) the ant can, in principle, sense its speed on the basis of efference copy of the muscle commands, afferent proprioception, and optic flow. Ronacher and Wehner (1995) have shown that optic flow does indeed play a role.

A second demonstration of return-to-origin behavior is described in a study of goslings of various species (Mittelstaedt & Mittelstaedt 1982; von Saint Paul 1982). At the outset, the goslings were imprinted on a human foster mother. They were then transported in a cart with open sides through a wooded and hilly region to a release point tens of meters from the origin of travel. When the foster mother disappeared from sight following the release, the goslings proceeded to walk quite directly to the origin. Under a critical manipulation, the cart was covered during portions of the outbound path to eliminate visual cues of travel. The goslings ignored these portions of occluded travel as if they were stationary. Their returns to the origin were those predicted by path integration during the periods of unoccluded transport, even if the return was in the direction opposite to the true origin (thus ruling out piloting). This study indicates that the goslings were doing path integration solely on the basis of optic information during the unoccluded portions of passive transport; that is, neither vestibular input nor proprioception was involved.

Quite the opposite basis for path integration is demonstrated in our third example: homing to the nest by the golden hamster (Etienne 1992; Etienne et al. 1988; Séguinot et al. 1993). In these experiments, the hamster is led by the experimenter using bait away from the nest, which is located within a large open arena. Upon receiving food, the hamster heads directly back to the nest, even when vision is excluded. This research shows that inertial cues and proprioception are sufficient for successful homing.

Another behavior implicating path integration is perceptually directed behavior—the animal senses something of interest (e.g., prey) and then attempts to locomote to it even though perceptual information about its location ceases to be available during the traverse. In an investigation of detour behavior of the toad, Collett (1982) conducted an experiment in which toads were allowed to view prey through a fence. When the toads attempted to detour around the fence, they encountered another barrier that occluded further view of the prey; even so, the toads continued orienting roughly in the direction of the unseen prey, indicating both path integration and updating of the internal representation of the prey. Similarly, Hill (1979) demonstrated path integration and spatial updating in a species of jumping spider. This spider moves along branches seeking insects as prey. Upon sighting prey, it begins traveling toward the prey while visually orienting along the branch so that the prey is no longer within its field of view. Every so often, the spider reorients in the expected direction of the prey in order to sight it. In a series of elegant experiments, some involving all three spatial dimensions, Hill demonstrated that the spider uses its movement to update its representation of the prey. In still another example, Regolin, Vallortigara, and Zanforlin (1995) found evidence of spatial updating in two-day-old chicks attempting to circumvent a barrier in order to reach an attractor.

MODELS OF PATH INTEGRATION

Although path integration is based on route information, the internal representation underlying path integration is very different from what is referred to as *route knowledge*. The latter is an internal representation of a route that has been traversed; as such, it contains information about the various path segments and turns as well as off-route landmarks that might be used in subsequent piloting. In contrast, the representation underlying path integration is a constantly updated abstraction derived from computations on the route information (e.g., travel velocity or the turns and segments of the path). At a minimum, the representation consists only of the traveler's updated location and orientation relative to the origin; consequently, using just this representation, there is no possibility that the traveler can either retrace the route just traveled or proceed directly to any point along the route other than its origin.

Two basic models of path integration can be distinguished by the frequency with which updating of the representation occurs. In moment-to-moment updating, the traveler is continually estimating current position and orientation on the basis of travel velocity and the prior estimate of position and orientation. In configural updating, the traveler maintains some more elaborate representation of the traveled path and every so often updates the estimate of current position and orientation based on the prior estimate and the stored representation of the path. An extreme example would be maintaining a complete representation of the path since the last position fix (e.g., from piloting) and, when one needs to proceed to some previously encountered location (e.g., the last fix or distinctive point along the route), computing the bearing and distance of that location on the basis of the stored representation. A less extreme example would be periodically updating the estimate of current position and orientation on the basis of the last several straight segments and intervening turns; after each update, the path representation on which it is based can be eliminated from memory. We discuss a particular configural model and the motivation behind it later in the chapter.

Almost all modeling of path integration has focused on moment-to-moment models; Benhamou and Séguinot (1995) and Maurer and Séguinot (1995) review most of the extant models. We focus on three normative models in connection with the return-to-origin task on the terrestrial plane.* In the Cartesian model proposed by Mittelstaedt (1985), the traveler maintains a representation only of current heading and position, the latter expressed as Cartesian coordinates along two cardinal axes (Mittelstaedt 1985) centered on the origin of locomotion. Polar models use polar (spherical) coordinates for position and have been proposed in two basic forms. The environment-centered form (Gallistel 1990; Müller & Wehner 1988) employs a representation consisting of the traveler's heading and the traveler's position, the latter of which is expressed in terms of distance and bearing from the origin of locomotion; heading and bearing are thus measured with respect to the environmental frame of reference. The traveler-centered form (Benhamou, Sauvé, & Bovet 1990; Fujita, Loomis, Klatzky, & Golledge 1990) represents the origin of locomotion within the traveler's coordinate frame and, thus, requires only two parameters: distance and relative bearing to the origin. Rieser (1989:1158) discusses the different functional implications of the two types of models. Being normative models, they fail to predict the errors observed in return-to-origin behavior. Various modifications of these models incorporating stochastic and systematic error, associated either with encoding of heading and velocity or computing of the updated position estimate, have been proposed (Benhamou et al. 1990; Benhamou & Séguinot 1995; Fujita et al. 1990; Maurer & Séguinot 1995; Müller & Wehner 1988). Other researchers have proposed mechanistic models of path integration (in the

---

*To keep the discussion simple, we consider translation only along the two spatial dimensions of a terrestrial plane and rotation of the traveler about a vertical axis (which determines the traveler's heading). Consequently, three degrees of freedom are sufficient to characterize the position and orientation of the traveler. We also assume that the traveler has information, perhaps imperfect, about travel velocity (speed and course) and rotation rate.

rat) based on research on hippocampal place and head-direction cells (McNaughton et al. 1996; McNaughton, Chen, & Markus 1991; Wan, Touretzky, & Redish 1994a).

HUMAN PATH INTEGRATION

SHORTCUTTING AND LEARNING OF SPATIAL LAYOUT

Much of the experimental research on human path integration has employed tasks in which the subject uses movement-based information to update the estimates of one or more locations encountered along the travel path. The first such task is a version of the return-to-origin task, typically involving completion of a path consisting of a small number of straight segments separated by turns (Beritoff 1965; Juurmaa & Suonio 1975; Klatzky, Beall, Loomis, Golledge, & Philbeck unpublished; Klatzky et al. 1990; Loomis, Beall, Klatzky, Golledge, & Philbeck 1995a; Loomis et al. 1993; Mittelstaedt & Glasauer 1991; Passini, Proulx, & Rainville 1990; Sauve 1989; Worchel 1951, 1952; Yamamoto 1991). In such a task, the subject is led along the outbound legs of the path and then asked to travel unaided to the origin in the absence of piloting information. In attempting to do so, the subject demonstrates his or her knowledge of the relationship between the origin and drop-off points on the basis of path integration along the intervening path. Other variants of path completion have had the subject indicate only the direction of the origin from the drop-off point, typically by pointing to it using a protractor (e.g., Able & Gergits 1985; Adler & Pelkie 1985; Baker 1985; Gould 1985; Klatzky, Loomis, Beall, Chance, & Golledge 1998; Rieser & Frymire 1995; Sadalla & Montello 1989; Sholl 1989).

The second such task requires the subject to indicate the spatial disposition of distinctive locations encountered during travel along a route. In some studies vision is excluded, whereas in others vision is restricted by occluding walls that permit vision of only the immediate environment. After route traversal, knowledge of the spatial disposition of the different locations is assessed, on the basis of either estimated distances or estimated bearings (Allen, Kirasic, Dobson, Long, & Beck 1996; Baker 1985; Chance, Gaunet, Beall, & Loomis submitted; Dodds, Howarth, & Carter 1982; Herman, Chatman, & Roth 1983; Lindberg & Gärling 1982; Ochaita & Huertas 1993; Passini et al. 1990).

The third such task involves learning the spatial disposition of a number of target locations experienced in relation to a common origin

(Juurmaa & Lehtinen-Railo 1994; Landau, Spelke, & Gleitman 1984; Lehtinen-Railo & Juurmaa 1994; Loomis et al. 1993; Rieser, Guth, & Hill 1986; Ungar, Blades, Spencer, & Morsley 1994). During the training phase, the subject is led from the home location to each of two or more target locations; sometimes, multiple traverses are allowed between home and each target location so as to allow better spatial learning of the targets relative to home. During the testing phase, the subject is led to one of the target locations, from which he or she either indicates the bearings of the other target locations or attempts to walk directly to one of them.

The motivation for much of this research derives from interest in whether congenitally blind individuals are deficient in spatial ability relative to sighted and adventitiously blind individuals. Although some of the studies cited previously found no differences between the groups, the evidence overall indicates that the congenitally blind perform more poorly, suggesting deficits in spatial ability (Beritoff 1965; Dodds et al. 1982; Herman et al. 1983; Juurmaa & Lehtinen-Railo 1994; Juurmaa & Suonio 1975; Landau et al. 1984; Lehtinen-Railo & Juurmaa 1994; Loomis et al. 1993; Ochaita & Huertas 1993; Passini et al. 1990; Rieser et al. 1986; Ungar et al. 1994; Worchel 1951); see Golledge, Klatzky, and Loomis (1996b) and Thinus-Blanc and Gaunet (1997) for reviews of much of this work. Despite this general finding, the lack of visual experience does not preclude the development of normal spatial ability, for there are congenitally blind individuals who perform as well on these tasks as the best individuals from the other two groups (e.g., Juurmaa & Lehtinen-Railo 1994; Loomis et al. 1993).

Of the three tasks mentioned, path completion is the most direct way of assessing path integration ability. A number of systematic laboratory studies have been reported. We discuss most of these in what follows; one we do not discuss (an unpublished thesis by Sauve 1989) is briefly described by Maurer and Séguinot (1995). In all of these studies, the subjects were guided by the experimenter over the outbound route (either walking or riding in a wheelchair) and were deprived of visual and auditory cues about their position and orientation.

Klatzky et al. (1990) studied completion of paths consisting of from one to three legs in the outbound portion. Twenty-one blindfolded sighted subjects completed each of the 12 paths twice at each of two scales. Subject trajectories were measured with a video tracking system. Figure 5.2 shows the two- and three-leg outbound paths and the results obtained at

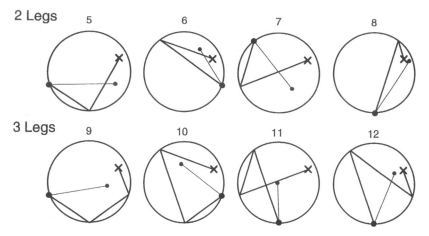

FIGURE 5.2. Performance of the average subject in an experiment on path completion (Klatzky et al. 1990). The eight panels show eight of the paths (identified by number) used in the experiment. The radius of each circle was 5 m. In each panel, X marks the origin of travel, the large dot represents the terminus of the outbound path (two or three legs), and the small dot represents the centroid of the stopping points of the 12 subjects. (Adapted from figure 5 of Klatzky et al. 1990 by permission of the Helen Dwight Reid Educational Foundation. Published by Heldref Publications, 1319 Eighteenth Street, N.W., Washington, D.C. 20036-1802. Copyright 1990.)

the larger scale (circle diameters equal to 10 m). The X represents the origin of locomotion and the larger filled circle represents the drop-off point. Subjects were attempting to return to the origin. The smaller filled circle represents the centroid of the subjects' stopping points.

As one would expect under the assumption that path integration is a noisy process, path completion performance worsened with number of legs. Mean absolute bearing error (the difference between return bearing and correct bearing to the origin, which also equals the turn error) was 26 degrees for the two-segment outbound paths and 35 degrees for the three-segment paths at the larger scale. Absolute distance errors for these same paths were 175 and 250 cm, respectively. Interestingly, there was little influence of scale on these two measures. Of particular interest is the poor performance exhibited in path 11, which involved a crossover during the outbound path. In addition to the large mean absolute errors, the mean signed errors were generally different from zero, indicating the presence of systematic biases in path integration.

Sholl (1989) completed an extensive path completion study comprising two experiments. One of these, employing 30 subjects, was similar to that of Klatzky et al. (1990), for subjects were guided while walking along the outbound paths. She used a total of six configurations, two each of two, three, and four legs in the outbound portion. Leg lengths were either 2.4 or 4.9 m. The two- and three-leg paths had one 4.9-m leg; the four-leg paths had two 4.9-m legs. All turns within a configuration were either 90 or 135 degrees. At the drop-off point, subjects pointed in the direction of the origin using a protractor. Mean absolute errors were not analyzed, but mean signed errors showed little influence of number of legs. On the whole, the average subject for each configuration performed well above chance, even on the four-leg paths.

The second experiment, employing 29 subjects, was identical to the first except that subjects were passively transported in a wheelchair. Passive transport resulted in much poorer performance than walking, and performance worsened with number of legs. Moreover, Sholl discovered an important fact relating to individual differences. Prior to the wheelchair experiment, she had classified her female subjects on the basis of performance on a paper-and-pencil task of spatial ability (the water-level task), which she hypothesized reflected vestibular sensitivity. Subjects classified as "poor" on the water-level task performed more poorly on the wheelchair path completion task than those classified as "good." In the walking experiment, no such group differences emerged. Sholl interpreted the wheelchair results in terms of differential sensitivity to otolith information, which is involved in sensing linear translations during passive transport.

The most extensive investigation of path completion was conducted by Loomis et al. (1993) as part of a larger study of spatial cognition in blind and sighted individuals. Twelve blindfolded sighted, 12 congenitally blind, and 13 adventitiously blind subjects were led outbound along two legs of a triangle and then attempted to return to the origin. Each subject completed 27 triangles, created by crossing three values of the first leg (2, 4, and 6 m), three values of the second leg (2, 4, and 6 m), and three values of the intervening turn (60, 90, and 120 degrees). Figure 5.3A depicts the origin (marked with an X), the outbound legs, and the drop-off points (marked by the filled circles). Subjects held onto a pair of handlebars carried by the experimenter and were guided at a normal walking speed over the two outbound legs, whereupon the subject attempted to walk unaided to the origin. Subject trajectories were measured by a video tracking system. None of the subjects exhibited good performance, and there was considerable variation between subjects though no discernible differences among the three groups (figures 2–5 in Loomis et al. 1993).

The mean absolute errors in the response turn toward the origin were 24, 22, and 24 degrees for the congenitally blind, adventitiously blind, and sighted subjects; the corresponding mean absolute errors in the return walking distance were 137, 107, and 168 cm, respectively.

Figure 5.3B gives the average return paths connecting the drop-off points with the centroids of the stopping points, averaged over the 37 subjects, for the 27 different triangles. Analysis of these responses (Loomis et al. 1993) revealed that the subjects made systematic errors that depended upon the path parameters. The average subject turned too little when large turns to the origin were called for and overturned when small turns were called for. This is manifest in figure 5.3B by the tendency for the different return paths to be directed toward points of convergence that lie beyond the origin. Less apparent in the figure is the other error pattern associated with the return distance: overshooting of short distances and undershooting of long distances. Both patterns of systematic error suggest the presence of systematic error in the underlying path integration process (either encoding the outbound path or computing the location of the origin).

Fujita, Klatzky, Loomis, and Golledge (1993) proposed an "encoding-error" model to account for the systematic errors in the Loomis et al. (1993)

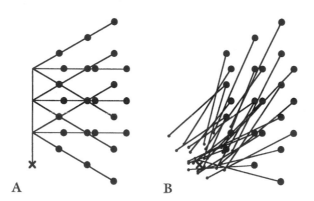

A   B

FIGURE 5.3. (**A**) Depiction of the outbound paths in an experiment on triangle completion (Loomis et al. 1993). The X represents the origin of travel, and each dot represents the terminus of a two-leg path. The first leg was 2, 4, or 6 m; the second leg was also 2, 4, or 6 m; and the turn was 60, 90, or 120 degrees. (**B**) Performance of the average subject. The large dots represent the termini from panel A and the small dots represent the centroids of the stopping points of the 37 subjects. If performance were perfect, the small dots would all coincide with the X. (Adapted from figures 2–5 of Loomis et al. 1993.)

data. The primary assumption of the model is that all of the systematic error in subjects' performance is the result of error in sensing the outbound path and representing that information in memory. Thus, it was assumed that when subjects attempted to return to the estimated position of the origin, they did so without systematic error. Justification for this assumption will be given later.

The encoding-error model is an example of a configural model, as defined earlier. The authors assumed that the values of the outbound turns and leg lengths were encoded upon completion of the turns and legs, respectively. The model fit to the data consists of encoding functions for turn and leg length. The turn encoding function provides the internally represented value for each stimulus turn value; the distance encoding function does the same for distance. The encoding functions took the following form:

$$t' = k_t \times t + i_t \qquad (1)$$
$$d' = k_d \times d + i_d \qquad (2)$$

where t and d are the stimulus values of turn and distance, t' and d' are the corresponding encoded values, k_t and k_d are slopes less than 1.0, and i_t and i_d are additive constants different from zero.

To give some appreciation of the form of these encoding functions, we consider the pattern of responses under two alternative models. In the first of these, we assume moment-to-moment updating with correct encoding of turn and underestimation of each footstep by a constant factor of 0.5. The resulting encoding functions for the legs and turns would be

$$t' = t \qquad (3)$$
$$d' = 0.5 \times d \qquad (4)$$

Figure 5.4A depicts the stimulus path (thick solid lines), internally represented triangle (dotted lines), and the predicted return leg (thin solid line). The model assumes no error in execution of the actively produced response (that is, the encoded path), in sharp contrast to the misencoding of the passively guided outbound legs. Under this assumption of no execution error, the predicted return leg has the same length as the internally represented return leg; similarly, the turns between the second and return legs are the same for the physical and internally represented triangles.

Figure 5.4C shows the pattern of predicted responses under this model. As can be seen, if there were a constant underestimation of each footstep during the outbound path but no error during execution of the

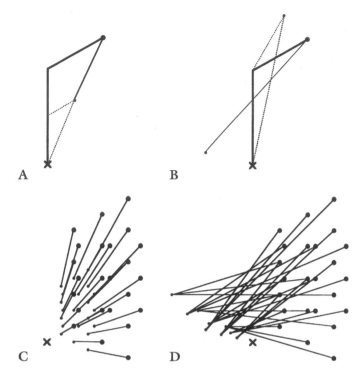

FIGURE 5.4. Predictions of triangle completion performance if subjects were consistently misperceiving the lengths of the outbound segments (panels A and C) or consistently misperceiving the stimulus turns (panels B and D). (**A**) The heavy solid lines represent the outbound path and the large filled dot represents the drop-off point. Under the assumption that the subject perceived the outbound segments to be half as large as they were, the short oblique dotted line represents the encoded second leg. Its upper terminus represents the subject's encoded location after the second leg. If the subject correctly perceived the length of the actively produced return, his or her perceived response is depicted by the longer oblique line while his or her actual response is depicted by the thin solid line. The small dot represents the stopping point of the subject's actual response. (**B**) The heavy solid lines represent the outbound path and the large filled dot represents the drop-off point. Under the assumption that the subject perceived the stimulus turn to be half as large as it was, the short oblique dotted line represents the subject's encoded location after the second leg. If the subject correctly perceived the actively produced response turn, his or her perceived response is depicted by the longer oblique line while his or her actual response is depicted by the thin solid line. The small dot represents the stopping point of the subject's actual response. (**C**) Predicted responses for underestimation of the outbound segments (panel A). The large dots represent the 27 drop-off points of the triangle completion experiment (see figure 5.3), the lines represent the predicted responses, and the small dots represent the corresponding predicted stopping points. (**D**) Predicted responses for underestimation of the stimulus turn (panel B). The large dots represent the 27 drop-off points of the triangle completion experiment (see figure 5.3), the lines represent the predicted responses, and the small dots represent the corresponding predicted stopping points.

response, the subject responses would converge in the direction of the origin, but the observed return legs would all undershoot by the same proportion. (If, instead, subjects were to continue doing path integration using the same encoding functions on the return leg, performance would be without error.)

In the second alternative model, we again assume moment-to-moment updating, this time with correct encoding of distance but constant underestimation of turn rate (with a proportionality constant of 0.5). The resulting encoding functions would be

$$t' = 0.5 \times t \qquad (5)$$
$$d' = d \qquad (6)$$

Figure 5.4B portrays one stimulus path and the corresponding internal representation. Figure 5.4D gives the predicted responses. In this case, all of the returns pass off to one side of the origin, and return legs for a constant stimulus turn terminate at a common point.

Fujita et al. (1993) found that nonzero intercepts in the encoding functions were required to account for the observed pattern of responses, both for the "average subject" as well as for most of the individual subjects for whom fits were made. The fits for the average subject were as follows:

$$t' = 0.48 \times t + 43.6 \qquad (7)$$
$$d' = 0.6 \ \times d + 1.2 \qquad (8)$$

with t measured in degrees and d measured in meters. In spite of the small number of parameters, the fits to the centroids of figure 5.4B were excellent.

Nonzero intercepts in the encoding functions have no obvious interpretation in terms of moment-to-moment updating. Other evidence that is inconsistent with moment-to-moment updating in human path completion comes from an analysis of other completion experiments (Klatzky et al. 1990; Loomis et al. 1993). This analysis showed that the subjects' latency to begin the return toward the origin increased with path complexity; the latencies for paths 9 and 11 (figure 5.3) were greater than those for paths 5 and 7. If subjects had been doing moment-to-moment updating, their current estimates of the origin should be independent of complexity.

If performance in the path completion experiment of Loomis et al. (1993) is incompatible with moment-to-moment updating, what utility might there be in configural updating, with which the data are consistent? It may be that when humans traverse paths with linear segments, they tend naturally to store the segment lengths and turns between them, in case the need arises to reverse course and retrace the configuration. If this should be true, however, such a strategy cannot succeed in the general case, for it does not seem to apply naturally to paths that are varying continuously in course.

The encoding functions fit to the average subject provide an accurate encoding of stimulus values near the stimulus mean. This "regression to the mean" might be explained in two very different ways. The first is that the encoding functions are immutable, with the correctly encoded values fortuitously coinciding with the middle (and mean) values used in the experiment. The second possibility is that the encoding functions reflect an adaptation process, whereby the encoding functions center on the mean of the stimulus distribution over some epoch of the subject's experience.

Klatzky et al. (unpublished observations) conducted a triangle completion experiment to distinguish between these two possibilities. They devised sets of triangles, differing in terms of the range of turn values (10–70 and 110–170) and in terms of the range of leg values (1–4 m and 4–7 m). The different stimulus ranges were blocked so that during a given session, subjects attempted to complete triangles having a limited range of values. It was expected that if the encoding functions center on the stimulus mean, the encoding functions would differ dramatically between the stimulus sets. Only small shifts in the encoding functions were observed. On the other hand, the encoding function parameters differed from those fit to the Loomis et al. (1993) data set. The authors concluded that the encoding functions are not immutable and do reflect the past experience of the subjects, but that the epoch (or number of trials) over which the adaptation occurs must be much longer than the sessions of this experiment.

This same experiment employed another manipulation of interest. Some subjects were tested blindfolded, whereas others were tested with a vision-limiting device. The device limited the subject's field of view to the lower forward quadrant, such that only that portion of the floor within 1.5 m was visible. This afforded general optic flow information about rotations and translations as well as information about the upcoming path, which was marked on the floor. The authors predicted that providing the subjects with more information about the outbound path ought to improve path completion performance. Performance did improve, thus supporting the basic premise of the encoding-error model. Large errors were still common, however, implying that optic flow is not sufficient to eliminate the errors in human path integration even for such short travel paths.

Although the four-parameter encoding-error model is modestly successful in accounting for triangle-completion performance, it does not account well for performance with more complex paths (Fujita et al. 1993). For a more detailed treatment of the model and the experiments used to test it, see Klatzky, Loomis, & Golledge (1997).

The data we have reviewed so far on human path completion clearly indicate that humans are incapable of navigating precisely by path integration alone. Before concluding that path integration is of little use to humans, we should consider that humans might do much better if permitted the sources of information available to other species, like the desert ant, that perform exceedingly well in return-to-origin tests. As have others that have been performed (Beritoff 1965; Juurmaa & Suonio 1975; Mittelstaedt & Glasauer 1991; Passini et al. 1990; Sauve 1989; Worchel 1951, 1952; Yamamoto 1991), three of the previously mentioned path completion studies took care to eliminate allothetic information about heading and translational velocity. The fourth (Klatzky et al. 1997) did include a condition with optic flow within a limited field of view, but significant errors were still common. Research must be done to assess the accuracy of human path integration under optimal conditions. These would include the presence of an azimuthal reference (providing allothetic heading information) and the availability of unrestricted optic or acoustic flow about self-rotation and self-translation.

Another condition likely to promote better return-to-origin performance would be allowing the subject actively to plan and execute the outbound path. In all of the human path completion studies, an experimenter guides the subject over the outbound path. Perhaps if the subject selected the outbound leg lengths and turns, return-to-origin performance would be much better.

Still another condition likely to facilitate performance is providing the subject with a mental representation of some environment. Rieser and Frymire (1995) had blindfolded subjects point to the origin after traversing an outbound path of several legs. Subjects in one condition traversed the outbound path without any mental image of the environment, whereas subjects in two other conditions traversed the path while imaginally updating objects that had either been viewed from the origin or had been instantiated by remembering a known environment. Subjects without any mental image did more poorly in pointing to the origin than those in the other two conditions. Thus, even though a mental image of the environment provides no information about current position and orientation, its presence in memory seems to facilitate path integration. The facilita tion of path integration by spatial updating of imaginary off-route objects is consistent with one hypothesis about the Micronesian navigators' use of "Etaks" (imaginary islands) as they navigate from one Pacific island to another (Gladwin 1970; Hutchins 1995).

HUMAN PATH INTEGRATION:
PERCEPTUALLY DIRECTED ACTION

Most of the research concerned specifically with understanding human path integration has employed the path completion (or return-to-origin) task. For this task the subject needs only to keep track of the origin with respect to self. Quite a different task, perceptually directed action, involves maintaining an internal representation of locations other than the origin; it, too, is directly informative about path integration. In a perceptually directed task, the subject is presented perceptual information (usually visual) about the location of a target and then attempts, in the absence of further perceptual input about the target, to indicate the target location using some motoric response; the most common motoric response is blind walking to the target. The aforementioned experiments dealing with detour behavior by the toad (Collett 1982) and reorienting behavior by the jumping spider (Hill 1979) are examples of perceptually directed action.

All perceptually directed walking tasks rely on three distinct processes: perception of the target location during stimulus exposure, updating of self-position and self-orientation by means of path integration, and spatial updating of the internal representation of the initially perceived target on the basis of the updated self-position and self-orientation (Böök & Gärling 1981b; Loomis, Da Silva, Fujita, & Fukusima 1992; Rieser & Rider 1991). Many experiments have been done on perceptually directed action. Some of these have been done to assess the accuracy of distance perception (e.g., Ashmead, Davis, & Northington 1995; Corlett, Patla, & Williams 1985; Elliott 1986, 1987; Fukusima, Loomis, & Da Silva 1997; Loomis et al. 1992; Loomis, Klatzky, Philbeck, & Golledge 1998; Philbeck & Loomis, 1997; Philbeck, Loomis, & Beall 1997; Rieser, Ashmead, Talor, & Youngquist 1990). More of the experiments have focused on the functional properties of path integration and spatial updating, which are frequently treated as a unitary process (Amorim, Glasauer, Corpinot, & Berthöz 1997; Böök & Gärling 1981b; Corlett, Byblow, & Taylor 1990; Corlett & Patla 1987; Easton & Sholl 1995; Farrell & Robertson submitted; Glasauer, Amorim, Vitte, & Berthöz 1994; Horn 1996; Israël, Chapuis, Glasauer, Charade, & Berthöz 1993; Laurent & Cavallo 1985; Loarer &

Savoyant 1991; May 1996; Mittelstaedt & Glasauer 1991; Pick & Rieser 1982; Presson & Montello 1994; Rider & Rieser 1988; Rieser 1989; Rieser & Frymire 1995; Rieser, Pick, Ashmead, & Garing 1995; Rieser & Rider 1991; Thomson 1980, 1983).

Here we review two experiments indicating the effectiveness of path integration in perceptually directed action. The first of these, by Philbeck et al. (1997), had subjects view targets and then walk to their locations without further visual input. The self-luminous targets were located at eye level 1.5, 3.1, or 6.0 m from the subject and viewed either in a well-lit room or in darkness. Under good lighting, the available distance cues permitted accurate perceptual localization of the targets, but in the darkened room, the eye-level targets were generally mislocalized. In a given trial, the subject binocularly viewed the target for several seconds, after which the subject attempted to walk without vision to the target along either a direct or an indirect path, depending on instructions. The first leg of the two indirect paths ran parallel to a wall that the subjects could feel. Upon feeling the appropriate marker (either at 1.5 m or 3.0 m from the origin), the subject turned and walked the rest of the way to the target. The question of interest here is whether the subjects walked to the same point regardless of the path taken. Figure 5.5 gives the results for nine subjects in the lights-on condition and nine subjects in the lights-off condition. For each target the centroids of the stopping points for the three paths are shown. In striking contrast to the pattern of return paths in triangle completion (figure 5.3B) that fail to converge within the workspace, the walking trajectories converge for each target under both conditions. As expected, the convergence points are nearly coincident with the physical target locations in the lights-on condition, indicating quite accurate perception of the targets from the viewing point; also, as expected, the convergence points deviate from the target locations in the lights-off condition because of the impoverished distance cues in that condition. The fact that the centroids are nearly coincident regardless of path indicates little systematic error in path integration and in spatial updating.

In the second experiment (Loomis et al. 1998, experiment 3), seven subjects either viewed a visual target or listened to an auditory target in a large open field and then attempted to walk to it without further perceptual input during the traverse. The target was either 3 or 10 m distant and had a bearing of one of four values: 30 degrees left, 80 degrees left, 30 degrees right, or 80 degrees right (relative to the subject's initial heading). On a given trial, the subject walked either directly to the target or along an indirect path, the first segment of which was 5 m straight

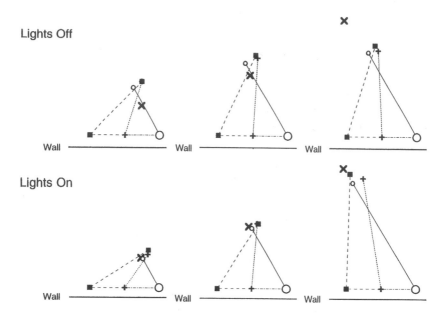

FIGURE 5.5. Performance of the average subject in an experiment on visually directed action (Philbeck et al. 1997). In each panel, the viewing position is indicated by the large open circle and the visual target by the large X. After viewing the target (with room lights either on or off), the subject attempted to walk without vision to the target's location, either along a direct path or along one of the two indirect paths, the first segments of which were parallel to a wall. The small circles, crosses, and squares nearest the target symbol represent the centroids of the stopping points of the nine subjects in each group (lights-on and lights-off). (Adapted from figures 3 and 5 of Philbeck et al. 1997.)

ahead. The two left panels of figure 5.6 give the centroids of the stopping locations for the auditory condition, and the two right panels give the same for the visual condition. The centroids were closer to the targets in the visual condition, indicating that visual distance perception was more accurate than auditory distance perception under these conditions. Of greater interest here is the congruence of the centroids of the stopping points for the direct and indirect paths for all targets for the two sensory modalities. This congruence, like that of the preceding experiment, suggests little systematic error associated with either path integration or spatial updating.

The excellent performance in both experiments, which are on a scale comparable to that of the triangle completion experiment of Loomis et al. (1993), contrasts with the relatively poor performance in triangle completion. The latter shows greater systematic error, as manifest in the poor convergence of the average return trajectories in figure 5.3B. We should, however, note one qualification about comparing the data of the two tasks— triangle completion involves two turns (one in the outbound path and one for the response) in contrast to a single turn in the indirect walking tasks. Incidentally, it is the absence of systematic error in most of the perceptually directed tasks cited above that motivated the critical assumption of the encoding-error model of no systematic error in executing the return to the origin.

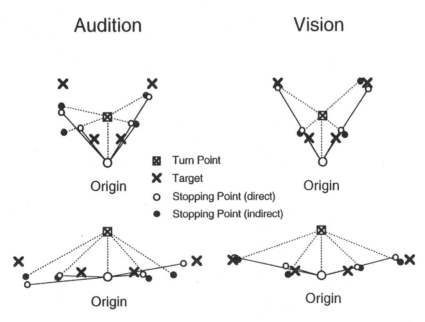

FIGURE 5.6. Performance of the average subject in an experiment on perceptually directed action (Loomis et al. 1998). In each panel, the viewing position is indicated by the large open circle and the visual or auditory target by the large X. For purposes of clarity, the data for the targets at 30 and 80 degrees azimuth relative to the origin are presented separately in the upper and lower panels, respectively. After listening to a sound source (*left panels*) or viewing a target (*right panels*), the subject attempted to walk without further visual or auditory input to the target's location, either along a direct route or along indirect paths; the small inscribed x represents the turn point for the indirect paths. The small open and filled circles represent the turn points for the indirect paths. The small open and filled circles represent the centroids of the stopping points of the seven subjects for the direct and indirect paths, respectively, in each of the two conditions (audition and vision). (Adapted from figure 7 of Loomis et al. 1998.)

One likely reason that path integration is more accurate in perceptually directed action is that the subject plays a more active role in planning the outbound path. Another possible reason might stem from the fact that in perceptually directed action the locomotion goal (the perceived target) is generally in the anterior hemifield, whereas in triangle completion the locomotion goal (the origin) is in the posterior hemifield during most of the outbound traverse. Thus, it is possible that spatial updating is more accurate for targets ahead than behind. Horn (1996) conducted an experiment to address this very question. On a given trial, the subject viewed a target off to one side, donned a blindfold, turned the body to position the imagined target in front or behind, and then sidestepped while imaginally updating the target. The subject then turned his or her body back to the original heading (to control for heading prior to the response) and then turned to face the updated target. Although there was a statistically significant effect of "spatial field" on the absolute error associated with facing the target, this effect was small and of no practical significance. Thus, it appears that the more accurate path integration of perceptually directed action is not explicable in terms of "spatial field."

INPUTS FOR SENSING ROTATION AND TRANSLATION

Path integration in humans potentially relies on a number of allothetic and idiothetic inputs, including optic flow, acoustic flow, efference copy of the muscle commands, afferent proprioception, and vestibular signals from the otoliths and semicircular canals. A number of studies have been conducted to assess the contributions of some of these potential inputs to human path integration.

One manipulation has been to assess whether proprioception contributes to path integration. As mentioned earlier, Sholl (1989) found that walking the outbound path resulted in much better path completion performance than did passive transport in a wheelchair. A similar experiment by Juurmaa and Suonio (1975) also found a difference, albeit a small one. Still another by Loomis et al. (1995a) found little effect of travel mode. Mittelstaedt and Glasauer (1991) compared walking and passive transport in performance of both triangle completion and visually directed tasks. Wheelchair performance was generally noisier (judging from their figures), but just as important is their finding that manipulating walking speed and wheelchair speed had opposite effects in both tasks, a finding for which they propose a mechanistic account.

The contribution of optic flow above and beyond those of idiothetic inputs has been assessed in two experiments. Klatzky et al. (1997) compared triangle completion in two conditions: without vision and with optic flow restricted to the lower frontal visual field. Performance was slightly better with the restricted visual input than with no vision at all. Loomis et al. (1995a) used a head-mounted virtual visual display with limited field-of-view as a means of providing optic flow information without landmark information. In their comparison of the vision and no vision conditions, there were no obvious differences in triangle completion performance. Clearly, more research is needed to determine the conditions under which optic flow contributes to path integration.

The study by Loomis et al. (1995a) using a virtual display included another condition of interest. Here the subjects remained seated in a stationary chair and received only optic flow about the outbound two legs of a triangle. The turns toward the origin exhibited considerably more error than in conditions where subjects walked under guidance or were transported in a wheelchair. Optic flow permitted above chance responding but was less effective than inertial cues alone or inertial cues together with proprioception.

Finally, a great deal of research has established that inertial information plays a role in path integration, broadly defined. Most of these studies, reported in a literature too voluminous to consider here, have shown that humans are somewhat sensitive to inertial information produced during passive rotations and translations. Although much of this research shows sensitivity to rotations and translations by way of the vestibulo-ocular reflex, other work has shown that humans update self-position and self-orientation on the basis of vestibular and other inertial cues (e.g., Berthöz, Israël, Georges-François, Grasso, & Tsuzuku 1995; Blouin, Gauthier, & Vercher 1995; Israël & Berthöz 1992; Israël et al. 1993; Ivanenko, Grasso, Israël, & Berthöz, 1997). The above-mentioned studies comparing walking and wheelchair transport indicate that inertial information results in above-chance performance of path completion. A question that remains is whether inertial information improves performance above and beyond what is obtained with other inputs. In a triangle completion task involving walking, Worchel (1952) found that the subjects with poor vestibular function actually performed slightly better than normals and argued for some compensation process. Glasauer et al. (1994) found that labyrinthine-defective subjects did as well as normal subjects in a task of visually directed action, judging on the basis of distance error. Based on the very limited evidence, it appears that if proprioceptive input is available, vestibular input is superfluous.

NATURAL SETTINGS

Given the significant performance errors in the path integration studies we have reviewed, it might appear that path integration plays little role in real-world human navigation. Concluding so would be premature, for, as we have argued, a human navigating by path integration in the real world often has access to sources of information (e.g., an azimuthal reference, unrestricted optic flow, and visual matching of local perspectives) not yet investigated in controlled studies.

In conclusion, we draw attention to some natural settings in which path integration must play a critical role. First, path integration in its broadest sense is involved whenever a person is exploring a novel environment without an external map and without distant landmarks to permit updating of position. The Micronesian navigators who travel over the open Pacific Ocean toward distant islands clearly use some form of path integration, facilitated by spatial updating of imaginary islands (Gladwin 1970; Hutchins 1995). So do hikers, orienteerers, and hunters traveling through dense forest or jungle; nomads traversing featureless expanses of desert; and sightseers exploring old cities with narrow streets and alleys (Baker 1981; Hutchins 1995; Lewis 1976). Even more dependent on path integration are blind individuals, for the limited travel range of sound denies them distant acoustic landmarks for the updating of position.

Other natural settings are less general, but no less a part of the "real world." Firefighters entering a smoke-filled building depend on path integration for positional awareness. Cave explorers (Montello & Lemberg 1995) and divers face the challenge of having to do path integration in all three spatial dimensions. Finally, pilots flying under instrument conditions while being vectored by ground controllers must engage in imaginal path integration to maintain an estimate of their relationship to their destination and any hazardous terrain.

Obviously, the controlled laboratory research we have reviewed on human path integration is greatly removed from navigation "in the wild" (Hutchins 1995). As such, there may be reason to question its representativeness. The crucial question, however, is whether the experimental tasks considered here tap the same basic processes as real-world navigation. Confident that they do, we believe that this research is the necessary starting point for understanding human path integration ability, for only through controlled experimentation can one begin to appreciate how stochastic and systematic errors within the path integration process depend upon the available sensory inputs and other aspects of the navigational setting. We hope that an improved understanding of real-world path integration would follow from the extension of controlled studies to include a wider range of allothetic sources of information. Better understanding of human navigation, of course, requires research both on the functional properties of piloting and on how piloting and path integration cooperate in determining navigation performance.

MICHEL-ANGE AMORIM

6

*A Neurocognitive
Approach
to Human Navigation*

The information processing approach of cognitive psychology is an effort to understand human performance as the consequence of mental operations (cognitive processes) executed on incoming information acquired from perception and on stored information (representations). Accordingly, human navigation involves several stages of information processing, for example, acquisition, coding, and integration of sensory information, as well as processing spatial information in order to localize oneself in relation to the environment and the objects it contains. Cognitive psychology makes extensive use of reaction times (Luce 1986) and errors as performance measurements in order to infer the elements of mental organization. For example, one basic postulate is that whenever a mental operation or information transformation is executed it takes some time, so reaction time increases with the amount of processing performed.

Golledge (1987) described how the information processing approach fits within environmental cognition. In this chapter, a neurocognitive approach to human navigation is illustrated based on the model of visuo-

The studies by Amorim and collaborators reported in this chapter were supported by a doctoral grant from the Centre National d'Etudes Spatiales to the author.

spatial cognition proposed by Kosslyn and colleagues (Kosslyn 1991; Kosslyn, Van Kleek, & Kirby 1990). Then, in order to validate this model, two studies are presented in some detail. One of these studies investigates the role of reference frames in computing locations in space; the other compares two different processing modes for the updating of an object's location and orientation. The usefulness of estimating the importance of the observed effects is highlighted. Finally, the neurocognitive approach is applied to the study of pathological cases of environmental cognition, that is, topographical disorientation.

A NEUROCOGNITIVE FRAMEWORK

In 1973, Downs and Stea claimed that *cognitive mapping* is the major process for environmental cognition. It is a process "composed of a series of psychological transformations by which an individual acquires, codes, stores, recalls and decodes information about the relative locations and attributes of phenomena in the everyday spatial environment" (Downs & Stea 1973a:9). Here, it is proposed that this concept be decomposed into the components described in Kosslyn's (1991; Kosslyn et al. 1990) model of visuo-spatial cognition. Briefly, according to this model (figure 6.1), the available multisensory information is first processed by a spatiotopic mapping subsystem. Then, information concerning real or imaginary locations can be specified relative either to the viewer (egocentric mapping, e.g., "The

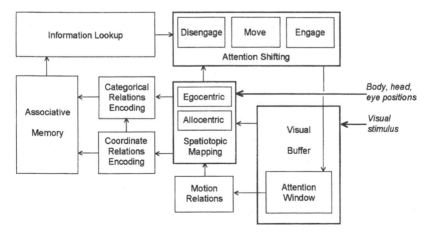

FIGURE 6.1. Model of visuo-spatial cognition. (Adapted from Kosslyn 1991 and reproduced from Amorim & Stucchi 1997, by permission of Elsevier Science NL.)

computer is 50 cm away relative to me") or to an object or place that is part of the environment (allocentric mapping, e.g., "The book is on the desk"). Afterwards, information on locations is encoded either in distance, direction, and orientation coordinates (coordinate relations encoding), or in categories such as "connected to," "left of," "under," and "above" (categorical relations encoding) (Carlson-Radvansky & Irwin 1993). This coded spatial information then goes into the associative memory, where it is matched to stored information about the location, name, and function of the other object(s) contained in the visual scene, and possibly organized in terms of subjective hierarchies (McNamara, Hardy, & Hirtle 1989). In order to access (look up) spatial information in an imaginary or a real environment, attention is shifted to the visual scene activated in working memory (visual buffer). This image scanning is realized by disengaging attention from the current location, shifting attention (moving) to a new location in space, and engaging attention at that new location (Kosslyn 1991). Particular attention is paid to the processing of motion relations coming from the visual inputs in order to pick up the world invariants (e.g., object-to-object distances and directions) within the changing perspective structure relating the navigator to an environment (Gibson 1979). The neural structures underlying each of these processes of high-level vision as well as their associated pathologies are rather well identified (for a review, see Kosslyn & Koenig 1992).

Landmarks are the bricks of our spatial knowledge (Golledge 1987). The manner in which objects and locations are processed in order to evoke a spatial representation is described next.

WHERE IS WHAT?

Studies on primates showed that the information on objects ("what" system) and their locations ("where" system) is processed by two distinct neurological pathways (Mishkin, Ungerleider, & Macko 1983). An occipital-temporal cortical pathway ("ventral" system) is specialized for the extraction of the visual features of objects, for example, shape and color, whereas an occipital-posterior parietal pathway ("dorsal" system) processes spatial properties such as location, size, and orientation. In humans, position emission tomography studies (Haxby et al. 1991), event-related brain potential (ERP) recordings (Mecklinger & Pfeifer 1996), and neuropsychological studies (Farah, Hammond, Levine, & Calvanio 1988; Levine, Warach, & Farah 1985) yielded additional evidence supporting the existence of those two distinct processing systems operating on spatial knowledge. In his chronometric study of the method of loci mnemonics, Lea

(1975) showed that access to location precedes access to its associated item. Similarly, using ERP techniques, Mecklinger and Pfeifer (1996) showed that rehearsal processes for spatial information start earlier than those for objects.

According to Kosslyn (1991), the fact that people can report from memory where furniture is placed in their living rooms suggests that the outputs from the "what" and "where" systems are conjoined downstream in the so-called associative memory (see figure 6.1). This memory system also associates the incoming structural information with other information referring to the name, meaning, or function of objects or places. In summary, the spatial knowledge acquired from navigation is elaborated little by little through a conjunction of structural, phonemic, and semantic "spread" of encoding (Craik & Tulving 1975).

LEVELS OF ENCODING

Spatial knowledge can be organized on the basis of metrical spatial information such as distance and direction (McNamara 1986; Stevens & Coupe 1978). Nonspatial information may also serve as an organizing principle. For example, people tend to group buildings by functions, that is, commercial buildings, with other commercial buildings, and university buildings together, despite the spatial scattering of the buildings (Hirtle & Jonides 1985). Similarly, McNamara, Halpin, and Hardy (1992) demonstrated that nonspatial facts about objects are integrated with spatial knowledge in memory and that they facilitate location judgments. Astonishingly, the functional or semantic relationship between two objects may also influence motor responses. Castiello, Scarpa, and Bennett (1995) reported the case of a patient suffering from bilateral, occipital calcarine sulcus lesions, who was unable to point simultaneously or put together pictures (objects) that were not functionally or categorically related (e.g., two fruits, but not one fruit and one animal). This effect of higher-level categorical encoding of objects upon movements is an additional finding consistent with Kosslyn's (1991) view that the brain maps the world either on a metrical base or by using a categorical type of encoding that may allow the organization of spatial knowledge in terms of hierarchies (McNamara 1986).

Kosslyn and colleagues (1990) suggested that in addition to this distinction between categorical and metrical encoding, the brain would map the world using either an egocentric or an allocentric reference frame (see figure 6.1). The ways in which these reference frames might affect environment scanning in order to "map" the world were the subject of the study by Amorim and Stucchi (1997) that is presented in the next section.

SPATIOTOPIC MAPPING AND REFERENCE FRAMES

NEUROCOGNITIVE PROCESSING PATHWAYS

The question of how reference frames are used for the representation of objects and spatial relations is central in visual cognition (for a review, see Pinker 1985). Briefly, in a viewer-centered frame of reference, objects or places are represented in a retinocentric, head-centered, or body-centered coordinate system based on the viewer's perspective of the world. In an object-centered frame, objects or places are represented with respect to their intrinsic axes (gravity and prominent landmarks are used in the case of places). Localizing and reaching are operations that require a viewer-centered processing mode, whereas identifying, grasping, and manipulating require an object-centered processing mode (Jeannerod 1994). According to Strong (1994), object-centered visual constraints may be processed in the occipital-temporal cortical pathway (the "what" pathway), whereas those relative to the viewer may be processed in the occipital-posterior parietal pathway (the "where" pathway). On the other hand, object-relative motor constraints may be processed in temporal-prefrontal associations and those relative to the subject may be processed in the posterior parietal-prefrontal pathway (see figure 6.2).

MENTAL SCANNING

Experimental evidence suggests that mental exploration, such as scanning a memorized visual scene or imagining a walk (Kosslyn 1978), shares some basic common processes with perception and locomotion respectively.

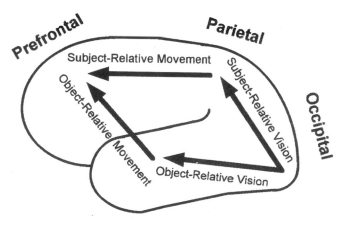

FIGURE 6.2. Four neurocognitive processing pathways. (Reproduced from Strong 1994, by permission of Cambridge University Press.)

Chronometric studies of visual scanning of mental images have demonstrated that reaction time is proportional to the scanned distance for two-dimensional maps (Kosslyn, Ball, & Reiser 1978), as well as for three-dimensional arrangements of objects (Pinker 1980). Studies on mental walking have also shown that the response time required for subjects to imagine reaching a location depends on distance, as in real walking (Decety 1991; Decety, Jeannerod, & Prablanc 1989; Kosslyn 1978).

The use of "you-are-here" maps implicitly suggests that *viewer-centered* (VC) frames of reference are easier to use than *object-centered* (OC) ones (Shepard & Hurwitz 1984). In Paris, a map indicating "the Eiffel tower is here" instead of "you are here" would probably make the task more difficult for the tourist to find his or her current position with respect to the monument. The purpose of the study conducted by Amorim and Stucchi (1997) was to investigate how the two exploration modes may differ in terms of processing time as a function of the explored distance. They induced their subjects to carry out either object- or viewer-centered mental exploration of an imagined environment using the same stimuli but different instructions. They also examined whether the time to locate an imagined position would increase with the size of the imaginary environment under VC and OC exploration conditions. The to-be-explored environment was that of an imagined clock drawn on the ground. An uppercase F portrayed on a computer screen was to be imagined standing up in the center of the clock (see figure 6.3). Clock representation is widely used by airplane instructors to indicate azimuth direction, because it is unequivocal. But unlike airplane pilots, who assume they stay at the center of the clock space, here the subject was supposed to be at the periphery of the clock (standing at

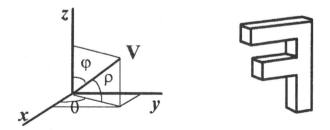

FIGURE 6.3. (*Left*) Spherical coordinates and coordinates of the subject's viewpoint *V* used to generate the object (Ammeraal 1986). (*Right*) Example of one orientation of the object, supposedly standing in the center of a clock drawn on the ground, with the viewer at the periphery of the clock looking toward the object. (Reproduced from Amorim & Stucchi 1997, by permission of Elsevier Science NL.)

1 of the 12 possible hours), looking toward its center to the F, which acted as a pointer toward hour locations. Previous work (Attneave & Pierce 1978) has shown that humans can accurately extrapolate directions and locations indicated by a pointer into perceived and imagined space, providing evidence that imagined space is functionally continuous with perceived space in the representational system.

In the Amorim and Stucchi (1997) study, subjects indicated either (1) the clock location pointed to by the F given their viewing position (VC condition) or (2) their location at the periphery of the clock given the location pointed to by the F (OC condition). For one group of subjects the imagined clock diameter was 3 m; for the other group it was 30 m. Results showed that response latencies were proportional to the explored imaginary distance and increased with the size of the imagined environment (3-m versus 30-m diameter). Furthermore, significantly higher mean processing times were found for the OC than the VC condition.

ESTIMATING THE MAGNITUDE OF THE EFFECTS

A quantitative estimation of the magnitude of effect in the parent population is important from a theoretical standpoint and goes beyond the fact of simply asserting the existence of an effect. The heuristic value of this procedure is its predictive approach, because it allows quantification of the parent effect with respect to the gathered data. In other words, it allows us to estimate the mean processing times required by each processing module postulated in models such as the one illustrated in figure 6.1. Using Bayesian ANOVA techniques (Rouanet & Lecoutre 1983), it is possible to build a distribution over the parent effect δ ($\delta = \mu_1 - \mu_2$) from the experimental data. The distribution is centered on the observed effect d ($d = m_1 - m_2$). Its dispersion—$(d/\sqrt{F})^2$, the observed effect d divided by the square root of its corresponding analysis of variance F ratio value, at the power of two—translates the potential of generalizability over δ, which is carried by the experimental information. The magnitude of the effect can be assessed through credibility limits obtained from this distribution (Bernard 1994; Rouanet 1996) as shown in figure 6.4.

Descriptively, the *frame of reference* effect observed in the Amorim and Stucchi (1997) study was 3.22 sec (mean of OC minus mean of VC). Its corresponding theoretical distribution is $t_{20}(3.22, 0.66^2)$, that is, a generalized t distribution centered on the observed effect and with a dispersion index $(d/\sqrt{F})^2$. From the usual t distribution tables, a Bayes-fiducial probability (guarantee) of 95 percent indicates that the effect in the population is greater than +2.08 sec. The same guarantee applies in regard to

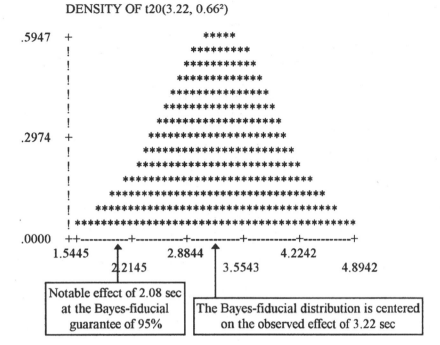

FIGURE 6.4. Output of the Bayes-fiducial inference program (Poitevineau & Lecoutre 1986), evaluating the magnitude of the *frame of reference* effect observed in the Amorim and Stucchi (1997) study.

the effect of the "clock size" factor being greater than 0.57 sec. In summary, it can confidently be stated that an additional processing time of at least 2.08 sec occurs during an object-centered exploration mode in comparison with the viewer-centered one, and that the time is 0.57 sec when imagining a larger environment.

Interestingly, the Bayesian approach allows the same guaranteed statement that the effect of frame of reference used to map the environment (at least 2.08 sec) is larger than the effect of the size of the explored environment itself (at least 0.57 sec), at least within the limits tested. This kind of conclusion is very important for the cognitive modeling approach, in order to find out to which spatial component information processing is mainly devoted when mapping the environment.

ATTENTION-SHIFTING PROCESS
Although response latencies were in general proportional to the explored *imaginary distance* (angular difference between the viewer position and

the location pointed by the object), a nonlinear trend between reaction time and angular difference was found among 25 percent of the subjects in the Amorim and Stucchi (1997) study. The imagined environment was explored in two possible ways, as illustrated in figure 6.5.

Most subjects displayed a pattern of performance (figure 6.5, right side) consistent with the interpretation that they shifted attention around the circle periphery from the initial attended location toward the targeted location, inducing a linear trend in exploration time. Other subjects displayed a pattern of performance (figure 6.5, left side) consistent with the interpretation that, for the larger angular disparities between object orientation and viewer position, subjects directly allocated their attention to the

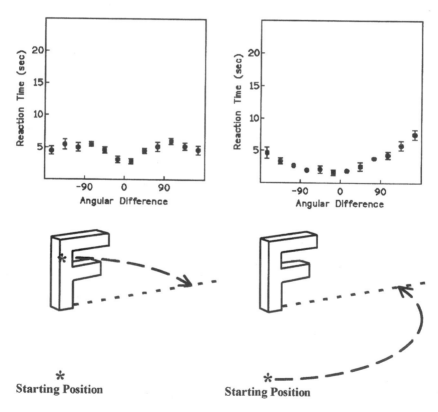

FIGURE 6.5. Example of discontinuous (*left*) versus continuous (*right*) modes of exploration for two different participants performing the VC task. The lower panels illustrate how attention was shifted along the imaginary clock in order to obtain the pattern of reaction times observed for each exploration mode. (The upper panels are reproduced from Amorim & Stucchi 1997, by permission of Elsevier Science NL.)

opposite location with respect to the starting position and then scanned the distance left toward the one for which they were looking. They thereby performed a kind of "perspective-change" further accompanied by a scanning process. This result suggests that several exploration modes may be adopted to map the environment. The issue of allocation of processing resources for updating the incoming spatial information during human navigation is discussed in the next section.

MODES OF SPATIAL UPDATING

Mentally updating the location and orientation of a previously seen distant person (e.g., a sniper) or object while walking in complete darkness is crucial and may be performed in two ways. The passing scene can be imagined continuously, paying less attention to the actual path along which one is walking (Loarer & Savoyant 1991; Rieser 1990). Or the path can be attended to—as if wanting to retrace exactly the steps taken—with a consequent failure to attend to the scene around the path except that at the terminus of the path, the new directions of the parts of the surrounding scene can be "reconstructed" or "inferred" (Huttenlocher & Presson 1973; Levine, Jankovic, & Palij 1982). Both kinds of computation involve a process called *path integration* (Mittelstaedt & Glasauer 1991; Mittelstaedt & Mittelstaedt 1980), which uses the available vestibular, kinesthetic, and motor-command information to maintain self-orientation and position when locomoting in the temporary absence of vision (Loomis et al. 1993). Previous studies used concurrent task paradigms to show that such a process demands central processing capacity (Böök & Gärling 1980; Corlett, Kozub, & Quick 1990; Lindberg & Gärling 1983); that is, navigation is effortful rather than automatic and is related to visuo-spatial imagery (Klatzky et al. 1990; Thorndyke & Hayes-Roth 1982). No study has addressed the specific issue of the effect of two different processing modes on nonvisual navigation, in terms of processing resources allocated differently to relevant spatial information in order to fulfill the navigation task, rather than in terms of interfering concurrent tasks that would never allow a correct response to emerge anyway.

Although there exists a literature on solution strategies for a few spatial tasks (see Lohman 1988 for details), such a study is certainly lacking in the field of nonvisual navigation. It was the goal of Amorim, Glasauer, Corpinot, and Berthöz (1997) to show that there are at least two different modes, or strategies, for processing information centrally in order to update spatial information while walking without vision, and to describe their effects in terms of processing time and error. In addition, in contrast to previous studies on blind-walking, which were mainly concerned with the updating of the location of single objects or configuration of objects (Amorim, Loomis, & Fukusima 1998; Easton & Sholl 1995; Loomis et al. 1993; Rieser 1989; Rieser & Rider 1991), Amorim and colleagues (1997) examined the updating of both the location *and* the orientation of a previously viewed object in geographic space.

In summary, the contention of the latter authors was that until now the stages of processing subsequent to path integration have been neglected in the literature on nonvisual navigation. Nevertheless, researchers agree that the processing of locational information demands central processing (Böök & Gärling 1980; Corlett et al. 1990; Lindberg & Gärling 1983). After a preliminary calibration test evaluating a subject's ability to memorize the orientation of a three-dimensional object in space, Amorim et al. (1997) compared the characteristics and effects of two different processing modes on the updating of the object's orientation and location during nonvisual navigation, as well as on the response latencies and locomotor activity.

Locating one's position on the basis of a directionally polarized object is a common task in everyday life. For example, it is typical to retrieve our position in a town by noting the direction of landmarks, such as signs, statues, buildings, and so on. The object used in the Amorim et al. (1997) study was a three-dimensional uppercase letter F. The selection of such an object was based on its peculiar symmetry properties, namely, its clear directional polarization. The attributes of this object inside the geographic space may be defined as follows: (1) its vertical bar indicates its *location* and (2) its two horizontal bars indicate the direction it points in space, that is, its *orientation*. Observers viewed and memorized the attributes of this to-be-updated object under reduced-cue condition (Philbeck & Loomis 1997): that is, the object borders glowed in complete darkness. Then the subjects were asked to close their eyes and the object was hidden. The experimenter led them to a new viewing position falling along an L-shaped trajectory around the object.

In one condition, using an OC updating mode, subjects mainly *rotated the object mentally* with respect to their current position while walking without vision. In another condition, the subjects adopted a *trajectory-centered* (TC) updating mode, in which they were only concerned with mentally *tracing the path they walk,* and with refreshing the initial view of the object in memory. Accordingly, during the walk, the TC subjects must assess the distance they walk and any angle turned. For the OC subjects, this updating of distance and angle must be accompanied by a continuous updating

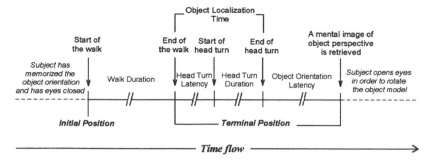

FIGURE 6.6. Experimental events flow chart. (Reproduced from Amorim et al. 1997, by permission of the Psychonomic Society.)

of the location and the orientation of the object. Therefore, during the walk, the OC task is more demanding than the TC task in terms of processing capacity or cognitive load. One would thus expect slower body movements for the OC subjects than for the TC subjects. To verify this effect, the experimenter who led the observers to the terminal position controlled *only* that they did not deviate from the imposed path. The subjects were explicitly encouraged to control the speed of their locomotor movement (i.e., slowing down or speeding up the movements when necessary) so that the task would be performed optimally.

At the end of the walk, observers were asked to face the (nonvisible) object in order to indicate where they thought it was located. Subjects were asked to align their body, head, and line of sight together toward where they thought the object (i.e., its vertical bar) was located. Once the object was faced, observers opened their eyes and gave the orientation of the object (i.e., its horizontal bars) by rotating a model of the object, inside a box fixed relative to the observer's head, to its estimated perspective relative to the current vantage point. Because in the OC task observers had updated this attribute along with their walk, shorter reaction times were expected than in the TC task. Indeed, in the latter condition, they had to perform additional computations to end up with the object's orientation response. The different processing times measured in the Amorim et al. (1997) study are illustrated in figure 6.6.

EFFECT OF TWO DIFFERENT PROCESSING MODES
ON LOCOMOTOR ACTIVITY AND RESPONSE LATENCIES
Remember that in the OC updating mode observers continuously kept track of the object's perspective during the walk, whereas in the TC mode,

observers deduced its perspective at a final viewing position from continuous trajectory-mapping and recall of the initial view on the object.
Results obtained by Amorim et al. (1997) showed that the spontaneous
locomotor velocity adopted by the subjects was significantly slower in the
OC mode than in the TC mode. Actually, in the latter processing mode,
subjects mainly had to process the trajectory of their walk, that is, they only
had to update the distance walked and the angle turned until the terminal position. In contrast, during the OC task, this updating of distance and
angle had to be accompanied by a continuous updating of the object orientation and location itself, which imposed a much larger cognitive load on
the subject and thus led to slower walking and turning velocities.

In addition, Amorim et al. (1997) showed that subjects take more time
to retrieve information on object orientation in the TC task than in the OC
one. Because observers have continuously updated the object's perspective
during their walk, in the OC task the response is permanently available in
the visuo-spatial buffer where image transformation is supposed to occur
(Kosslyn 1981, 1991). On the contrary, in the TC task, subjects were only
concerned with mentally tracing the path they walked and refreshing the
initial view of the object in memory. Consequently, from the terminal position, additional computations were required in order to determine the object's
orientation response. Using Bayesian procedures, this supplementary
spatial inference process that subjects performed in the TC mode in order
to reconstruct the object's perspective was estimated to add at least 2.43 sec
to the object orientation latency as compared with the OC task. Results also
showed larger absolute errors in the updating of the object's orientation
in the TC than in the OC task, but no difference between the tasks in terms
of signed errors. The larger absolute errors in the TC task reflected the presence of misalignment and "mirror image" reversal errors typical of the use
of directional information from a vantage point different from the one in
which it was learned (Levine, Jankovic, & Palij 1982; Rossano & Warren
1989). In summary, careful examination of the processing times shows that,
in the OC mode, most of the processing occurs during the walk, whereas
in the TC mode it is supplied at the end of the walk (see figure 6.7).

Instead of viewing each of the two different processing modes investigated by Amorim et al. (1997) as being distinct, it is possible to consider
them as complementary for the control of human navigation. Along these
lines, congruent pieces of evidence in favor of a cooperation between
both processing modes are supplied by research on multisensor integration modeling (Bernstein 1967; Droulez & Berthöz 1991; Durrant-White
1990) and the navigation of autonomous mobile robots (Brown, Durrant-

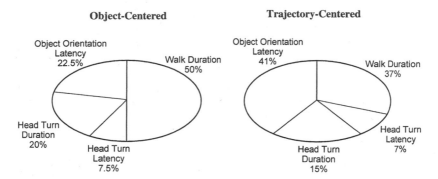

FIGURE 6.7. Proportion of processing time allocated to each component of the object orientation updating task while walking in the temporary absence of vision, for the two processing modes investigated by Amorim and collaborators (1997).

White, Leonard, Rao, & Steer 1989). The evidence suggests that when information from one sensor (e.g., vision) is missing, cooperative information is supplied by other sensors in order to update an internal model of the missing sensor input. Due to errors that increase over time, however, new visual input will be necessary periodically for resetting the internal model of the sensor input (Brown et al. 1989). Similarly, one may suggest that during nonvisual navigation available sensors cooperate to allow *partial* re-creation of the missing visual information through mental imagery. At certain steps, however, spatial inference (computational) processes would be necessary to allow reconstruction of the *global* perspective structure representation. During locomotion in everyday life, therefore, only part of the spatial layout may be monitored during displacement, and then the remainder can be regenerated from the new vantage points. Such "updating-reconstruction-updating" cycles would constitute the basis of perspective structure updating during nonvisual navigation.

TOPOGRAPHICAL DISORIENTATION

Although in theory humans can correctly update both their position and orientation relative to the world's invariant structure, spatial disorientation may happen for various reasons. Distortions and fluctuations in topographic memory may constitute one major cause of disorientation when navigating in the environment (Giraudo & Pailhous 1994; Golledge 1987). The other source of disorientation is a problem linked to perception while encoding environment features during one's travel. Those problems become more obvious when studying topographical disorders subsequent to brain lesions. De Renzi (1982) classified neurological problems relating to perception of environmental features at either a local (e.g., inability to make sense of a landmark) or a global (e.g., inability to relate together the features of a visual scene) level as *topographical agnosia*. On the other hand, he suggested the label *topographical amnesia* for the disorders arising from either an inability to access the memory for places or a loss of the memory content.

Farah (1984) used the componential information processing model of imagery provided by Kosslyn (1980) in order to interpret the patterns of deficits and preserved abilities in the neurological reports of loss of mental imagery following brain damage. Because human navigation makes use of visuo-spatial imagery (Klatzky et al. 1990; Thorndyke & Hayes-Roth 1982), the approach taken by Farah (1984) may also be applied to the understanding of topographical disorientation. Accordingly, some disorders may be due to information-bearing structures such as the long-term visual memory as well as the medium in which images occur, that is, the "visual buffer." Other disorders may be due to information-manipulating processes such as those that "generate," "inspect," and "transform" the image in the visual buffer. Visual scenes are activated in the visual buffer either by the generation of an image from long-term memory or by a perceptual encoding process. Through careful examination of the performance of patients at different tasks, it may be possible to infer the components of imagery that were individually susceptible to brain damage. Indeed, the visual scene may be simply detected or inspected for further processing: description (question-answering tasks), copying (drawing or construction tasks), or matching with long-term visual memories (recognition tasks).

Guariglia, Padovani, Pantano, and Pizzamiglio (1993) reported the case of a patient with a unilateral neglect restricted to visual imagery subsequent to a lesion involving the right frontal and anterior temporal lobes. Their patient did not show any visual neglect for stimuli located in a far or near space or on his own body. But when asked to describe a familiar piazza or well-inspected room from memory, from two opposite perspectives, he consistently omitted half of the elements on the left and none on the right side. This result suggests that long-term memory is not impaired itself, because by combining the left-side views taken from two opposite perspectives, in theory one should be able to reconstruct the entire room. On the other hand, the patient showed no neglect for visually presented items, which suggests that the visual buffer is also intact. Therefore, if we apply Farah's (1984) componential analysis to this case study, we arrive at the conclusion that the deficit shown by this patient concerned the image-generation process.

Neurological patients may also display syndromes suggestive of a dissociation between modes of processing spatial information centered on objects or landmarks versus the walked trajectory. Persons suffering from constructive topographic agnosia (Grüsser & Landis 1991) due to lesions in the lateral temporo-parieto-occipital lobe orient themselves well with respect to landmarks, but they are unable to combine the different structures located in the extrapersonal space in the right order. In light of the findings of Amorim et al. (1997) such patients have a deficit in implementing a trajectory-centered mode of processing, and instead benefit from using an object-centered mode of processing of their spatial representations, despite the higher on-line cognitive cost of this last processing mode.

As a general conclusion, this chapter is a definite invitation to consider human navigation performance with respect to the neurocognitive functions using approaches similar to that proposed by Kosslyn and colleagues (Kosslyn 1991; Kosslyn et al. 1990). Finally, by investigating the processing modes used to perform the navigation tasks and estimating the importance of the different processes tapped by experimental manipulations, as well as examining brain dysfunctions related to navigation, we may confidently hope to figure out the basis of human wayfinding and improve spatial behavior.

JOHN J. RIESER

7

Dynamic Spatial Orientation and the Coupling of Representation and Action

Spatial orientation—knowing where one is relative to objects in the surrounding environment—is a central feature of situational awareness and readiness to act. Spatial orientation is dynamic. When a person moves, his or her perspective *rotates* and *translates* in directions and at rates that are dependably related to the geometry and rate of self-movement. When there is light, people can see some of the changing self-to-object distances and directions directly, but other objects go out of view, for example, when the person turns away from them, walks past them, or walks past a wall that occludes them. The need to maintain awareness of spatial orientation relative to things that are out of view is particularly acute when people walk without vision in the dark or when, like pilots, firefighters, and deepwater divers, they operate in situations that are rendered visually unreliable and impoverished by clouds, smoke, or turbulence.

Research shows that while walking with vision, people readily maintain their spatial orientation relative to places they have passed. For example, Attneave and Farrar (1977) noted that when adults viewed a row of seven objects and then turned in place to face away from them (resulting in reversal of the objects' left-right positions relative to the subject's new facing direction), they rapidly and accurately judged the objects' changed left-right directions. Everyday experiences with children's games like pin the tail on the donkey and blindman's buff show that people can maintain awareness of dynamic orientation relative to remembered places when walking without vision, although their accuracy is sometimes less than they would hope. Studies have quantified the accuracy of paths including simple rotations like those used by Attneave and Farrar (Pick, Rieser, Wagner, & Garing in press; Rieser 1989), simple translations (Rieser, Ashmead, Talor, & Youngquist 1990; Steenhuis & Goodale 1988; Thomson 1983), and complex paths consisting of turns and translations (Böök & Gärling 1981b; Rieser, Guth, & Hill 1986). Research on the early development of dynamic spatial orientation when walking without vision indicates that it emerges by about one year of age and then continues to increase in its precision and ranges of distance (Lepecq 1989; Rider & Rieser 1988; Rieser & Rider 1991).

My purpose here is to present a perceptual-action learning theory to account for dynamic spatial orientation while walking without vision or access to other useful environmental reference information, to deduce some of the implications of the theory, and to summarize three lines of experimental evidence from my laboratory that converge on the theory. Two points should be noted at the outset. First, a systematic review of the literature is not included in this chapter, which is more narrowly focused. Second, whereas the narrative tells the hypothetico-deductive story of thesis, implications, and experimental tests, the research process often consisted of descriptive studies: a search for a few good facts to constrain the theoretical possibilities.

THE GENERAL MODEL AND SPECIFIC THESIS

How is it that people maintain awareness of spatial orientation when walking without vision and without access to nonvisual sources of input about the surroundings? The thesis to explain the psychological processing and the general model within which the thesis exists are depicted in figure 7.1. The elements of the functional model consist of the terms *perception, representation,* and *action;* lines linking them, to stand for how

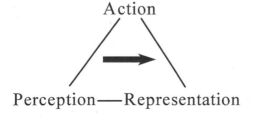

FIGURE 7.1. The system consists of three coupled subsystems: action-representation, action-perception, and perception-representation.

they are coupled together in pairwise fashion; and a single causal arrow, standing for the claim that the coupling of perception and action accounts for the coupling of representation and action.

ELEMENTS OF THE MODEL

Consider what the terms mean and how the model works in the context of the Rieser, Guth, and Hill (1982, 1986) task to assess dynamic spatial orientation. People were asked to learn the locations of six objects in a large room, walk without vision to a new observation point, and aim a pointer to localize their bearing to each of the remembered targets. (Some of the experiments involved tests with participants who were blind. These are discussed later in this chapter.) During the learning phase in one condition, participants learned about the targets by viewing them from a single home-base position; in another condition they learned about them nonvisually, following a sighted guide from a single home-base position to a target and back to home-base, then from home-base to the next target, and so forth. The walk to the new observation point occurred without vision or access to other environmental reference information; and the new observation point differed from the observation point at home-base in both location and heading. The result is that most of the participants pointed toward the targets well, averaging about 20 degrees of error. Their localizations were reliably better than would be expected by chance under conditions like these. Although the responses involved heading judgments, computer simulations showed that the levels of accuracy obtained were possible only if people updated their position as well as their heading relative to the remembered set of targets.

In terms of the present model, participants were asked to perceive the targets from home-base, either by looking at them or by walking to them nonvisually. In addition they were asked to keep the targets in mind during the test procedures, representing them in working memory. Finally, they were asked to walk to a new observation point, an action that required them to update their spatial orientation relative to the remembered targets.

Consider the meaning of the links in the model, coupling the elements together in pairwise fashion. The perception-action link stands for the fact that an actor-observer's actions result in corresponding changes in perspective. James J. Gibson (1958, 1966, 1979), together with others (e.g., Lee 1980; Neisser 1976; Pick 1990), developed the implications of this coupling across a broad range of perceptual skills. In the context of spatial orientation while walking without vision, when people act by turning in place, the perceptible result is that their perspectives rotate; that is, the network of self-to-object distances remains the same and the self-to-object directions all change at the same rate, which is equal to the rate of self-turning and opposite in its direction. When people act by walking along any straight path without changing their headings, the perceptible result is that the network of self-to-object distances and directions translate. The rates of change vary from object to object, depending on an object's initial distance from the actor-observer together with its initial direction relative to the direction of walking. What is meant by the link coupling perception and action in the present model is that afferent and efferent motor input on the action side covary in consistent ways with the perceptible changes in perspective. So, for example, the biomechanics of turning in place consistently results in rotations (not translations) in perspective, whereas the biomechanics of walking a straight path consistently results in translations (not rotations) in perspective.

The perception-representation link stands for the fact that actor-observers remember much of what they learn about their surroundings. Roger Shepard and others (Kosslyn 1980; Shepard & Cooper 1982; Shepard & Metzler 1971) investigated many of the possible similarities between perception and the resulting spatial representations and imagery. Consider the perception-representation link in the context of spatial orientation while walking without vision. Knowing about one's spatial orientation from the home-base position is a matter of remembering the distances and directions that one perceived visually or nonvisually. There are legitimate issues about the degree to which what is remembered is the same as what was perceived. For example, with visual perception distances and directions might be systematically distorted when encoded in memory (Haber, Haber, Penningroth, Novak, & Radgowski 1993), although data indicate there is no systematic distortion in memory of straightline distances that were viewed or walked (Rieser et al. 1990; Rieser & Rider 1991). As another example, Thomson (1983) suggested that the relation of what was perceived and remembered changes over time, such that the memory representation is of very high fidelity for about 8 sec and then decays rapidly; however, others have found little or no short-term memory decay (Elliott 1987; Rieser et al. 1990; Steenhuis & Goodale 1988). The same issues apply in principle to the perception of walking without vision, although they have just begun to be investigated.

The key phenomenon here is knowing one's dynamic spatial orientation while walking without vision, that is, being aware of the changing distances and directions relative to remembered objects in the surroundings. It is not obvious how this awareness should be labeled. It involves remembering objects in the surroundings, but it is not "memory" in the usual sense of rote memory. Likewise, it involves perceiving one's dynamic spatial orientation. The perceiving is driven by real-time afferent (proprioceptive) and efferent motor information specifying the walk without vision, but it is coupled with memory for objects in the surroundings. This real-time integration of motor input and spatial representation reflects the coupling of action with representation.

CENTRAL THESIS

What processes might account for this coupling? The central thesis is that the coupling of perception and action is the basis for the coupling of representation and action. While walking with access to visual and nonvisual information about the surroundings, people can learn directly about the covariation of environmental flow and the afferent-efferent motor input associated with the biomechanical actions of walking. This learning about the reliable relation of perception and action, in turn, is the basis for the coupling of representation and action, which accounts for awareness of changing spatial orientation relative to the surroundings while walking without environmental information. The next section presents some basic concepts about the perception-representation-action system.

BASIC CONCEPTS

The phrase *losing one's way* typically means that one does not know the directions and distances needed to get back to a starting point or other useful destination while one is under way. The meaning is important, because it highlights four concepts central to this chapter: Spatial orientation is something that can be known. What is known involves dynamic changes in geometrical relations, as self-to-observer distances and directions change whenever one is underway. The knowledge is relational: what is known needs to specify the frame of reference selected to fit the actor-observer's goals. Finally, the perception of one's own actions is a central feature of the model. Each of these concepts is discussed in turn.

KNOWLEDGE OF SPATIAL ORIENTATION

There are different ways to come to know one's own spatial orientation. In order to find out, one can, for example, solicit aid, look it up in an atlas, figure it out by reasoning about initial propositions like one's path of travel and landmarks in the local neighborhood, or perceive it. These ways of finding out involve different types of input information, knowledge, and processing. For example, in order to solicit aid, one must know one is lost, know how to communicate to solicit aid, and coordinate one's communication with one's destination and travel limitations. To use an atlas, one must know one's present location, destination, and orientation, and generally know how to read an atlas. There are many ways to figure something out by reasoning, including if-then statements like "If I am at the courthouse facing the theater, then I need to turn right" (this type of reasoning depends on detailed knowledge of the destination in terms of the immediate surroundings) or reasoning like "If my immediate right turn for a block was preceded by a left turn and so forth, then my destination must be straight ahead."

The emphasis in this chapter is on perceiving one's dynamic spatial orientation while walking without vision. Some take this statement to be self-contradictory by definition. The reason is that, on the one hand, their definition of "perception" involves processing sensory input in real time, and on the other hand, spatial orientation while walking without vision means maintaining knowledge of changes in self-to-object distances and directions relative to things in the remembered surroundings. According to such reasoning, directly perceiving dynamic spatial orientation while walking without vision is impossible for two reasons. First, it involves the integration of knowledge of the surroundings with real-time motor input for self-movement. Second, because the self-to-object distances and directions change dynamically, they must be calculated, not perceived.

Concerning the first reason, perception is typically informed by knowledge, whether what is perceived be a phrase of music, an object such as a particular person's face, or a particular perspective on a particular neighborhood. Concerning the second reason, events occur over time, and the perception of events involves the integration of input over time. Just as this is the case for perceiving the distance to an object via parallax or optical flow, for perceiving a melody, and for perceiving a facial expression, so too is it the case for perceiving one's dynamic spatial orientation while walking without vision. There is nothing definitionally incorrect about it. The perception-representation-action model, together with the perception-action learning thesis, explains how dynamic spatial orientation when walking without vision might be perceived instead of "figured out."

DYNAMIC AND GEOMETRICAL PROPERTIES
OF SPATIAL ORIENTATION

Spatial orientation is dynamic, because self-movement results in changes in self-to-object distances or directions. This chapter does not focus on what can be known about spatial orientation from a static position, but on processes involved in keeping up-to-date on the changes caused by loco-motion and other forms of self-movement. Spatial orientation involves know-ing about the geometry of motions in space, and it is useful to distinguish *rotations* from *translations*. *Rotation* refers either to a type of movement (a change in heading without a change in location) or to a change in per-spective (self-to-object directions all change at the same rate, with the rate equal to the rate of self-turning but opposite in its direction). The word *translation* refers either to a type of movement (a change in location, by moving in any direction(s) along any straight or curved path, without any change in heading) or to a change in perspective (self-to-object distances and directions change at different rates, and the rates depend jointly on an object's distance from the actor-observer and an object's direction relative to the direction of translation).

FRAMES OF REFERENCE

Spatial orientation and object localization are flip sides of the same conceptual coin. Whereas *spatial orientation* is defined as knowing one's distance and heading relative to one or more objects, *object localization* is defined as know-ing the distance and direction to objects in the surroundings. It makes sense to distinguish environment-centered from viewer-centered frames of ref-erence for what is known. Viewer-centered knowledge consists of knowing the distance and bearing to objects. Environment-centered frames include many alternatives, for example, magnetic compass directions, knowledge of the relative positions of other objects to use as a frame, and knowl-edge of single landmarks to use as a beacon.

PERCEPTION OF SELF-MOVEMENT

Self-movement occurs in different ways, and the ways result in different types of afferent and efferent information. Many involve the fully active motoric participation of the actor-observer, for example, walking, skating, or swimming. Others involve varying degrees of active motor participation—for example, driving an automobile, following a sighted guide while unable to see, riding the rear seat on a tandem bicycle, or being a passenger on a train. But even the most passive instance of being a passenger involves some motor participation, namely, postural adjustments for the angular and linear accelerations occurring during the movement.

Consider two ways to classify different types of information that specify self-movement. First, consider efference (feed-forward from the intended and soon-to-be-produced act) and afference (feedback from the recently produced act). In principle, the availability of efference depends on the nature of the action. So, for example, when walking for the purpose of reaching a given destination via a particular route, it makes sense to suppose that the efference includes higher levels that specify the entire route, lower levels that specify the next step, and so forth. On an afternoon stroll with no particular destination, however, only the lower level might be specified. For the purposes of this chapter, it is relevant to consider the efference available when following a sighted guide: The guided partner grasps the guide's upper arm and follows half a step behind (Hill & Ponder 1976). In the context of skillful guided partnerships, the guided person readily anticipates the next step and can anticipate forthcoming features like steps and doorways and obstacles, so there would be the chance for efference. In an unskilled context, however, the guided person could anticipate and plan for less. The sources of afference specifying self-movement include joint and muscle input signaling limb movements, pressures exerted on the guided partner's hand while grasping the guiding partner's upper arm, vestibular input from the semicircular canals (specifying angular accelerations from turning) and otoliths (signaling linear accelerations from motion and gravity), and optical flow. The availability of the different types depends on the type of action and on the situation.

Second, consider exteroception, proprioception, and exproprioception (Gibson 1958; Lee 1978). By *exteroception* is meant perception of features of the surroundings and relations among those features (including spatial relations) as they are often known by looking and listening, and sometimes by sniffing and feeling wind or temperature. *Proprioception* signifies perception of the positions of body parts relative to one another as well as changes in their relative positions. *Exproprioception* means perception of body position relative to the surrounding environment as well as change in position.

By definition, awareness of dynamic spatial orientation is exproprioceptive. To assess it one needs to ask people to judge their position, heading, or movements relative to the surroundings. Contrary to this, studies assessing the perception of self-movement have been mainly proprioceptive, including studies investigating biomechanical inputs as well as optical flow input (e.g., see Wertheim & Warren 1990).

TWO-STEP AND ONE-STEP PROCESSES
FOR THE INTEGRATION OF ACTION AND REPRESENTATION
Awareness of dynamic spatial orientation when walking without vision implies that people integrate their knowledge of the surroundings as viewed from an initial observation point with the afferent (proprioceptive) or efferent motor information associated with their perception of self-movement.

Most research about the perception of self-movement has assumed that self-movement is perceived relative to a body-centered frame of reference, whereby people perceive changes over time in their body positions, which are not themselves linked with the environment. Because people maintain their spatial orientation when walking without vision or nonvisual access to the surrounding environment, this assumption leads, in turn, to the need for a two-step process whereby people first perceive the velocities and accelerations from instant to instant, and then integrate those body-centered perceptions with their knowledge of the surrounding environment.

A two-step process such as this is plausible but inefficient, first deriving proprioceptive awareness as a step toward the goal of achieving the exproprioceptive awareness needed for subsequent action. People's awareness does not fit two-stage models. Instead, people report perceiving their self-movement directly in environmental terms, maintaining their awareness of their movements in terms of their changing distances and directions relative to objects fixed in the remembered surroundings (Garing 1998). More efficient would be a one-step process, in which exproprioceptive awareness of self-movement is perceived more directly, without first figuring the body-centered perceptions.

SITUATIONAL VARIATIONS
IN THE COUPLING OF PERCEPTION AND ACTION

The relation of rates of self-movement specified by proprioception and rates of change in perspective is locally dependable in given situations and circumstances. But it varies from situation to situation, especially when people use tools and vehicles to aid their locomotion. So, for example, during forward walking a stride of a given length typically results in environmental translation of the corresponding distance. But there are tool-related variations—for example, when one skates, skis, walks on stilts, or walks on shipboard—and there are environment-related variations in the relation of effort and environmental flow rate—when walking uphill or downhill, with the wind or against it.

The relation of the direction of self-movement and of the corresponding environmental flow (however it is specified in sensory input) is less variable. It is almost always the case that turning in place results in rotations in perspective (that is, self-to-object distances all change at the same rate and distances stay the same), whereas forward walking (or walking along any straight line) results in translations in perspective (self-to-object distances and directions change at varying rates, depending on the rate of locomotion, self-to-object distance, self-to-object elevation, and direction of walking relative to target positions). Forward walking (with eyes forward) typically results in forward translations in perspective (not rotations, not left-to-right translations), and turning clockwise typically results in a counterclockwise rotation in perspective (not a clockwise rotation, not a translation). Exceptions to this direction-direction specificity occur in the context of vehicular travel, for example, while walking on board a moving ship, plane, or train.

Variations in the direction of self-movement relative to the direction of optical flow vary whenever one turns one's head. So for example, forward walking with eyes forward co-occurs with radially expanding optical flow, whereas with eyes and head turned to the left it co-occurs with right-left optical flow. While turning in place in the clockwise direction, one can turn head and eyes in the counterclockwise direction, resulting in variations ranging up to brief periods of optical flow in either the clockwise or the counterclockwise direction.

ASSESSING SPATIAL ORIENTATION WHILE WALKING WITHOUT VISION

The general idea is to assess how well people know their dynamic spatial orientation in situations in which they do not have access to environmental information. Across all the studies discussed here, access to visual input was eliminated by blindfolds that completely occluded central and peripheral vision. Access to auditory input was restricted so that subjects could hear the instructions but could not localize the instructions or ambient sounds, because they were input over an FM broadcasting system and earphones. The outdoor studies were conducted in situations in which thermal cues from the sun were not useful and on flat, grassy fields or paved fields where there were no distinctive features on the ground surface.

People were asked to walk without vision. Sometimes the walking was part of the input to which they were asked to respond; sometimes it was the method of responding in order to localize the position of a remembered target. When walking was used as the method of responding, subjects typically walked on their own; when walking was used as a part of the input, they typically were guided by the sighted-guide method described earlier. Whether independent or guided, subjects needed to practice walking without vision to have confidence in their physical safety. Many reported that when they lacked confidence in their safety, their attention shifted away from exproprioceptive awareness of their position relative to the remembered surroundings to proprioceptive awareness of the body surfaces they wished to protect. Ten minutes or so of practice at walking without vision under test conditions seemed effective at assisting subjects in gaining confidence; the evidence for this is simply that subjects who have practiced for 10 min or so perform with as much accuracy as subjects who are highly practiced at walking while blindfolded (Rieser et al. 1990).

The guided walking typically has been arranged with the "sighted-guide" (Hill & Ponder 1976) method used to assist persons with visual impairment. The method consists of asking a person to grasp the sighted guide's upper arm and lag slightly behind the guide. With relatively little practice, guided partners report it is easy to perceive the guiding partner's gait and then actively regulate their own gait accordingly.

DIFFERENT TASKS TO ASSESS WALKING WITHOUT VISION

Different tasks have been used in different studies, motivated in part by human constraints involved in testing people who are young or who have visual impairment, and in part by the logic of what we have tried to learn. The main tasks and logic are listed here.

Path Integration

In a path integration task subjects are asked to note when they begin walking a path, which can be varied in number of turns and lengths of legs, and then asked to localize the distance or direction toward the starting point. All the logically necessary information is intrinsic to the path walked; there is no need to integrate knowledge of the surroundings with the biomechanical activities of walking. Knowing the heading and the distance back to the starting point at the end of the path depends on subjects perceiving the angles turned and distances of the path segments walked, and integrating the angles and distances over space and time. Geometrically, accurate judgments of the heading back to the start depend on accurate knowledge of the distances walked as well as the angles turned, and accurate judgments of the distance back to the start depend on accurate knowledge of the angles as well as the distances. Worchel (1951) reported the early use of path integration tasks with persons who were blind.

Walk to Target

In a walk-to-target task, subjects are asked to localize a single target object by looking (or listening) to it, then to attempt walking to its remembered location without vision. Logically, precision of walking depends on the precision of the initial visual perception, the precision of the memory representation, and the precision with which the walking is perceived relative to the remembered target. Thomson (1983) reported a relatively early version of a walk-to-target task.

Localize Remembered Targets

In localizing remembered targets, subjects are asked to study one or more targets arrayed around them in the surroundings, walk without vision to a novel observation point, and then localize the direction or distance from the novel observation point to one or more of the remembered targets. Logically, precision of localization depends on the precision of visual perception, the precision of the memory representation, and the precision with which the walking is perceived relative to the knowledge representation. Rieser et al. (1982) reported an early use of this task.

ACCESS TO KNOWLEDGE
OF THE SURROUNDING ENVIRONMENT

In walking-without-vision tasks, subjects typically begin by looking around their surroundings, walk, and judge their spatial orientation relative to targets in their actual remembered surroundings. Young (1989) and Rieser, Garing, and Young (1994) created a *virtual surroundings* method. While in a laboratory, participants were asked to call to mind some remote place that was familiar to them, imagine standing in a particular point of observation there, physically walk to a new observation point while keeping in mind their remote surroundings, and judge their changing spatial orientation relative to the remote surroundings. The results were that even three-year-old children, like adults, were able to judge their dynamic spatial orientation relative to virtual surroundings in conditions like these.

Mark Frymire (Rieser, Frymire, & Berry 1997) adapted the methods to manipulate access to knowledge of any surroundings when walking without vision. In a *virtual ganzfeld* condition (a *ganzfeld* being an unstructured field of view), subjects were equipped with a sound system, blindfolded, and guided around campus along tortuously circuitous paths for 10 min or so, until they said they did not know their location. During the tests, subjects did not know their surroundings and typically reported walking without awareness of any particular surroundings. In a *virtual environment* condition, subjects were disoriented to their actual surroundings. Then they were asked to generate an image of some familiar surroundings and to keep those imagined surroundings in mind during the tests of walking without vision. Finally, in the *real world* condition, subjects viewed their actual surroundings at the start of each trial and were asked to keep them in mind during the tests of walking without vision.

EVIDENCE TO SUPPORT THE THESIS

In this section three implications deduced from the perception-action learning thesis are stated, three experimental methods used to examine different implications of the thesis are summarized, and the ways in which the results of the three methods converge on the thesis are shown.

The first implication of the thesis is that changes in the coupling of perception and action should cause corresponding changes in the coupling of representation and action, as assessed by walking-without-vision tasks. The method involves a brief experimental manipulation of the perception-action coupling and a test for the predicted changes in the representation-action coupling, as assessed by walking without vision.

The second implication is that if the coupling of representation and action is based on experience with the coupling of perception and action, then walking without vision should show some of the same geometrical constraints as walking with vision and access to other sources of environmental flow. The approach involves reasoning by analogy to identify features of spatial orientation while walking with vision that should also apply when walking without vision. The approach involves path integration tests to see how well the analogous features predict walking-without-vision performance.

The third implication is that visual experience might influence the developmental course of the representation-action coupling, because vision specifies the perception-action coupling with more precision than other sources of information specifying environmental flow. According to this, persons born without vision are at a disadvantage in learning the perception-action coupling with as much precision as persons who can see. The method employs naturally occurring group quasi-experiments, comparing the spatial orientation of persons who were born without vision to that of persons who have life experience with vision.

EXPERIMENTAL MANIPULATIONS
OF THE PERCEPTION-ACTION RELATION

The basic idea is that while walking with access to visual and nonvisual sources of environmental information, people learn how the world looks (and sounds and feels) while walking at different rates and in different directions. Then while walking without vision, people know the dynamic changes in how the world looks-sounds-feels, draw on this knowledge to perceive their self-movement in environment-centered terms, and thus are aware of their dynamic spatial orientation. If the central thesis is correct—that those experiences with the coupling of perception and action account for the coupling of representation and action—then manipulations of the one should lead to corresponding changes in the other, as assessed by predictable changes in performance on walking-without-vision tasks. A long series of experiments that support the theory for forward walking (Rieser, Pick, Ashmead, & Garing 1995) and for turning in place (Pick et al. in press) was recently reported.

Forward Walking

The forward-walking method consisted of pretests, a learning intervention, and post-tests. During the pre- and post-tests, subjects stood in an open field, viewed a target located about 8 m straight ahead, then attempted to walk without vision to the remembered target location. During the learning intervention phases, subjects walked either in a *biomechanically faster* condition, in which the rate of walking was paired with a slower-than-normal rate of movement through the surroundings, or in a *biomechanically slower* condition, in which it was paired with a faster-than-normal rate. To arrange these conditions, subjects walked atop a motor-driven treadmill (the treadmill's motor controlled the biomechanical rate of walking), and treadmill and subject were towed on a trailer behind a tractor (the tractor's motor controlled the rate of movement relative to the surrounding environment). After walking in the biomechanically slower condition, subjects tended to stop short of the remembered target during the post-tests, as if perceiving their rates of walking as faster than their actual rates. After walking in the biomechanically faster condition, on the other hand, subjects tended to walk past the target, as if perceiving their rates as slower relative to the remembered objects than their actual rates.

The findings were replicated across ten experiments. The results showed that the change in calibration was not due to a simple motor aftereffect of walking alone (without discrepant visual input) nor a simple sensory aftereffect of optical flow alone (without walking activity to cause it). In addition, the results demonstrated response generalization such that changes in the calibration of forward walking generalized to the functionally similar action of side-stepping, but did not generalize to functionally different actions like throwing and turning in place.

Turning in Place

The calibration of turning in place was investigated with the same pre-test, intervention, and post-test methods but with a human turntable instead of a tractor and trailer. The turntable consisted of a waist-high T-bar centered within a 122-cm disc. Subjects stood atop the disc while grasping the T-bar. When the T-bar turned, driven by an electric motor, subjects needed to step in order to keep it centered at their waist. Thus, it controlled their rates of turning relative to the surroundings. When the disc turned (it was driven by a different electric motor), subjects needed to step to compensate for its turning rate. The two motors could work in conjunction to vary the biomechanical rate relative to the environmental flow rate. For example, when the T-bar turned 5 rpm in the clockwise direction and the disc turned 5 rpm in the counterclockwise direction, the result was a 5-rpm rate of environmental flow and a 10-rpm rate of leg activity. As others have observed (Bles 1981; Garing 1998; Lackner & DiZio 1988), stepping in place simply to compensate for the rotation of the disc gives rise to compelling perceptions of self-movement.

To induce changes in calibration, subjects stepped biomechanically for 10 min at one rate while physically turning with eyes open at a faster or slower rate. During the pre- and post-tests, subjects were asked to view a target object, close their eyes, and then turn in the clockwise or counterclockwise direction to face the remembered object. After walking in the biomechanically faster condition, subjects tended to turn too far in order to face the target, averaging 85 degrees of overshoot. In the biomechanically slower condition, on the other hand, subjects tended to turn too short a distance, averaging 35 degrees of undershoot. The change in calibration of turning was directionally specific: subjects showed little or no error during the post-tests when turning in the direction opposite to that in which they had turned during the learning phase.

The main result, the one relevant here, is that the overshoot-versus-undershoot direction of the post-test errors matched the biomechanically faster versus slower learning condition. This shows that the change in the perception-action relation caused a change in how people coordinated their turning with the knowledge of the surroundings. Note that the magnitude of the overshoot was significantly larger than the magnitude of the undershoot. Pick et al. (in press) reported a series of studies aimed at understanding this and found an additional motor aftereffect. It resulted from simply turning in any constant direction; this aftereffect added to the change due to the new perception-action relation.

STUDIES OF PATH INTEGRATION
WHEN WALKING WITHOUT VISION

An implication of the perception-representation-action model is that walking without vision is more precisely perceived in environment-centered terms (relative to the remembered surroundings) than in body-centered terms (relative to the temporal stream of changing body positions). It is easy to show that this fits people's subjective impressions. For example, when asked to walk a path with many turns with or without vision and then tell about how they moved, people typically describe walking from place to place in the remembered surroundings, only occasionally describing the intrinsic shape of the path walked. They are skillful at keeping track of the places visited but poor at drawing or verbally describing the shape of the path walked.

An implication of the theory is that distances translated without vision are fine-tuned in environmental terms. When walking with vision, people use optical flow, both to situate themselves in the surrounding environment and to fine-tune the calibration of distances walked. The present hypothesis is that people use their knowledge of the surrounding environment in analogous ways.

Remembered Surroundings versus Virtual Ganzfelds

One implication is that even when walking without vision, people typically have their immediate surroundings in mind, and thus can calibrate their walking relative to their remembered surroundings. In studies involving the virtual surroundings discussed earlier, Rieser et al. (1994) found that when children as young as three years were asked to walk without vision, they perceived their movements relative to objects in the remembered surroundings. Although these earlier studies showed that adults and young children alike tend spontaneously to perceive their walking without vision in environment-centered terms, they do not show anything about the relative precision of environment-centered versus body-centered perception of walking without vision.

Mark Frymire and I conducted an experiment to test the hypothesis that the perception of walking without vision is more precisely accomplished in environment-centered than in body-centered terms (Frymire 1997; Rieser et al. 1997). Subjects were equipped with blindfolds and sound systems and tested across a series of path integration trials. During each trial, sighted-guide techniques were used to guide subjects along a four-segment route. At the end of each route they were asked to judge the heading toward the remembered starting position by turning to face it (headings were assessed with a hand-held electronic compass) and asked to judge the distance by walking to the starting position (distances walked were recorded with electronic tape measures).

One condition, the virtual ganzfeld condition, was contrived so that a participant's only option would be to use a body-centered strategy. The point was to disorient participants so they could not guess their actual location, by blindfolding them and guiding them for 10 min around campus along circuitous paths while conversing with them. Participants were unable to judge their location on campus. After finishing the path integration task, all reported that while performing it they imagined themselves to be in a large, featureless empty field. The actual surroundings condition was contrived so that subjects would use an environment-centered strategy. At the start of each path integration trial, subjects viewed their actual surroundings and were asked to keep several landmarks in mind while walking.

In terms of task complexity and information processing load, one would predict better performance in the virtual ganzfeld condition than in the actual surroundings condition, because knowledge of the surroundings was not logically needed to perform the task and the instruction to keep potentially irrelevant details about the surroundings in mind might interfere with paying attention to the path and remembering the starting point. Despite this, the results were that the 16-degree average error in the actual surroundings condition was significantly smaller than the 38-degree average in the virtual ganzfeld condition.

The finding fits with the theory that self-movement is fine-tuned in environmental terms, even in conditions under which the environmental terms are mediated by knowledge, not by real-time input. Subjects underwent a third condition to help control for the possible distractions caused by walking while disoriented. The third, imagined surroundings, condition consisted of a combination of elements of the other two. As in the virtual ganzfeld condition, subjects were thoroughly disoriented before the block of trials, so they were unable to guess their test location. After reaching the test site, however, they were asked to imagine that they were standing in the open area facing the student center on campus and to generate an image of the surroundings there. Then, as in the actual surroundings condition, subjects were asked to keep in mind several of the landmarks in that imagined place while walking the test paths. The results were that subjects performed significantly better in these imagined surroundings (the errors averaged 25 degrees) than in the virtual ganzfeld condition.

Two things are important to note. One is that, as predicted, the findings support the view that the perception of self-movement is fine-tuned in environmental terms. But the other is that subjects performed ably in the virtual ganzfeld condition, presumably based on afferent and efferent motor inputs, without fine-tuning them in environmental terms.

Nearby versus Far-Away Targets
According to the theory, the better performance in the actual surroundings and imagined surroundings conditions than in the virtual ganzfeld condition suggests that the perception of self-movement is fine-tuned in

terms of flow relative to remembered features of the environment. If this is the case, then the geometry of environmental flow should constrain the benefits of remembering the surroundings. A main feature of environmental flow is that the rate of change in the self-to-environment angles and proportional distances varies as a function of the self-to-object distance, with nearby objects showing a faster rate of change than far-away objects. Because an implication of the model is that, even when walking without vision, these rates of change calibrate the perception of distances walked, it follows that path integration when walking relative to remembered nearby objects would be more precise than when walking relative to remembered faraway objects. Rieser et al. (1997) tested this implication. The tests were conducted in an open city park. Path integration tests were conducted as before, with subjects standing near the center of the park. In the faraway-landmark condition, subjects were asked to study the locations of five of the buildings surrounding the park and to keep several of them in mind while being tested. The landmarks varied from 120 to 160 m in distance. In the nearby-landmark condition, five chairs were placed around the subject, each in line with one of the five buildings, and each about 12 m from the subject. Subjects were asked to study the locations of the chairs and to keep several of them in mind while being tested. As predicted, errors on the path integration tests were significantly smaller in the nearby- than the faraway-landmark condition.

THE ROLE OF VISUAL EXPERIENCE

The third implication involves the role that visual experience might play in the developmental course of the representation-action coupling. The general idea is that experience with environmental flow fine-tunes the perception of walking without vision and adds to the precision of dynamic spatial orientation while walking without vision. It is access to information specifying generic environmental flow, not necessarily optical information, that is thought to fine-tune dynamic spatial orientation. Nevertheless, visual input specifics environmental flow with more precision and across a wider range of distances or radial directions than the other senses, so it makes sense to hypothesize that the precision of nonvisual orientation is limited by the precision with which environmental flow has been experienced. If this is the case, then the dynamic spatial orientation of congenitally blind individuals would be less precise than that of individuals who were blinded after childhood and than that of sighted individuals walking while blindfolded.

The hypothesis is that people acquire a general model for the covariation of environmental flow and the biomechanical activities of walking. The specific covariation is not fixed. Indeed, from our experimental manipulations of the calibration of walking without vision, we know that the calibration is flexible, adapted to temporary variations in how people move through their surroundings. But what could be acquired from visual experience are limits on the amount of precision with which the covariation can be fine-tuned.

Precision of Dynamic Spatial Orientation as a Function of Onset of Total Blindness

Rieser et al. (1982, 1986) tested the dynamic spatial orientation of blind and blindfolded sighted adults. The subjects were six sighted individuals tested while blindfolded, six individuals who were blinded after eight years of age, and six who were blinded before two years of age (the people who were blind ranged from being totally blind to having at most light perception). We assessed dynamic spatial orientation with a three-phase procedure. In the learning phase, subjects were shown the target locations by walking to each from a constant home-base position. They repeated the paths until they felt that they knew the remembered target locations and showed they could aim a pointer accurately at each while standing at the home-base. Then subjects were guided and walked along a J-shaped path to a novel location and facing direction. Finally, they were asked to aim a pointer toward the remembered targets from the new observation point. In the locomotion condition they were asked to aim the pointer at the remembered targets while they stood at the novel observation point. In the imagination-only condition they were guided back from the novel observation point to the home-base, asked to imagine standing again at the novel observation point, and aim the pointer at the remembered targets.

In the locomotion condition the sighted and late-blinded subjects responded similarly. Each subject reported that the task was easy, and each responded rapidly (averaging 2 sec) and accurately (averaging 18-degree error). The early-blinded subjects showed a different pattern. All reported that the task was difficult, and they responded more slowly (averaging 5 sec) and less accurately (averaging 55-degree error) than the other subjects. This finding fits the theory, namely, that for the early-blinded participants the walk to the novel observation point was not as precisely coupled with their knowledge of the target objects as it was for the other subjects.

It is important to consider alternatives. For example, given the small samples, perhaps we inadvertently selected a sample of relatively unskilled early-blinded individuals compared to our other samples, or perhaps early-blind individuals globally have less spatial experience and so perform worse at a whole range of spatial tasks. Performance in the imagination-only condition provides a method to evaluate these alternatives. In the imagination-only condition, subjects needed to imagine walking back to the new observation point. The sighted and late-blinded subjects rated this task to be moderately difficult, averaging 5 sec and 56-degree error. The early-blinded subjects found it similarly difficult, and their latencies and errors did not significantly differ from those of the other subjects.

Given that all six of the early-blinded subjects in this study performed worse in the locomotion condition than any of the other 12 participants, the findings are consistent with the possibility that visual experience plays a necessary role in fine-tuning dynamic spatial orientation. But later research has shown that this is a probabilistic affair. Several more recent studies have failed to find differences. For example, in their studies of blind individuals who were selected in part because of their good spatial skills, Loomis et al. (1993) asked early- and late-blinded subjects to perform a triangle completion task, in which they were guided along two segments of an L-shaped path and asked to walk to the beginning point. Their early-blinded subjects, selected as having very good spatial skills, performed as well as their late-blinded subjects. Haber et al. (1993) reported similar findings with a somewhat different task. These studies with highly selected early-blinded participants serve as demonstrations that the levels of dynamic orientation that typify late-blinded and blindfolded sighted individuals can be achieved without visual experience early in life.

Life Experience with Broad Visual Fields

An implication of the thesis is that individuals who were born with small visual fields would be deficient at dynamic spatial orientation whereas those born with poor acuities but broad visual fields would not be deficient. Again, this prediction depends on analysis of the geometry of translation movements, like forward walking. For a given rate of forward walking, the rates of change of the distances and directions relative to different objects in the surroundings vary, as a function of both the particular object's distance from the observer (nearby objects change at faster rates than faraway ones) and the object's direction relative to the direction of locomotion. For example, the object in line with the direction of locomotion changes its relative distance faster than the peripheral objects, the direction toward in-line objects changes not at all, and the peripheral ones undergo rapid rates of angular change. The basis for the prediction is that people with congenitally small visual fields have less opportunity than others to see how these rates vary across the field.

This prediction was tested in a study of people with varying types of visual impairment (Rieser, Hill, Talor, Bradfield, & Rosen 1992). The participants varied in the age of onset of abnormal vision (early-onset was defined as beginning by two years of age; late-onset was defined as beginning after five years of age). In addition, they varied in terms of their residual vision (field deficits were defined as having visual fields of less than 20 degrees; acuity deficits were defined as having visual acuity of less than 20/300).

Because the tests were conducted in field settings throughout the eastern and midwestern United States, we did not have methods of controlling perceptual information. So instead of assessing dynamic orientation, we assessed a product of it, namely, people's knowledge of the object-to-object relations among the landmarks in familiar places. For each individual subject, we selected a region of town that they traveled regularly—typically, their home, shopping, or work neighborhood. In each neighborhood, we selected eight landmarks that were well known and often traversed by the individual subjects. During the tests, subjects were asked to suppose they were standing at particular locations in their familiar neighborhood and then to judge the directions to the different landmarks from that imagined observation point.

The results closely fit the predictions. The errors of five of the groups—the early-onset acuity loss, late-onset acuity loss, late-onset field loss, late-onset blind, and blindfolded sighted groups—did not significantly differ from one another. As would be predicted, however, the errors of the individuals in the early-onset small-field group and in the early-onset blindness group were both significantly worse than those of the individuals in the other five groups.

CONCLUSIONS AND IMPLICATIONS

This chapter is about dynamic spatial orientation while people walk without vision, keeping up to date on their changing distance and direction relative to objects in the remembered surroundings while walking without access to environmental information. Spatial orientation was placed in the context of a functional model relating perception, representation, and action. The major features of the model are the pairwise links coupling perception with action, perception with representation, and action with representation. Included, too, is a causal relation denoting that opportunities to observe and learn about the coupling of perception and action account for people's knowledge of the coupling of action and representation, which in turn mediates spatial orientation when walking without vision under conditions similar to those reviewed here.

Three sets of testable implications were deduced from the model, together with different methods to test each set. The results of the three different methods all converge on the model. One implication is that experimental changes in the relative rates of the perception-action coupling should lead to predictable changes in the action-representation coupling, assessed through tests of walking without vision. To assess this for forward walking, tractors and treadmills were used to manipulate the relation of walking rate and environmental flow rate; to assess this for turning in place, a human turntable was used. The results in both cases demonstrated the predicted functional relation between the changes. After people walked or turned with vision in a situation in which a faster-than-normal rate of environmental flow was paired with their locomotion, they tended to walk or turn too short a distance when walking without vision to reach remembered targets. Similarly, after walking or turning was paired with a slower-than-normal rate of environmental flow, they tended to walk or turn too long a distance.

A second implication is that dynamic spatial orientation is fine-tuned in environmental terms, even when walking in the absence of environmental input. To determine this people were assessed in a path integration task, walking without vision along a four-segment path and then asked to judge the heading back toward the start of the path. In one experiment path integration was assessed in conditions in which people walked without vision in a virtual ganzfeld, without any way to know about objects in their surroundings, compared with walking in conditions in which people were aware of their actual surroundings or imagined remote surroundings. As predicted, more accurate path integration performance occurred in the latter two conditions than in the former condition. In another experiment path integration was assessed in conditions where people walked without vision relative to remembered objects that were nearby versus remembered objects that were far away. The idea was that if knowledge of environmental flow fine-tuned dynamic spatial orientation, then path integration would be more accurate when walking relative to remembered objects that were nearby, since a given unit of translation results in a faster environmental flow rate relative to nearby objects than faraway ones.

Finally, a third implication is that visual experience may play a role in the early development of dynamic spatial orientation. The implication stems from the observation that environmental flow is most precisely specified by access to information from a broad field of vision. The prediction is that persons with a history of broad field experience have the best opportunity to learn the perception-action coupling and thus should have the most accurate knowledge of the action-representation coupling. Consistent with this, studies indicate that persons who lose their vision later in life tend to show more accurate dynamic orientation than those who lose their vision early in life. In addition, persons who were born with small but usable visual fields tend to show poor levels of accuracy, similar to those of people who were born without vision.

In principle, the model is general across age-related variations in the couplings, and across different actions in addition to locomotion. Consider both in turn. The coupling of perception and action varies with ontogenesis. For example, the rate of environmental flow depends on one's stride length. As children's limbs grow, the environmental flow rate resulting from a given stride rate increases. Observations showing that even infants know about their dynamic spatial orientation while walking or moving without vision indicate that the basic system comes on line early in life. The increasing precision in orientation with age indicates that the system is updated to take growth into account.

The model can, in principle, be applied to any action. Rieser and Schwartz (see Rieser 1993) investigated its relevance to object manipulation and dynamic object perception. Adults were asked to study a complex object from a particular perspective and then decide whether a comparison object was the "same" shape or a mirror image of the original one. After they had studied the object, a shelf was placed over hand and object so that the subjects could not see them, and a visual comparison object was placed atop the shelf. The visual comparison object was either aligned with the studied object or rotated 90 or 180 degrees relative to it. In the imagination-only condition the subject was simply asked to make the decision. In the action-with-imagination condition the subject was asked to rotate his or her hand (together with the object, which was also under the shelf) either 90 or 180 degrees. The results indicated that the participants were aware of the perspective into which they had rotated the object, not the perspective from which the object was originally viewed.

Whatever its generality across ontogenesis and across actions and action systems, the converging lines of evidence indicate that dynamic spatial orientation when walking without vision under conditions similar to those discussed here is mediated by the coupling of representation and action, and that this coupling in turn is based on knowledge of the coupling of perception and action.

III

Wayfinding and Cognitive Maps in Nonhuman Species

A primary concern of this book is to explore questions of internal representation, such as that implied in the work on cognitive maps, and the processes of wayfinding. In previous sections we focused on humans, in whom cognitive processing is well established but the tie to navigation and wayfinding is not strongly defined. In this section the authors focus on navigation and wayfinding by nonhuman species, in which the presence of cognitive maps is being strongly debated. In these chapters, biological and ecological scientists examine wayfinding and discuss the possibility that different species have and use cognitive maps.

Etienne, Maurer, Georgakopoulos, and Griffin begin (in chapter 8) with a review of the significance of dead reckoning or path integration and landmark use in the representation of space. In many ways this provides a view that complements chapter 5 by Loomis et al., which presents a human navigator's view of the same process. In particular they examine suggestions that dead reckoning (which does not involve learning an environment) seems more dominant in nonhumans, whereas landmark-guided movement may be more dominant in humans. The problem of how different species combine the systems in wayfinding is examined in great detail. Drawing on examples from their group's work with small mammals, Etienne et al. suggest that animals may well have a simple cognitive map that helps their memory for routes and places (such as sources of food or food storage areas). But not all animals may have such cognitive maps. Judd, Dale, and Collett follow with a chapter on the use of landmarks for finding places by insects and for examining the microstructure of insect behavior. Their thesis is that landmarks are important for all phases of insect movement behavior and that landmarks must therefore be represented symbolically in insect memory. Roswitha and Wolfgang Wiltschko then offer a chapter on compass orientation as a basic element in avian orientation and navigation. The Wiltschkos draw several parallels between the development of navigational aptitude in young birds and the development of spatial knowledge acquisition as children grow to adulthood. The similarities of information processing, storage, and use among birds and humans give credence to the existence of different types of route and grid cognitive maps in birds as well as humans. Catherine Thinus-Blanc and Florence Gaunet present the final chapter in this section (chapter 11), focusing on spatial processing in animals and humans. Immediately we are faced with questions of the universality of cognitive maps and the degree to which similar processing operations appear to take place in human and nonhuman navigational and wayfinding activities.

In chapter 8, Etienne, Maurer, Georgakopoulos, and Griffin begin from the viewpoint that spatial representation as defined originally by Tolman (1948) and more recently

by O'Keefe and Nadel (1978) refers to a high level of spatial information processing. They use the term *cognitive map* to imply that a subject organizes the familiar environment as a system of interconnected places and that it applies a set of transformation rules to this system, which may consist of a limited number of complementary operations (such as hypothesized by Piaget 1937) or that optimize goal-directed movements. Thus whether human or nonhuman, a subject must be able to pilot and perform new route selection before being credited with possessing a cognitive map. The authors define piloting in terms of planning and performing a goal-directed path by deducing an itinerary from the memorized spatial relations between a goal and a traveler's current position, while new route performance implies an ability to select the most economical alternative path (including shortest path and shortcuts) in both familiar and unfamiliar settings. If a cognitive map alone is used, then piloting and path following must take place without either the use of beacons or reference to external landmarks. Etienne et al. argue that the general literature has yet to yield convincing evidence that spatial knowledge reaches this degree of coherence in species other than primates. They suggest using the term *spatial representation* or, more precisely, the *representation of locomotive space,* for their work with nonprimate animal species. Thus their chapter directly addresses the question of the universality of cognitive maps by suggesting that whereas spatial representation may be universal, cognitive maps may develop only in a limited number of species.

Etienne et al. then point out that the attribution of specific systems of representation to different species poses severe problems. They argue that if one ascribes to an animal or a young child particular forms of spatial representation, inevitably one begins by analyzing subjects' behavior in specific functional contexts to see how observed behaviors fit certain aspects of the environment. The authors make a strong statement that all sedentary species adapt their locomotor behavior to relevant features in the spatial environment in order to reach their goals without getting lost. Thus the observed correspondence between behavior and functionally meaningful aspects of the environment gives insights into what the traveler knows about the environment and thus how the external world is represented or modeled.

The authors then examine the process of dead reckoning, with and without the possible use of ancillary landmarks. They report that many theories of navigation emphasize that dead reckoning (path integration) plays a significant role in spatial representation and wayfinding across the entire animal kingdom from insects and other invertebrates to mammals (Gallistel 1990). Then, building on this fascinating introduction, the authors examine the role of dead reckoning in the representation of space in a comparative perspective, including hymenopterans and

rodents. They describe how insects and mammals use dead reckoning as current route-based information and how they use landmark-place associations as long-term location-based references. They then consider how the species previously mentioned represents space on the basis of route-based and location-based information, and on the interaction between these two categories of references. Their conclusion is that at this stage we have no clear-cut behavioral results to indicate that spatial representation in rodents and other nonprimate species satisfies the criterion for true cognitive maps. They argue that there is no common agreement whether rats are capable of piloting without being guided by immediately accessible landmarks, and that it remains questionable whether rodents can plan detours or take shortcuts without resorting to external cues or using dead reckoning to determine a vector from the start to the goal.

In chapter 9, Judd, Dale, and Collett examine the fine structure of view-based navigation in insects. They begin by asserting that insects learn landmarks as two-dimensional views. These views are highly dependent on vantage points, so that even over a relatively short section of a foraging trip, the insect's view of a nearby landmark will change appreciably. Insects simplify the problems of using such retinotopic views for navigation in a number of ways. For example, bees and wasps restrict the range of directions in which they approach a familiar place so that they capture roughly the same sequence of retinal images from visit to visit (i.e., approach from the same perspective view). They are guided into the vicinity of the goal by aiming at a nearby beacon landmark. Because of changes in image size and shape, a single stored view of the beacon is unlikely to allow the insect to recognize it over the whole range of possible approaches. In addition, the authors claim that wood ants are shown to take several "snapshots" of a beacon at different distances in the early stages of learning a new environment. Once close to a beacon, the insect relinquishes fixation either to approach another beacon or to approach the goal. This transition is achieved by linking a stereotyped action to a frontally stored view of the beacon. By this means the insect can acquire a standard view of the next beacon or arrive at a point close enough to the goal to allow image matching of the goal itself or the nearest landmark. The goal is then pinpointed by moving so that the image on the retina matches the view of nearby landmarks. The authors go on to suggest that there is surprising similarity in the motor constraints and landmark strategies of real insects and those of simple simulated "creatures" that have "evolved" artificially. Again, the parallel between human and non-human species stands out. In earlier chapters, Golledge (chapter 1) and Allen (chapter 2) commented on the importance of perceptual viewpoint for human landmark recognition and reviewed the work of Cornell and Hay (1984), which emphasized the significance of

changing perspectives (i.e., their "look-back" strategy) for the development of route knowledge, landmark recognition, and origin and destination recognition by children traveling through unfamiliar environments.

Moving from ground-based animals and low-flying insects to birds, the internationally acclaimed team of Roswitha and Wolfgang Wiltschko (chapter 10) discuss compass orientation and basic elements in avian orientation and navigation. Birds face orientation tasks in two behavioral contexts: homing and migration. Because of the long distances involved in migration, birds must establish contact to their goal indirectly via an external reference. Three such mechanisms have been described: a magnetic compass based on the field lines of the geomagnetic field and two compass mechanisms based on celestial cues, namely a sun compass and a star compass.

In order to use a compass, birds must first determine the compass course leading to their destination. For homing, experimental evidence indicates that experienced pigeons can derive the home course from site-specific information obtained at the starting point of the return flight. Their ability to do this even at distant, unfamiliar sites has led to the concept of the navigational "map," which is a directionally oriented representation of the distribution of environmental gradients within the home region. It can be extrapolated beyond the range of direct experience. Birds determine their home course by comparing local values of these gradients with the home values.

The "map" is based on individual experience. During an early phase in life, young pigeons derive their home course from directional information collected during the outward journey. On spontaneous flights, they record prominent landmarks and changes in navigational factors and combine this information with the direction flown to form the navigational map. Once the map is established, it is preferentially used, because it permits the correction of errors. The navigational map is a cognitive map because it allows novel routes; it differs from cognitive maps discussed for other animals by the size of the area covered and by including continuous factors like gradients. One cannot help but be impressed, at least superficially, by the similarities between their descriptions of learning behavior of these young pigeons and the developmental processes in young children that Allen describes in chapter 2.

In migration, birds must reach a distant region of the world. The course leading to this goal area is constant; the birds possess genetically coded information on their migratory direction. The conversion of this information into an actual compass course requires external references, which are provided by celestial rotation and by the geomagnetic field. Celestial rotation indicates a reference direction away from the celestial pole, whereas the magnetic field defines a specific deviation from this

course, resulting in the population-specific migration course. Both types of cues continue to interact during migratory flights. Depending on the nature of the orientation tasks, birds make use of innate information or of individual learning processes. In both strategies, however, external references provided by compass mechanisms are essential components.

Thinus-Blanc and Gaunet (chapter 11) acknowledge that many studies of spatial cognition in animals and humans refer to the notion of a cognitive map. They define this as an internal representation of an environment where places and their spatial relationships (such as angles and distances) are charted. This notion of a *cognitive map* has been extensively criticized in the past by Thinus-Blanc because the expression is antinomic and can easily lead to misunderstanding. The authors point out that "cognitive" refers to dynamic processes and "map" refers to a static picture of the real world. To this extent, the term *cognitive mapping* is functionally more correct. Internal spatial representations are said to be useful for orienting in a given environment just as they contribute to the organization of new spatial information as it is accrued. Thus Thinus-Blanc and Gaunet argue that spatial representation may be viewed as maps of the environment but more appropriately should be viewed as cognitive or active information-seeking structures. They draw on data from animal and human studies and related theoretical work to support this hypothesis.

Do animals have cognitive maps? Are there differences among humans and other species with regard to how environmental information is perceived, encoded, stored, manipulated, decoded, and used? Certainly some ants use the same "look back" strategies that appear to help humans excel in route learning; regardless of point of view, landmark recognition is common to all species. Birds appear to use both mosaic and grid maps, the latter particularly in long distance migration, whereas humans appear to have a larger geocentered representation in which local egocentered data can be embedded as a travel need arises. Young homing pigeons learn about their environments in a way that parallels the way young children learn about theirs. Ants, bees, and wasps use landmarks to guide route-following activities and their return to a home base, and itinerant monarch butterflies return to places late in the year that their dead grandparents left at the beginning of the year. Honey bees "dance" to produce a three-dimensional map of the location of nectar sites. Chimpanzees learn routes and do exact route retraces to find hidden food supplies, and squirrels know where to dig through a snow cover in winter to find an acorn buried in late summer. Sahara ants take food directly home after irregular wanderings. Do they use cognitive maps? Are their spatial representations retinal snapshots or hardwired response patterns? Or is their behavior a result of continuous spatial updating so that no spatial cognitive processing is needed to know where they are or where to go? Do the earth's magnetic field, the sun, or celestial compasses provide auxiliary wayfinding and location aids comparable to the simple tools (e.g., magnetic compass and chronometer) or complicated tools (such as satellite-based global positioning systems) that now dominate human activity? For answers to these questions, read on.

ARIANE S. ETIENNE
ROLAND MAURER
JOSEPHINE GEORGAKOPOULOS
ANDREA GRIFFIN

8

Dead Reckoning (Path Integration), Landmarks, and Representation of Space in a Comparative Perspective

Wayfinding naturally calls to mind cognitive maps, a seminal concept since the blossoming of the cognitive sciences in the last few decades. Like most cognitive terms, cognitive maps can be defined in different ways depending on the conceptual frame of analysis, the use of operational criteria, and the range of animals to which we wish to apply the term.

In this chapter, we start from the viewpoint that spatial representation as defined by Tolman (1948) and more recently by O'Keefe and Nadel (1978) refers to a high level of spatial information processing. The term *cognitive map* implies not only that the subject organizes the familiar environment as a system of interconnected places but also that it applies to this system a set of transformation rules. Following Poincaré (1904) and Piaget (1937, 1956), these rules may be formalized in terms of a limited number of complementary operations (the "group of displacements") that optimize goal-directed movements. To be credited with a cognitive map a subject therefore must be able to perform adequate new routes, that is, choose the most economical alternative path (such as a shortcut or a

Our research is supported by Swiss NSF grant 31-39311.93. Our warmest thanks to T. S. Collett, T. Gärling, and R. Golledge for revising the manuscript.

detour) under new conditions. In its broadest sense, the cognitive map concept implies the capacity for piloting. This means that the navigator can plan a goal-directed path at his current location by deducing a new itinerary from the memorized spatial relationships between different locations. The choice and subsequent performance of the path have to occur without the help of beacons or of continuous landmark guidance.

Comparative studies have yet to bring converging evidence that spatial knowledge reaches the required degree of coherence in species other than primates. We therefore focus this chapter on the less-loaded term of *spatial representation* or, more precisely, on the *representation of locomotor space,* that is, the space in which an organism changes the location or orientation of its entire body (as opposed to perceptual, postural, and manipulatory space). In animals, this representation includes all forms of spatial cognition which are expressed by locomotor behavior that helps the subject explore and exploit the environment in an adaptive manner, so as to satisfy his biological needs. A foraging bee, for instance, may represent locomotor space in general as an area over which it can apply hardwired vector navigation, the home range as a set of loosely interconnected, fixed routes labeled with sequentially arranged landmarks, and privileged places within this environment as the origin (nest) or endpoint (foraging site) of vectors associated to landmarks.

The attribution of specific systems of representation to human or nonhuman species poses severe difficulties, which we avoid by adopting a deliberately realistic epistemological attitude. To ascribe to an animal (Lorenz 1973; von Uexküll 1909) or a young child (Piaget 1937) particular forms of representation one starts by analyzing the subject's behavior in a given functional context and by examining how well the observed behavior is adapted to certain aspects of the environment. With respect to space, all territorial species adapt their locomotor behavior to relevant features of the spatial environment in order to reach their goals without getting lost. In general terms, this correspondence between the observed behavior and functionally meaningful aspects of the environment expresses what the subject knows about these aspects and therefore how he represents or models the external world (Palmer 1978; Roitblat 1982). Evidently, subtle experiments are needed to tease out what a subject really knows about the world.

Dead reckoning, also called *path integration,* is the process by which a subject continuously monitors the route taken from a particular reference position so that he or she always knows his or her distance and direction from the reference point. For instance, desert ants are capable of exploring their fairly homogeneous environment without relying on landmarks. Instead, they measure, combine, and "integrate" the angular and linear components of their locomotion so as to remain constantly informed of their distance and direction from the nest. Navigating predominantly by dead reckoning and usually commuting between the nest and only one foraging site during a given time-span, ants may represent their foraging range by one vector that is anchored to the nest and points to a particular foraging site.

Current theories of navigation (Gallistel 1990) emphasize the role of dead reckoning for spatial representation across species, from insects and other invertebrates to mammals. For humans, this possibility raises the question of how information of which we are little aware influences our perception and representation of space. Consciously, we perceive the spatial environment mainly as the layout of landscape features. It is, however, very likely that our general representation of locomotor space owes as much to the indirect apprehension of space through continuous internal feedback and external reafferences (such as visual flow) originating from locomotion as it does to the direct perception of discrete spatial cues from the environment. This is particularly the case with respect to the representation of space as a continuum. Locomotion determines an individual's current position in a continuous manner and therefore contributes in an essential way to his "position sense" (Gallistel 1994), that is, the representation of where he or she is located in familiar space. At the same time, locomotion interlinks particular locations, which the subject singles out as discrete places according to their functional signification.

Our aim here is to examine the role of dead reckoning and landmarks in the representation of space through a comparative account of hymenopterans and rodents, that is, of those species in which dead reckoning and landmark learning has been studied most extensively (see also Benhamou & Poucet 1996). We describe how these insects and mammals use path integration to obtain current information about position and landmarks to provide long-term location-based references. This is followed by a discussion of how these species represent space using route-based and location-based information, emphasizing particularly the interrelations between the two kinds of spatial information. From a more general cognitive viewpoint we wish to show how insects (whose particular subsystems of the central nervous system function in a relatively autonomous manner) and rodents (which possess a more telencephalized mammalian brain), use the same spatial strategies either in relative functional segregation or as an integrated system.

DEAD RECKONING

GENERAL REQUIREMENTS

A convenient way to study dead reckoning in animals is to analyze homing behavior in central-place foragers, that is, in species that commute between their nest and particular feeding sites. Dead reckoning has already been defined as a process by which a subject remains constantly informed of his position relative to a particular point of reference by means of signals generated during locomotion, that is, of route-based information. For a central-space forager the major point of reference is the location of its nest or retreat, where the animal initiates and ends each foraging excursion. Thus, dead reckoning enables the animal to keep track of its position with respect to its home and therefore to return home directly at any moment and from any point of each particular foraging excursion.

Figure 8.1 illustrates dead reckoning in two central-place foragers during a round trip with an L-shaped outward journey. After having collected food, honey bees and golden hamsters home directly, without reproducing the outward detour, as if they knew the (approximate) return vector. Back in the hive, bees communicate the direction and distance to the feeding site to their nestmates through their dance axis, which may be described as a miniaturized outward vector to the feeding site.

Updating the current position vector imposes the following requirements on the subject. Whenever the animal moves it measures the angular and linear components of its locomotion through signals generated by locomotion. The signals that originate from rotations and translations are combined with one another to estimate momentary changes in location and head direction. The summation of moment-to-moment changes in position after each step yields a moment-to-moment indication of the animal's current position relative to home, or, in other terms, the current position vector.

After L-shaped outward journeys, bees and hamsters tend to commit systematic homing errors. In the examples shown in figure 8.1, the two species chose a homing direction with an inward error, which for hamsters at least, expresses mainly the overcompensation of the clockwise or counterclockwise rotation of the outward journey (Séguinot, Maurer, & Etienne 1993). The same biases are also shown by other species including ants, spiders, dogs, and humans (e.g., see Etienne, Berlie, Georgakopoulos, & Maurer 1998a; Maurer & Séguinot 1995; Wehner 1991). This formal similarity in the behavior of unrelated species suggests that path integration may involve a similar algorithm for processing route-based information across different taxa.

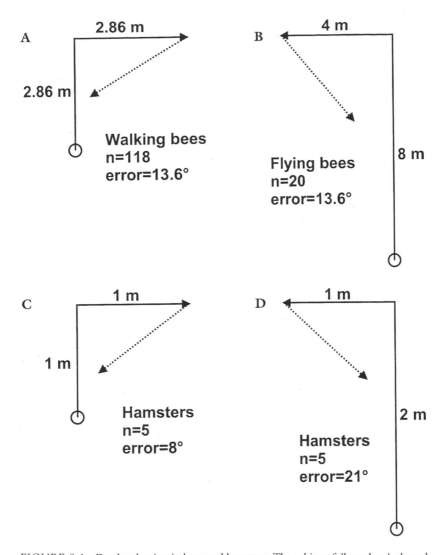

FIGURE 8.1. Dead reckoning in bees and hamsters. The subjects followed an L-shaped outward journey (solid line with an arrow) from their nest (small circle) to a foraging place. The direction of the estimated return vector to the nest (hatched arrow) was deduced from the axis of the figure-eight dance of bees and measured directly for hamsters. Bees could use the sun as a directional reference during the entire foraging trip. Hamsters were tested under infrared light and auditory, olfactory, and tactile cues were eliminated or masked. (Data from Bisetzky 1957 on walking bees, Lindauer 1963 on flying bees, and Séguinot et al. 1993 on golden hamsters.)

The analysis of systematic errors has been used to model this algorithm; by now, it is generally admitted that this algorithm leads to a very approximate form of path integration (Benhamou, Sauvé, & Bovet 1990; Benhamou & Séguinot 1995; Fujita, Klatzky, Loomis, & Golledge 1993; Maurer 1998; Maurer & Séguinot 1995; Müller & Wehner 1988, 1994).

Whatever the interspecific similarity in the formal processing of route-based information to update position, the input variables into the path integration system may vary among different taxa. The most important differences in the measurement of rotations and translations depend on whether the two components of locomotion are estimated with or without the help of external references. Bees and ants measure rotations and therefore the current direction of their progression with the help of the time-compensated sun azimuth (the sun's position in the horizontal plane; see von Frisch 1965), and, according to recent findings, translations by means of optical flow and possibly proprioceptive information (Esch & Burns 1996; Srinivasan, Zhang, Lehrer, & Collett 1996; Wehner, Michel, & Antonsen 1996).

Dusk- and night-active rodents, on the other hand, rely primarily on dead reckoning when visual cues are not available and do not use any external compass. This was confirmed by experiments on hamsters in which the animals were able to choose the correct homing direction while being deprived of any directional reference from the environment, including the geomagnetic field (Etienne, Mauer, Saucy, & Teroni 1986). Rather, the animals assess direction and distance through inertial signals from the vestibular system, somatosensory feedback, and efference copies (copies of the motor signals that specify speed of locomotion and changes in direction), with vestibular signals playing the main role in the estimation of rotations, but not in the assessment of translations (see Etienne, Maurer, & Séguinot 1996).

IMPORTANCE OF AN EXTERNAL REFERENCE
FOR MEASURING DIRECTION

The use of external references greatly enhances the precision in the measurement of direction and distance. This is particularly important with respect to the estimation of direction. When bees and ants advance in a given direction, they can measure this direction continuously with respect to the sun azimuth. By contrast, relying entirely on evanescent signals from within the organism, rodents measure changes in direction only at the very moment of their occurrence, during the stimulation of the corresponding sensor systems. Dead reckoning is a recurrent process in which changes in position and orientation are calculated at every step on the basis of the estimated position and bearing at the end of the preceding step. It is therefore open to cumulative errors. The less precise the input variables, the faster errors accumulate.

As shown by extensive research on bees and desert ants (Rossel & Wehner 1982; Wehner 1994), the insects derive compass information from the polarization pattern of blue and ultraviolet (UV) sky light, a pattern that is anchored to and therefore moves with the sun. While walking, the ants maintain the local sky in the fixation plane of their uppermost dorsal eye region. In this "dorsal rim region," the insects are equipped with ultraviolet receptors, each of which has a peak sensitivity for UV light that is specifically polarized. Thanks to their spatial distribution on the dorsal rim, the UV receptors form a two-dimensional sensory template that is roughly complementary to the polarization pattern of UV light on the sky. By rotating its head around the vertical axis the insect increases the stimulation of its receptor system until it is in register with the symmetry line of the skylight pattern, which corresponds to the solar and antisolar meridian. To distinguish between the two meridians and therefore to use the celestial polarization pattern as a true compass, the insect can rely on chromatic differences in skylight between the solar and antisolar hemisphere. Thus, the insect determines the current sun azimuth by using a few simple rules that allow it to match its hardwired representation of the sky with the real sky.

As is the case for all species that derive compass information from the sun, hymenopterous insects need to compensate for the daily east to west movement of the sun. Recent experiments on bees (Dyer 1996; Dyer & Dickinson 1994) and ants (Wehner & Müller 1993) have solved the longstanding puzzle of how these insects acquire a relatively precise representation of the (nonlinear) movement of the sun's azimuth. Foragers who have been exposed to the sky only during the early morning or late afternoon behave as if they imagine the sun's horizontal displacement as a step function, with the sun rising and setting in opposite directions and changing its position abruptly by 180 degrees around midday. In natural conditions the insects adjust this innate representation to the real displacement of the sun azimuth, using salient landscape features and their internal clock as spatial and temporal references. Thus the insect learns where the sun is located with respect to the edge of a wood or the horizon skyline at different hours of the day. In bees, this learning process culminates in the capacity to locate the sun's current azimuth with respect to the landscape even under complete overcast (Dyer & Gould 1981). This unique performance resembles the "re-representation" of external entities that are out of sight; it may, however, be explained through a rigid mechanism that relates the sun's successive positions relative to the horizon skyline to the current state of the animal's internal clock.

Benefiting from an efficient path integration system, bees and ants navigate over relatively long distances by depending on dead reckoning mainly, although their flight vector can be influenced by landmarks encountered en route (Chittka, Kunze, Shipman, & Buchmann 1995). Furthermore, these insects can keep vector information based on dead reckoning in memory. Thus, after landing at a foraging site, bees may store their flight vector and recall it at a later moment by making reference to the sun azimuth. They may then fly home along the inverted outward vector. As already mentioned, bees also represent through their figure-eight dance a miniature version of the outward vector to direct new foragers to the same location. On the vertical dance floor of the visually completely shielded nest, they use gravity as a directional reference instead of the sun and are able to repeat their dance session after considerable time intervals, indicating the direction and distance to the same feeding site without themselves leaving the nest again.

Path integration not assisted by external references may play an important role in conjunction with other orientation mechanisms but by itself it has limited functional value. During homing behavior in the context of hoarding, golden hamsters estimate the rotational component of the outward journey to a feeding place with limited accuracy (see Etienne, Maurer, & Saucy 1988; Etienne et al. 1996). Furthermore, the animals cannot refer a currently calculated vector to a stable external compass and therefore are unable to recall its direction. This means that hamsters have to update their position during the entire foraging excursion, updating also being needed when they are moving around a feeding place without changing their location.

In rodents (Alyan & Jander 1994; Mittelstaedt & Mittelstaedt 1982; Potegal 1987), dead reckoning may be considered mainly as a back-up strategy that allows an animal to explore unfamiliar parts of its environment over limited distances when familiar landmarks are not available. We shall see, however, that in mammals dead reckoning may also be the prewired spatial frame that incorporates landmark information and, in particular, the spatial relationships between landmarks. Dead reckoning may, therefore, underlie spatial representation in a fundamental manner if we consider this process part of an integrated system.

DOES DEAD RECKONING OCCUR IN AN EGOCENTERED
OR GEOCENTERED FRAME OF REFERENCE?

An important issue is whether the representation of space is egocentered (subject-centered) or geocentered (earthbound). In an egocentered representation, the frame of reference is anchored to the animal's head, which, by definition, is at the origin of the coordinate system, whereas in a geocentered representation, the frame of reference is anchored to the earth (and its origin, in the simplest case, can be assumed to lie at the starting point of the excursion).

As illustrated by our examples of homing in animals, dead reckoning in its simplest form involves two points in space: the fixed, earthbound point from where the subject departs, and the animal's current position, which changes from step to step. Dead reckoning does not, however, require a general frame of reference such as a geocentered Cartesian or polar coordinate system to which the animal would relate its own position. Consequently, whether we assume the positional variables to be represented egocentrically or geocentrically would seem to be mostly a matter of personal belief (but for a discussion of the possible roles and interplay between the two modes of representation, see Gallistel 1990; Gallistel & Cramer 1996).

The first important model of dead reckoning, proposed by Mittelstaedt and Mittelstaedt (1973), described this process by means of strictly geocentered variables. In contrast, subsequent authors (most notably, Benhamou et al. 1990; Fujita et al. 1993; Müller & Wehner 1988) believed that a geocentered frame of reference was an unnecessary assumption for the simplest realization of path integration, whereby only one point is represented besides the origin of the coordinate system. Although the issue seems simple, it has been clouded by the absence of explicit definitions of what distinguishes an egocentered from a geocentered reference system. An example of this kind of confusion is apparent in Wehner's description of his own model of dead reckoning in desert ants (Müller & Wehner 1988; Wehner 1996; Wehner et al. 1996). The model is claimed to be egocentered, but actually the positional variables it uses are inscribed within an earthbound polar-coordinate system whose origin lies at the ants' nest and whose reference axis is fixed relative to the sun azimuth (time-adjusted to allow for the rotation of the earth).

In a purely egocentered frame of reference, a compass can only be used to estimate the changes in direction during the excursion (see Benhamou et al. 1990); on the contrary, in a geocentered frame of reference, the animal has to know its heading (as given by a compass) in order to compute its position after the current step: just knowing the instantaneous changes in direction will not do.

Our initial experiments on dead reckoning in golden hamsters revealed the existence of an *internal compass* that receives vestibular inputs. In certain homing experiments occurring in darkness and with a passive outward journey from the nest to the food source, the animals compensated for the outward rotations, but not for the outward translations, and therefore homed in a *constant direction,* irrespective of the true location of the nest (Etienne 1980; Etienne et al. 1986, 1988). Furthermore, one of the most striking recent findings of the neurobiology of rodents is the existence, in various structures of the rat's brain, of populations of visually and vestibularly driven neurons that code for geocentered directions. Because such a compass is mandatory only for dead reckoning in a geocentered frame of reference, it may well turn out that Mittelstaedt and Mittelstaedt (1973) were right after all.

LANDMARK-PLACE ASSOCIATIONS

Familiar landmarks that are associated with specific locations are most appropriate for helping the subject to return to particular locations on a long-term basis. The association of landmarks with particular places may be based on dead reckoning, which informs the subject of its own position within its home range and therefore also of the position of nearby landmarks. Once a landmark-place system has been established, however, it may also function by itself. This is suggested by experiments where the subject finds its way to a particular goal after having been displaced to a familiar release site (provided that the familiar release site does not trigger previously computed route-based vectors).

Stable landmark-place associations—the discrete units of spatial representation—are required not only for organizing familiar space on a long-term basis but also for pinpointing specific goal sites such as a nest entrance occluded from sight. Thus they are used by all territorial species. On the other hand, the mechanisms for learning to identify specific landmarks and associate them with particular places may vary considerably among different taxa. The processes by which animals learn to identify landmarks and to associate them with a given place are best revealed by careful experiments with insects, but knowledge of these processes still remains rather speculative with respect to mammals.

INSECTS

Hymenopterous insects like wasps, ants, and bees memorize and use landmarks according to their strategic function. Landmarks close to the goal play a crucial role in helping the insect to aim at the goal location and are therefore registered and recognized precisely, from a constant position. More distant landmarks may either form part of the general landmark panorama against which close landmarks are identified or be used for route guidance; these landmarks seem to be memorized more globally and from more variable positions.

To locate a goal such as the nest or a food source, honey bees (Cheng, Collett, Pichard, & Wehner 1987; Lauer & Lindauer 1971) and wasps (Tinbergen 1972; Zeil 1993b) *select landmarks* that are close to the goal and provide them with precise positional information. Motion parallax (image motion induced by the subject's own motion) allows the insects to filter out these landmarks independently of their angular size (Lehrer, Srinivasan, Zhang, & Horridge 1988). At the same time, the insects prefer relatively prominent objects that stick out from the ground and are visible from a distance to smaller and less conspicuous configurations (Cheng et al. 1987; Lauer & Lindauer 1971).

According to the "snapshot" model (Cartwright & Collett 1983), honeybees and other hymenopterans *memorize* particular landmarks by fixating the visual configurations from a given location and, most importantly, from a particular direction, which they choose and maintain with the help of a magnetic or visual compass (Collett & Baron 1994). This allows the insect to "photograph" the landmarks retinotopically, that is, according to their projection on a constant region of their compound eye.

When ants and wasps leave a goal place to which they will later return for the first time, they perform elaborate orientation flights following a sequence of fixed geometrical patterns (for a recent review see Collett & Zeil 1996). During these flights, the insects register the landmark objects from different vantage points. When revisiting the same site, they may then rely on landmark guidance to reach the spot from which they took a snapshot of the immediate goal region. To face the relevant landmarks from the same direction as during their first visit to the goal, they use their biological compass (Zeil 1993b; for further references see also Collett 1996a). It is under these conditions only that the animals can *recognize* the landmarks near the goal through image matching, that is, by establishing an accurate retinotopic correspondence between the actually perceived landmarks and their memorized snapshot. This means that in order to be recognized, the landmark object has to be projected on the same region of the retina as during its registration. Thus, once the insect has achieved accurate image matching, it has only to turn and/or advance by a given amount to reach the goal. Consequently, the learning process for identifying particular landmarks automatically implies the association of these landmarks with specific places.

To make sure they retrieve the correct snapshot memory while approaching the goal, the insects rely on contextual cues. Prominent features from the more distant landscape panorama play an important role in this respect, as they necessarily remain correlated with the feeding place. Furthermore, distant references give less accurate positional information than close ones, but present a more constant image to the approaching insect. Thus, in conflict situations between landmarks specifying the precise food location and information from the more distant surroundings indicating their own location within the general test space, bees base their search of the goal on distant rather than close landmarks (Collett & Kelber 1988).

En route to the goal, flying hymenopterans may be guided by single, prominent configurations or by sequences of landmarks. In experiments on sequence learning in bees, the insects linked specific motor instructions to each particular visual stimulus and expected to find particular configurations at specific points along a familiar route. Thus, bees link together particular path segments by relying on visual stimuli that appear at the appropriate time (Collett, Fry, & Wehner 1993). According to recent experiments by Dyer (1996), the landmarks along a route can be recognized from new vantage points, provided the latter are not too different from the view points under which the landmarks have been registered (Collett, personal communication).

After a passive displacement to a familiar feeding site, walking desert ants return home by following a sequence of landmarks (Wehner et al. 1996). But so far there is no clear evidence that ants go beyond recognizing each particular landmark, that is, that they memorize the order in which they encounter the familiar visual references. The guidance by particular landmarks also helps the ants to commute between the nest and feeding sites over shrubby areas, where the plants represent both obstacles *and* visual references. Particular individuals choose constant routes to walk through a test area with controlled "landmark obstacles," recognizing each configuration independently of its distance from the starting point and bypassing it in a constant direction, that is to the right or left (Collett, Dillmann, Giger, & Wehner 1992).

RODENTS

Although little is known about the way in which rodents *select* landmarks in the wild, laboratory studies show that rats and other species rely mainly on visual spatial cues to navigate in controlled conditions, priority being given to distal (extramaze) over proximal (intramaze) visual configurations (for references and neurophysiological data see Cressant, Muller, & Poucet 1997; Etienne, Joris-Lambert, Maurer, Reverdin, & Sitbon 1995; Etienne, Joris-Lambert, Reverdin, & Teroni 1993; Etienne, Teroni, Hurni, & Portenier 1990). Distant landmarks are related to the general shape of the test environment, which represents the major frame for perceiving and representing space according to recent studies (Cheng 1986; Gallistel 1990). Olfaction also plays an important role for these predominantly dusk- and night-active species (e.g. Benhamou 1989; Jamon 1994; Tomlinson & Johnston 1991). Recent experiments on rats show that reliance on olfactory cues depends on a number of different variables, in particular on the visual test conditions. Furthermore, when both visual and olfactory stimuli are presented, they interact with one another in complex ways (Lavenex & Schenk 1995, 1996, 1998).

Within an open experimental area, distal visual landmarks can be used for locating specific goal sites and as general directional references. Experiments on golden hamsters under controlled optical conditions (Etienne et al. 1995) show that the animals select the surrounding visual landscape as the main spatial reference for homing, provided the panorama combines a continuous, low frequency (Field 1987) background pattern with at least one superimposed vertical two- or three-dimensional structure. Hamsters seem to choose distant visual references according to general principles; these principles, however, cannot be related to known functional properties of the animals' visual system.

A major problem remains to be explained: how do rodents and other mammals *learn* to label places with particular landmarks? This process leads not only to the identification of specific landmarks through their intrinsic features, but also to the representation of their spatial relationship (distance and bearing) to the relevant site. In recent experiments (Biegler & Morris 1993, 1996), rats were presented with an array of two landmarks and a food source that remained in a fixed position with respect to one of these landmarks. The animals used this landmark as a predictor of the food location only if the two landmarks remained in a fixed position within the experimental arena. By contrast, the rats always learned to distinguish the two landmarks of the array from each other and therefore to recognize them independently of their absolute and relative positions within the experimental area.

This and other results point to the important fact that the identification of a landmark and its use for locating a goal depend on different learning processes, and that the general rules that govern associative learning (such as relations of contingency and predictability between the landmark as conditioned stimulus and the food source as unconditioned stimulus) do not suffice to explain how the animals link landmarks to a rewarded goal site. As recent data on the organization of the primate visual system suggest, information on object identity and object location is processed along different pathways, both in the sensory processing cortical areas (Ungerleider & Mishkin 1982) and in the prefrontal cortex that controls cognitive and executive functions (Wilson, Scalaidhe, & Goldman-Rakic 1992).

Classical as well as recent data (for reviews see O'Keefe & Nadel 1978; Poucet 1993) show that mammals use—and therefore perceive and store—landmarks in a configurational manner. Thus, the general pattern formed by distal cues plays a greater role in specifying a particular maze arm than does each cue taken by itself (Suzuki, Augerinos, & Black 1980; O'Keefe & Conway 1980). As for proximal references, rats tend to use them in a less integrated manner for simple cue guidance (e.g., Diez-Chamizo, Sterio, & Mackintosh 1985). Under appropriate conditions, however, the animals can also learn to rely on the configuration of visual and tactile cues on the floor and walls of a maze (Hughey & Koppenaal 1987).

Unlike insects, rodents and other mammals *recognize* a goal site through landmarks independently of the direction of their approach to the goal. The representation of a single place must therefore involve the combination of different "local views." It may be that the subject perceives, processes, and registers these views by performing rotatory movements at the goal location (Sharp 1991). Thereupon, the animal may identify a given place from any direction; approaching, for instance, the goal location from the south (and therefore looking north), the subject recognizes the local view it had previously registered at the goal location while looking north (Poucet 1993).

Furthermore, recent experiments show that rats can learn to identify a place by processing positional information from three identical dim light sources even when these cues are never presented together: One of the lights is turned on only when the subject is in the central part of the test apparatus, while the other two lights are presented when the subject is in the peripheral region (Rossier, Grobéty, & Schenk 1996). These data suggest that rodents can recognize particular places without relying on simple image matching.

An alternative conception of place recognition with the help of close landmarks was proposed by Collett, Cartwright, and Smith (1986) for

Mongolian gerbils (*Meriones unguiculatus*). The animals had to locate a feeding place with respect to one or several three-dimensional landmarks in a large open arena. They behaved as if they were able to compute new trajectories to the goal with the help of a single landmark, by subtracting the stored landmark-to-goal vector from the currently computed vector that links their own position to the landmark. Planning the goal-directed path through the relatively simple principle of vector addition requires a preliminary matching process between the actually perceived landmark array and its internal representation. This matching process is rather demanding; it requires not only the recognition of each landmark through its own properties and relative position in the array, but also a directional reference for aligning the represented (that is, memorized) and the real array with each other.

Here again the crucial question of a general directional reference for rodents and mammals in general arises. Only a few ethological studies point to the use of a universal external reference such as the sun (see Teroni, Portenier, & Etienne 1987) or the geomagnetic compass (Burda, Marhold, Westenberger, Wiltschko, & Wiltschko 1990; Mather & Baker 1980) in spatial orientation by rodents. Other, less stringent candidates for setting an overall reference direction are prominent extramaze cues or distant landscape features (O'Keefe & Nadel 1978), the geometric slope of the environment that the animal derives from the general cue distribution (O'Keefe 1991b), or the preferred axis of movement, which depends on gross asymmetries in the environment (Poucet 1993). Since the discovery of head direction cells (Taube, Muller, & Ranck 1990a; for recent reviews see Muller, Ranck, & Taube 1996; Taube, Goodridge, Golob, Dudchenko, & Stackman 1996), an internal compass which is driven by angular acceleration and reset by local views of the visual landscape represents an essential component of mammalian navigation in the physiological and neural network literature (McNaughton, Knierim, & Wilson 1994; McNaughton et al. 1996; Redish & Touretzky 1997; Skaggs, Knierim, Kudrimoti, & McNaughton 1995; Touretzky & Redish 1996; Wan, Touretzky, & Redish 1994b; Zhang 1996).

SPATIAL REPRESENTATION

So far we have described how certain insects and rodents find their way by relying either on dead reckoning or visual landmarks. A complete account of how these animals know or represent locomotor space would also include the contribution of other types of sensory information.

Olfactory information, for instance, plays a major role in guiding insects as well as mammals to a goal. But planning a trajectory in advance, which is the cognitively most interesting aspect of navigation, depends on dead reckoning and on vision, with other kinds of input controlling mainly the execution of the path.

The final part of this chapter aims at deducing from behavioral facts how dead reckoning and landmarks influence the representation of loco-motor space in hymenopterans and rodents. The focus of this section is on the interaction between route-based and site-dependent information and their mutual contribution to a so-called map. We use this term in its broadest meaning, referring to the question of whether, to what degree, and how the subject interlinks different locations of his environment and therefore represents familiar space as a more or less interwoven system. Thus, the concept of *cognitive map* refers to maps of the highest level: it implies a coherent representation of the spatial environment that generates the free combination of path segments so as to reach a goal in the most appropriate manner, whatever the point of departure or the availability of particular landmarks.

INSECTS

We have seen that bees and desert ants perform path integration with a considerable precision by using the sun azimuth as a directional reference. This requires the capacity to extract the sun's position in the horizontal plane from its position in three-dimensional space by relying on sky light patterns that are correlated with the current position of the sun. Further, the insects have to take into account the time course of the solar azimuth's daily movement through the sky. The way in which the insects achieve both conditions demonstrates nicely how their behavior is adjusted to complex features of celestial space, and therefore reflects an implicit knowledge or representation of these features. On the insect level, this representation is based on hardwired receptor systems, innate time-dependent response functions, and preprogrammed learning.

Thus, searching for food over unfamiliar terrain, desert ants walk and honeybees fly over considerable distances by relying exclusively on dead reck-oning. In conflict situations between dead reckoning and landmark-place associations, furthermore, the insects give a strong precedence to dead reck-oning and tend to ignore visual input (Wehner 1996). Finally, messenger bees have no means of recruiting newcomers to a profitable food source by informing them of landmark linked routes; instead, they literally dance their flight vector in the midst of their sisters, on the colony's dance floor. We may therefore conclude that the insects' representation of locomotor space in general is mainly based on route-based vector navigation.

Despite their secondary function for the navigation system of ants and bees, landmarks are absolutely necessary to stabilize the long-term representation of the home range. Landmarks close to crucial places are represented by two-dimensional snapshots that control site recognition through image matching. Being based on the direct projection of the landmark configurations on the retina, snapshots exclude form and size constancy and therefore do not allow the animals to recognize landmarks from different places or under different angles. By contrast, more distant landmarks that inform the insect of its general location or characterize a particular route can be recognized under new perspectives, possibly because their appearance changes less rapidly with the insect's displacement than the sight of close landmarks (Collett, personal communication).

Let us now consider how insects interlink route-based vector information and location-based landmark information. We have already seen that compass-aided path integration allows the insects to store vectors on a longer-term basis. When bees trained to forage at two different sites are displaced from the hive to one of these places, they tend to home directly (Menzel et al. 1996; Wehner, Bleuler, Nievergelt, & Shah 1990), as if the landmark-based recognition of the release site induced the retrieval of the corresponding return vector. According to the model of Cartwright and Collett (1987), bees may associate a homing vector with the landmarks around a feeding site and store at least two such vectors. Interestingly, the retrieval of a particular vector at a given site depends not only on the specific landmark panorama and on its perceptual salience, but also on internal factors linked to the insect's motivational and attentional state (Menzel et al. 1996).

The coupling between dead reckoning and landmark information can also be observed at the end of a homing trip. When ants have covered the correct bee-line distance known through dead reckoning, they search for the landmarks that match snapshots taken in the nest region (Wehner 1996). The expectation of seeing specific landmarks after having covered a given distance occurs only near the nest, when the path integrator approaches a zero state. Further away from home, when the insect is still en route, landmark memories are retrieved independently of the distance the animal has covered between the feeding place and the nest (Collett et al. 1992). Thus, the insects link landmarks to the current length of the homing vector only at the most crucial site of the familiar environment, namely, the nest region.

The main question pertaining to spatial representation concerns the general organization of space as a maplike system. Bees have been credited with route-based vector maps that generate new routes through vector addition and with topographic maps that relate particular landmark-place associations to each other through a common coordinate system.

That species that excel in path integration may possess a kind of vector map has been considered repeatedly in the literature. Recent experiments show, for instance, that bees that have been trained to fly to two foraging sites and are then released at a third feeding site halfway between the familiar training sites home directly, along the shortest path (Menzel et al. 1996). It has also been suggested that it might be possible to plan more elaborate shortcuts through the combination of vectors associated with landmarks that the insects could retrieve at a familiar releasing site (Cartwright & Collett 1987). Further experiments are needed to exclude the use of landmark guidance at the release site. At the moment it is unclear whether hymenopterans can acquire a more general vector map and therefore the capacity to plan and execute new routes by adding different vectors to each other (Wehner et al. 1996). But the very principle of dead reckoning implies that the insects can commute between particular feeding sites and the nest along the shortest path as well as through detours (see figure 8.2A).

In a seminal but controversial study, bees have been credited with map representation endowing them with a high level of flexibility in information processing (see figure 8.2C). After being trained to fly to a foraging site (F2), bees were captured at the hive and released at another site (F1) within their home range, from which they were not supposed to see any landmark associated with site F2. The bees flew directly from F1 to F2 and therefore seemed to perform a shortcut without landmark guidance (Gould 1986). However, further experiments under better-controlled conditions showed that bees take shortcuts only if, at the new releasing site F1, they can perceive distant landmarks learned on previous trips to F2 (Dyer 1991). It seems therefore that bees, rather than forming a representation of the geometrical relationships between different places and different routes, memorize landmark constellations while proceeding to a given goal, and can recognize these landmarks from new vantage points. Thus, the insects build up a limited, one-dimensional system of route maps, which imply landmark guidance (Dyer 1996).

Taken together, current data do not support the hypothesis that insects form true maps, that is, represent locomotor space within a general earthbound frame of reference. When it comes to acquiring a representation that

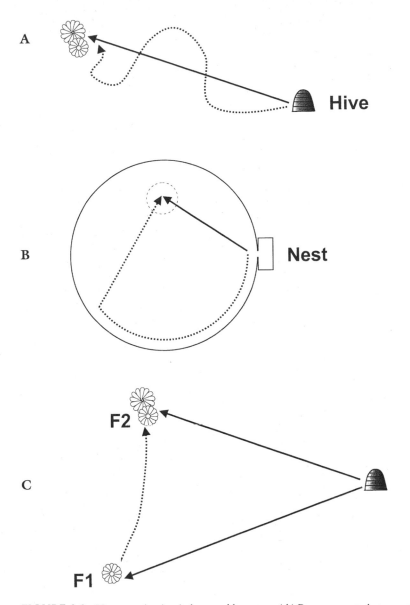

FIGURE 8.2. Vector navigation in bees and hamsters. (**A**) Bees commute between their nest and different feeding sites along stored vectors (solid arrow). (**B**) Likewise, hamsters can process and memorize a nest-to-food site vector (solid arrow). (**A** and **B**) From their nest, both species can reach a familiar feeding site either directly (solid arrow) or through detours (dotted path), by subtracting their current position vector from the stored nest-to-food site vector. Both species, furthermore, associate the end points of the memorized vectors with external cues. (**C**) The capacity to *plan* shortcuts between different, independently memorized feeding sites (F1, F2) without landmark guidance and after passive transport to F1 would suggest that the animals possess a true cognitive map (here represented as a vector map). This level of information processing does not seem to be reached by bees and has still to be examined for hamsters.

goes beyond the encoding of particular places in the form of snapshots, the insects are at most capable of recognizing sequences of local images when proceeding from one location to another (Wehner 1996) and of identifying landmarks under a new perspective. Ants and bees travel along novel goal-directed routes only with the help of currently perceived beacons, landmarks, or landmark sequences, which they have associated with particular routes and goal sites during previous foraging trips.

In conclusion, ants and bees owe their astonishing navigational capacities to specialized subsystems (for further data, see Wehner 1991). Some of these subsystems are entirely preprogrammed; others reach their final form through the (preprogrammed) registration of specific experience. The overall picture is a remarkable orchestration of precisely working mechanisms that are activated one at a time by either route-based or site-dependent inputs and thus cooperate sequentially rather than synchronously. Speaking of spatial representation in these species, we must keep these principles in mind. It is unlikely that insects know their home range (or parts of it) as an integrated system, where vectors and landmark-place associations are completely interlinked. In modular terms, the animals possess not one spatial module but at least two. We have mainly dealt here with the path integration and the snapshot subsystem. Both of these systems imply an astonishing degree of precision, in particular for matching internal representations with their external models. Their capacity is limited, however, when it comes to extracting more general features of the "objective" spatial environment.

RODENTS

We have seen that rodents and other mammals depend heavily on stable landmarks to navigate within their home range or in familiar laboratory conditions. As shown by experiments where the animals have been disoriented before being introduced into a familiar test apparatus, piloting can occur without any reliance on dead reckoning. It is therefore generally assumed that mammals represent a familiar environment independently of dead reckoning, as a system of interconnected places in which each place can be identified through associated landmarks and through its relations to other places.

However, homing experiments in conflict situations between visual references and signals deriving from active (Teroni et al. 1987) or passive (Teroni, Portenier, & Etienne 1990) motion testify that dead reckoning and inertial compass information continue to play a role during navigation in a familiar environment. In minor conflict situations pitting dead reckoning against a rich visual background by rotating an experimental circular arena (with a nest fixed to its wall) by 90 degrees, golden hamsters return from a food source at the arena center to their (rotated) nest at the periphery along compromise directions. Distal visual cues predominate over dead reckoning, but the influence of dead reckoning always remains noticeable (Teroni et al. 1987).

For the majority of subjects this continues to be the case when the visual information consists of only a weak light spot (see figure 8.3, experiment A). The hamsters lived before and throughout the test period in the experimental arena with a peripheral nest, where they established their granary. During the dark phase of the dark-light cycle, the light spot was always presented at a constant angular location at the arena periphery, so that the subjects could associate the single visual cue with the nest entrance. During the experiments, the animals followed a bait from the nest to a food source at the arena center under infrared light, that is, in complete darkness. The light spot was turned on as soon as the animals started to pouch food. In the control trials, the spot appeared in its usual position. In the experimental trials, it was shifted by 90 degrees, opposing directional information derived from dead reckoning and from vision by 90 degrees. Under these conditions, the animals chose compromise directions, most of them depending on the single visual cue more than on dead reckoning.

In further experiments based on the same general procedures, the conflict was increased during the experimental trials, either by opposing dead reckoning and the light spot by 180 degrees or by shifting the spot several times during the same trial (see figure 8.3, experiment B). Under these conditions, the homing directions were no longer influenced by the visual cue. Instead, a majority of animals returned toward the standard nest location and therefore relied on dead reckoning.

On the whole, research on golden hamsters and mice (Alyan & Jander 1994) points toward the priority of extramaze visual references over dead reckoning once the animal is completely familiar with the test environment. This is plausible if we consider the stability of prominent distal features in the animals' natural environment on the one hand, and the cumulative errors that always affect dead reckoning on the other hand. By contrast, if the mismatch among continuously integrated route-based signals and more episodically observed visual references exceeds certain limits, the animals may no longer rely on the visual world and therefore regress to the exclusive reliance on dead reckoning. Further conflict experiments also showed the importance of the correlation between different sets of cues. If dead reckoning and a prominent cue card at the periphery of the experimental arena were in conflict with the distal visual environment, the distal environment

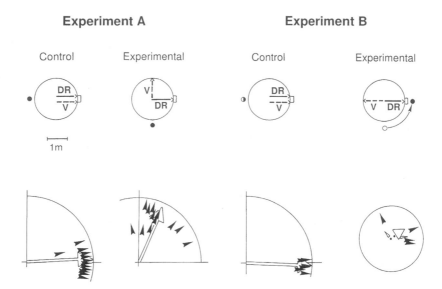

FIGURE 8.3. Homing directions of hamsters tested in conflict situations between visual information and dead reckoning. Upper row: The large circles represent the experimental arena with the nest (small rectangle) and the small circles the light spot at the arena periphery. The arrows indicate the expected homing direction if the subject depended either on the visual cue (V) or on dead reckoning (DR). In control trials, the spot appeared in its standard position, opposite the nest, as soon as the subject started to pouch food. In the experimental trials of experiment A, the spot appeared in a new position, diverging by 90 degrees from its standard position, when the animal started to pouch food. In those of experiment B, the spot was presented under a 90-degree shift during the outward journey, and was shifted by a further 90 degrees when the animal started to pouch food. Lower row: In the quadrants, the small arrows indicate the mean homing direction of particular subjects, each subject being tested 20 times in the control situation and 10 times in each experimental situation (empty arrowhead: $p > 0.05$; filled arrowhead: $p < 0.05$ or 0.01; Rayleigh test). The large arrows indicate the second-order vectors for the whole experimental group (dotted arrow: $p > 0.05$; double line arrow: $p < 0.01$; Moore's test). (Data from Etienne et al. 1990.)

still exerted a dominant influence on the subjects' homing direction. At the same time, however, dead reckoning and the cue card reinforced each other at the expense of the visual background (Etienne et al. 1993).

The behavioral results from conflict experiments agree with neurophysiological data on the activation of neural systems that implement spatial representation. During a small mismatch between a salient, strongly polarizing visual cue and vestibular information due to a passive rotation of the experimental apparatus, the place field of hippocampal place cells and the preferred direction of thalamic head direction cells follow the visual cue. If the mismatch is increased by quick rotations of the apparatus, the cells are increasingly influenced by vestibular signals and their activity remains aligned with the extra-maze environment, that is, the inertial frame of reference (Knierim, Kudrimoti, Skaggs, & McNaughton 1996). Further conflict experiments confirm the simultaneous influence of inertial and visual input on place and head-direction cells, visual cues predominating in general over vestibular signals unless the conflict exceeds certain limits and is induced in an abrupt rather than in a stepwise manner (Rotenberg, Kubie, & Muller 1993; Taube, Muller, & Ranck 1990b). The relation between the influence of visual and of vestibular input may change according to a number of experimental factors (e.g. Chen, Lin, Barnes, & McNaughton 1994; Wiener, Korshunov, Garcia, & Berthöz 1995).

Evidently, in contrast to conflict situations, the availability of correlated internal and external inputs enhances the accuracy of navigation. Thus, golden hamsters choose more precise homing directions (Teroni, unpublished results) when visual and inertial cues yield the same information than when they are set at variance.

This leads us to the issue of how dead reckoning and visual cues cooperate with one another. In natural conditions where different categories of spatial references tend to remain correlated, the cooperation between two completely different and therefore complementary strategies not only optimizes navigation, but constitutes a basic prerequisite for constructing and using a general representation or map of the environment. A number of supporters of the cognitive map theory have adopted this viewpoint, emphasizing the role of dead reckoning to a lesser (O'Keefe & Nadel 1978) or major (Gallistel 1990; McNaughton et al. 1996) extent.

Dead reckoning, a strategy that does not involve learning, is immediately functional in a new environment (Etienne et al. 1998a). It therefore allows the subject to keep track of its position during the exploration of unfamiliar space, cumulative errors being avoided by frequent returns to the point of reference where the animal resets the path integrator. Dead reckoning may be considered as a basic prerequisite for landmark learning, each landmark being automatically assigned a particular position with respect to the subject's own position. As such, dead reckoning may be the most fundamental component of the representation of space—a conception that has been implemented at the level of neurophysiological models (McNaughton et al. 1996). In this view, the hippocampus contains a synaptic matrix for path integration, to which viewpoint-specific landmark information becomes secondarily bound by associative learning. Behavioral data are still missing to support this viewpoint. On the other hand, neurophysiological studies show that in conflict situations between vestibular and visual information, a salient landmark starts to control place and head direction cells only after the animal has been exposed to the test apparatus during a given time span. By contrast, in an unfamiliar environment the cell activity is not yet driven by visual input and tends to remain correlated with vestibular information deriving from (active or passive) motion (McNaughton 1996; McNaughton et al. 1994). Furthermore, visual cues gain control over place and head direction cells only if the rats remain oriented with respect to an inertial frame of reference during their initial exploration of a new environment (Knierim, Kudrimoti, & McNaughton 1995). This result again points to the important fact that mammals, like bees, need the help of a compass for memorizing and later on for using landmark-place associations. But whereas bees rely on external compasses, current data suggest that mammals primarily use an internal compass based on vestibular signals and proprioceptive feedback from locomotion (Etienne et al. 1986, 1988, 1996; McNaughton et al. 1994, 1996).

Further neurophysiological data illustrate in an exemplary manner how head direction cells, which implement the internal compass, are driven by self-generated information in a new environment, whereas visual information tends to override (but not to erase) motion-dependent cues in familiar space. When rats leave a familiar cylinder with a salient cue card and enter a new alley leading to an unfamiliar rectangular box, postsubicular and thalamic head direction cells maintain their preferred direction under the control of signals derived from locomotion. In the next recording phase, the rats are reintroduced into the familiar cylinder where the cue card has been rotated by 90 degrees. The cells' preferred direction tends to follow the rotation of the visual stimulus. When the rat again enters the (now familiar) alley and box, the cells shift back to directional preferences consistent with the landmarks of the new environment (Taube & Burton 1995).

The possibility that dead reckoning sets the basic reference frame for assigning landmarks to particular locations has been considered so far. How landmarks cooperate with dead reckoning to safeguard its functional importance in a familiar environment is now examined. Current experiments on golden hamsters investigate to what extent landmarks correct the state of the path integrator and facilitate the planning and execution of trajectories that are based on route-based vector information and in particular allow the subject to pinpoint the goal.

As stated repeatedly, dead reckoning is necessarily affected by cumulative errors. Without being cleared of these errors, dead reckoning remains functional during short excursions only. Theoretically, the path integrator can be reset through frequent returns to a reference point, or, much more economically, by the sight of familiar landmarks. This second possibility plays a fundamental role in Gallistel's conception of navigation. In this perspective, the subject is supposed to correct the errors which degrade path integration through "episodic fixes" on the landmark panorama (Gallistel 1989).

So far, positive results with regard to the episodic visual fix hypothesis were obtained in a relatively simple situation (see figure 8.4). The homing direction of golden hamsters was again tested during hoarding excursions in a large experimental arena with a nest at the arena wall. All hoarding excursions took place under infrared light and therefore depended on dead reckoning alone, except for the experimental trials with the opportunity

Outward trip   **Return trip**

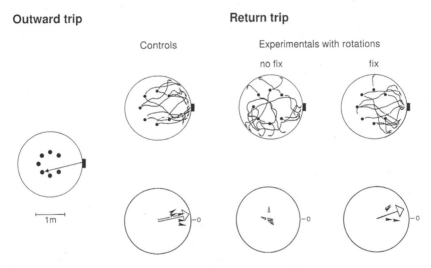

FIGURE 8.4. Testing the role of an "episodic visual fix" in hamsters. *Outward trip:* In all trials, the subject was led from its nest (small rectangle) to a platform (diameter = 25 cm) with a food source. The platform was always located at a radial distance of 50 cm from the arena center; its angular location was shifted between trials, corresponding in each trial to one of seven different positions. *Return trip:* The upper row shows the return paths of one subject from the platform to the arena periphery in the three experimental situations. In each situation, the animal was tested twice with the platform in a given angular position. The lower row shows the subjects' mean homing directions at a distance of 35 cm from the center of the platform. The small arrowheads indicate the mean homing direction for each subject (filled arrowheads: $p < 0.05$ or 0.01; empty arrowheads: $p > 0.05$; Rayleigh test). The large open arrows indicate a significant mean homing direction ($p < 0.05$ or 0.001; Moore's test) for the whole experimental group (Griffin & Etienne in press).

for a visual fix. In the control trials, the hamsters collected food on a stationary platform during about 30 seconds and then returned home. In the experimental trials without the opportunity for a fix, the animals were submitted to more than 10 full rotations (occurring in both directions) at the beginning of the food uptake, and then continued to fill their cheek-pouches on the stationary platform during a further interval of approximately 20 seconds. Former results have shown that after having experienced two to three full passive rotations, hamsters are no longer able to orient toward the nest (Etienne et al. 1988). As expected, this was also the case in the present experimental situation, where positional information from dead reckoning must have been completely washed out. In the experimental trials with the opportunity for a fix, the animals were again rotated more than ten times on the platform, and then presented with the room environment for about 10 seconds; after this presentation, they continued to take up food during about 10 seconds. The brief sight of the landscape re-established the animals' capacity to choose an approximately correct homing direction. Note that after the presentation of the experimental landscape, the animals continued to collect food and move around the food source. In consequence, they had to update their orientation with respect to the nest after the uptake of visual information.

Similar experiments with a two-legged outward journey, where the animals were rotated and then given the opportunity for a visual fix after having walked along the first outward leg, yielded heterogeneous results. In these conditions, the animals had to deduce their position after the first outward leg in order to continue path integration with respect to the nest and to integrate linear as well as angular movements during the second outward leg. In contrast, the hamsters' capacity to benefit from the brief presentation of the room environment in the previous experiment may have been due to the fact that directional information was enough to re-establish an approximately correct homing direction. This interpretation agrees with the fact that after the presentation of room cues the hamsters tended to proceed in the same general direction (that is, followed parallel and not convergent return paths) toward the nest region, irrespective of the (changing) location of the platform where they initiated the return path.

Recent experiments with rats showed that familiar visual cues may reset drifting postsubicular and thalamic head direction cells. After a first recording session with the cue card in a constant position in the experimental apparatus, the card was removed and a second recording session was run. If the preferred direction of a particular head direction cell rotated by more than 30 degrees in the absence of the card, the card was reintroduced in its former location. In general, the cell's preferred direction shifted immediately back to its original value. If the preferred direction of a cell did not shift in absence of the cue card, the card was reintroduced in a new angular position. In these conditions, the cell's preferred direction rotated to remain aligned with the cue card (Goodridge & Taube 1995). Similarly, the firing direction of thalamic head direction cells starts to drift if the rats move on an eight-arm maze in darkness during a given time span, but it is restored when the lights are turned on again (Mizumori & Williams 1993). Thus, within certain limits, visual information resets the internal compass.

In our initial experiments on homing during hoarding excursions under infrared light (Etienne et al. 1986, 1988), golden hamsters chose the approximately correct homing direction without relying on any external cue, dead reckoning informing them of their current position with respect to the nest. But having walked from the food source at the center to the periphery, the animals explored the wall in search of the nest door and therefore resorted to olfactory and tactile cues to find the nest entrance. Current experiments (Georgakopoulos, unpublished) examine two related questions: Can hamsters learn to locate a food source outside the nest through long-term vector information? That is, can they assess the location of the feeding place in terms of its direction and distance from the nest through dead reckoning and memorize the nest-to-food vector on a long-term basis? Should this be the case, do the animals again need external cues to back up the memorized vector and to pinpoint the exact feeding location?

Golden hamsters learn to proceed in darkness from their peripheral nest to a food source that is buried at a constant location in an open circular arena. In test trials, they are led (by a bait) from their nest to a particular point at the arena periphery; the end point of this guided trip along the wall varies between different test trials. From the peripheral end point the subjects must reach the goal location without being helped by any external cue. In these conditions, the hamsters are capable of choosing the correct goal direction from any point at the arena periphery by processing route-dependent vector information, that is, by subtracting their current-position vector from the stored nest-to-goal vector. But when they cross the goal zone, they stop only if food is buried there. By contrast, in experimental trials with an empty goal site, they walk over the goal zone without searching for food. Therefore it seems that in order to interrupt locomotion and perform the correct instrumental response, the hamsters need an external confirmation that they have arrived at the goal (Etienne et al. 1998b).

In a second version of the same experiment, the food source is hidden in a goal cylinder that belongs to a symmetrical array of four identical cylinders. The cylinders being evenly spaced around the arena center with no other landmarks (from within or without the arena) available, the goal cylinder can only be identified through positional information from dead reckoning. Test trials occur either in continuous darkness or with the presentation of the landmark array during the last phase of the trial, the arena being lit from the moment the animal has executed the forced detour along the periphery until the end of the trial. In both light conditions, the hamsters proceed to the cylinder at the correct location without exploring another cylinder in at least 89 percent of the trials. In continuous darkness, however, the animals frequently refuse to leave the periphery to proceed to the goal and instead return to the nest. This response is seldom observed when the subjects can see the landmark array from the periphery (Etienne et al. 1998b).

The conclusion from these results is that hamsters are indeed able to plan and to execute a goal-directed path through self-generated vector information only. However, landmarks play an important role in confirming vector information at two stages. At the end of the forced detour around the periphery, the animals have already processed the path that leads them to the goal through vector summation; yet the sight of a landmark at the goal location confirms the planned trajectory and may facilitate its execution. Furthermore, upon approaching the goal zone, feedback from locomotion is not enough to inform the subject that it has arrived precisely at the goal location. To start searching for the hidden food, the subjects need external cues previously associated with the goal site.

These arguments bring us back to landmark-vector associations, a concept used in connection with insect navigation. We may again ask to what extent rodents go beyond processing and using single vectors (see figures 8.2A and 8.2B) that are associated with landmarks near or at the goal site and are tied to a single point of reference, that is, the point of departure of the foraging trips at the nest. No behavioral data have been collected that may inform us on this issue, but neurophysiological research on rats suggests that this is so. Within a given environment with mobile landmarks and associated goal sites, different subgroups of hippocampal cells encode location with respect to different (behaviorally relevant) reference frames (Gothard, Skaggs, Moore, & McNaugton 1996). If dead reckoning represents the basic navigational process to which landmarks are bound secondarily, then these results imply that dead reckoning in rats occurs with respect to different reference points depending on the functional context of behavior (McNaughton et al. 1996).

Recent neuronal models link the hippocampal place code either to the geometrical properties of the environment (e.g., O'Keefe & Burgess 1996) or to the interaction between locomotor and visual cues. According to the last-mentioned class of models, dead reckoning is implemented by the hippocampal formation itself (McNaughton et al. 1996) or it occurs outside this structure (e.g., Wan et al. 1994b). Whatever the detailed circuitry of the hypothetical networks, motion signals and visual information converge on head direction and place cells and determine jointly their firing patterns in a known environment. Through feedback loops, the path integrator is checked and if necessary corrected by the sight of landmarks—a fundamental requirement that, as we have seen (see figure 8.4), still awaits full confirmation at the behavioral level.

It is beyond the scope of these models to predict the full range of spatial information processing in rodents. Neither do we know through clear-cut behavioral results whether spatial representation in rodents and other nonprimate species is flexible enough to satisfy the criteria for true cognitive maps. In the behavioral literature, there is no common agreement whether rats are capable of piloting without being guided by immediately accessible landmarks, for instance in a visually disconnected environment (e.g., Benhamou 1996; Schenk, Grobéty, & Gafner 1997). Likewise, it remains questionable whether rodents can plan detours (e.g., Tolman & Honzik 1930; Vauclair 1987) or shortcuts (e.g., Kendler & Gasser 1948; Tolman, Ritchie, & Kalish 1946) without resorting to external cue guidance, or without following a vector established through dead reckoning (see also Bennett 1996).

In our own research, we plan to examine whether golden hamsters are capable of planning and executing adequate new routes without any landmark guidance between different food goals. After having learned to reach (at least) two different goal sites from their nest, the subjects will be tested as to their capacity for proceeding directly from one goal to another through vector addition. In contrast to our previous experiments implying the performance of detours to reach a single goal (see figure 8.2B), success under these new conditions would testify that the animals combine several route-based vectors with each other and therefore use a true cognitive map (as in figure 8.2C, where it is represented as a vector map). So far, positive results of this nature have been obtained only in humans (Landau, Gleitman, & Spelke 1981).

SUMMARY AND CONCLUSIONS

Central-place foragers such as hymenopterous insects and rodents navigate within their home range by relying on two different strategies. Dead reckoning, which is based on feedback information from locomotion, informs the animals in a continuous manner on their current position with respect to their home. On the other hand, landmarks that are associated with particular places provide the animals with long-lasting, site-dependent information on the location of particular goals.

Flying honey bees and walking desert ants explore their environment over considerable distances by depending on a sun compass–aided and therefore relatively precise dead reckoning system. In the vicinity of the goal, the insects fixate nearby landmarks under particular viewpoints and store them as snapshots; subsequently, they recognize these landmarks from the same vantage point through retinotopic image matching. En route to a goal, the insects can also rely on landmarks, although route-based vector information tends to override landmark guidance. In conflict situations between positional information from dead reckoning and familiar landmarks, the animals stick to dead reckoning. Landmark information takes over only when the animals' path integrator approaches a zero state. In natural conditions, this is the case when the animal has returned to the point of departure of its current excursion, that is, the site of its colony, where it expects to find familiar landmarks that indicate the nest entrance.

In general, bees and ants show remarkably precise navigation strategies. The latter are based on hardwired information processing mechanisms and on specific predispositions for acquiring information about variable features of the spatial environment, knowledge of which cannot be coded genetically. On the other hand, the insects' route-based and location-based navigation strategies do not belong to a fully integrated system and therefore function sequentially rather than synchronously. Along the same line, insects do not seem to organize their home range as a set of interconnected vectors or places, that is, as a map. Rather, they represent familiar space in terms of single nest-to-foodsite vectors, site-specific landmark-place associations or landmark sequences along familiar routes where each landmark unit triggers off specific motor instructions. General landscape features and contextual cues resulting from the animals' former displacements seem to endow the insects' spatial representation with some degree of continuity, and therefore prevent the animals from getting lost in a mosaic of fragmentary knowledge about the spatial layout of their home range.

Like bees and desert ants, rodents commute between their nests, and feeding sites by relying on dead reckoning and on landmarks. Unlike these insects, rodents and other mammals do not use external references for measuring rotations (and translations) and consequently rapidly accumulate errors during path integration. In a familiar environment, these animals rely heavily on familiar cues, vision playing a predominant role, particularly for planning a given path. Spatial orientation in rodents is always controlled polymodally, however, as shown by the behavior of place and head direction cells in central nervous structures that are driven by a number of different external and internal stimuli.

In minor conflict situations between dead reckoning or internal compass information and visual cues, hamsters and other rodents show compromise behavior, with vision playing the leading role. This is no longer the case when visual and self-generated signals indicate completely opposed directions, or when visual references are shifted repeatedly during a particular foraging excursion. Thus, the animals trust visual landmarks stored in long-term memory more than feedback information from locomotion, provided that the conflict between the two categories of information does not exceed certain limits.

Compromise behavior between conflicting visual cues and feedback signals from locomotion show that the two categories of information interact with one another. Being completely different, but at the same time complementary, it is likely that route-based and location-based information also cooperate with one another to help the animals organize the spatial environment and optimize navigation in familiar space. Behavioral experiments showed that hamsters are disoriented after having been rotated in darkness, but that they reset their internal compass through the brief presentation of the visual surroundings. Hamsters can also be trained to go from the nest to a food source using route-based vector information. At the same time, however, the animals use visual cues to plan and execute the nest-to-goal vector; in darkness, they rely on nonvisual cues at the goal site to stop their progression and look for the food reward.

Neurobehavioral data on the firing of hippocampal place and head direction cells in the postsubicular and thalamic areas of rats yield analogous results in situations where vestibular and visual signals yield divergent information. These data also show that familiar visual cues reset the internal compass and suggest that dead reckoning may provide the animal with basic positional information on which landmark-place associations are grafted secondarily.

Taken together, rodents seem to possess all the prerequisites to represent space as an integrated system, based on the cooperative interaction between two deeply complementary navigation strategies. However, behavioral results have not yet shown in a convincing manner whether rodents possess a cognitive map. The main criteria for cognitive maps being the anticipation of the optimal path in a new situation, it may be that neuro-physiological data may solve this long-standing question by showing whether certain cells fire in anticipation of changes in location or direction before the animal starts to execute the path and therefore without guidance by perceivable cues at new locations.

SIMON P. D. JUDD
KYRAN DALE
THOMAS S. COLLETT

9

On the Fine Structure of View-Based Navigation in Insects

Insects can be impressive navigators. A honeybee, for example, will fly regularly between its hive and a rich patch of forage up to 10 km away (Vischer & Seeley 1982). A desert ant, *Cataglyphis fortis,* will search for dead insects 100 meters from its nest and, after finding one, will run straight home with it (Wehner 1987). Numerous navigational strategies and mechanisms contribute to these skilled performances. We focus here on just the end of such a journey and the way in which an ant, bee, or wasp uses landmarks to locate its goal. Our main concern is with one particular aspect of this problem. Hymenopterans learn landmarks as views from specific vantage points when facing in specific directions. What specializations are needed to make view-based navigation work?

The major problem to be confronted when devising a successful view-based navigation system is that views change as the insect moves. In figure 9.1 are shown several views of a scene as a camera approaches a target from two directions that are 60 degrees apart. The scene in the print, or on an insect's retina, alters greatly during the approach and even more

Research described in this chapter was funded by the BBSRC. Simon Judd and Kyran Dale are supported by BBSRC research studentships.

when viewed from different directions. An animal that stores scenes as views may find it difficult to recognize the same scene from an unfamiliar vantage point. Little can be done to avoid the transformations that must occur when an observer approaches a target, and we will see that insects cope with this difficulty by learning multiple views of the scene. But the observer can adopt simple motor strategies to limit changes in viewing direction.

The other side of the coin is that the process of landmark navigation relies upon changing views. An animal that has buried a seed at a position relative to the tree stump in figure 9.1 can retrieve the seed by moving until the view of the tree stump matches the view that the animal had when hiding the seed. And, in fact, the precision with which the seed can be located depends upon how much the view changes as the animal moves toward the hiding place.

Views seen from a single vantage point also change their appearance. Wind, rain, and large animals can have dramatic effects on the small-scale features insects use to pinpoint a goal. These changes are essentially irreversible and can only be accommodated by storing new views. During the course of a day, movements of the sun and the clouds alter the position and distribution of cast shadows. To cope with such changes, insects need ways of distinguishing edges formed by the boundaries of objects from those generated by cast shadows.

This chapter begins with evidence that insects learn two dimensional patterns in retinotopic coordinates and that familiar scenes are also stored as two-dimensional views. Next is described the way in which two-dimensional views guide an insect's path and the details of those motor strategies, which simplify the task of view-based navigation and reduce the number of views that must be learned. Rapidly changing views mean that insects have to memorize several views of the scene immediately around a feeding site, and the central part of the chapter is a discussion of the use and acquisition of multiple views by wood ants that have been trained to approach a cone to collect sucrose at its base. The final section is concerned with the motor strategies of simple simulated creatures that have been selected through a process of artificial evolution solely for their ability to return to a position marked by a landmark. There is remarkable convergence in the paths and motor strategies of real and simulated creatures suggesting that the motor strategies found in insects are indeed a response to the problems of navigating with retinotopically organized views.

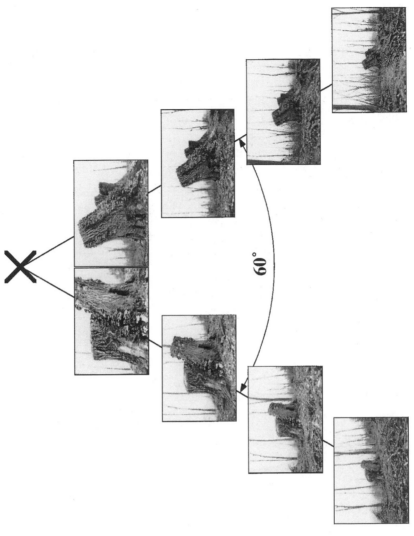

FIGURE 9.1. Views of a tree stump from two directions 60 degrees apart taken at four different distances to demonstrate how the scene changes with vantage point. Position of the tree stump is marked by the cross.

TWO-DIMENSIONAL PATTERNS
STORED IN RETINOTOPIC COORDINATES

Studies by Wehner (e.g., Wehner 1981) some 30 years ago first suggested that honeybees learn visual patterns in retinotopic coordinates. In one of his experiments, bees were trained to enter a horizontal tube behind which was a large disk divided alternately into black and white equiangular sectors. Different bees were trained to radial gratings of different spatial frequencies and then given a choice between two tubes with a similar sectored disk behind each. One disk was an exact match of the training pattern, the other was a similar disk rotated through half a period so that, viewed from the entrance to the tube, the positions of the black and white sectors on the retina were reversed. Bees hovered just in front of each tube before entering one. They had a strong preference for the tube that led to the disk in the training orientation. The preference broke down when bees were trained to fine radial gratings with individual sectors smaller than about 10 degrees (figure 9.2).

The ability to distinguish among such patterns argues that the rewarded pattern has been stored retinotopically and that it is only recognized when its elements fall on the same region of retina that viewed it during learning. More recently, Dill, Wolf, and Heisenberg (1993) have measured the amount of vertical translational shift on the retina that *Drosophila* spp. will tolerate before the flies fail to recognize a pattern. Recognition breaks down once the pattern has been displaced by more than about 3 degrees, which is the angular distance between the optic axes of adjacent *Drosophila* ommatidia.

VIEWING DIRECTIONS FOR ACCESSING
RETINOTOPICALLY STORED PATTERNS

For an insect to recognize such retinotopically stored patterns, it must recover the viewing direction and vantage point that it adopted during learning. In the two studies just described, the insect was in a fixed position in space. The bee hovered immediately in front of the tube before entering it and film analysis showed that the insect faced the tube in an orientation that remained constant in pitch, roll, and yaw (Wehner & Flatt 1977). Maintaining this fixed position relative to the tube meant that the pattern would fall automatically on the same retinal location. In the Dill and Heisenberg experiment, a fly was attached rigidly to a torque meter that measured the torque developed about a vertical axis. The fly's turning force was used to

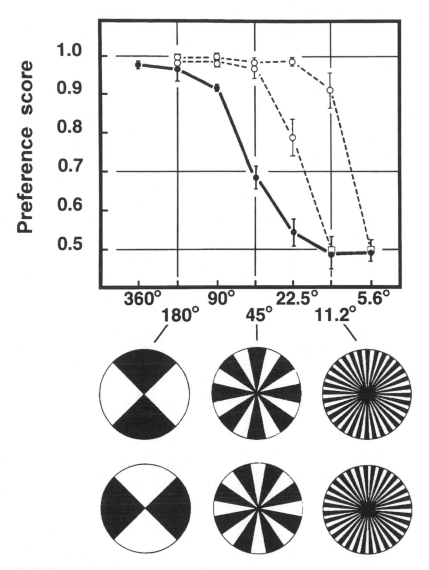

FIGURE 9.2. Retinotopic memories. Honeybees were trained to enter a horizontal perspex tube in order to collect sugar water. The tube was placed in front and at the center of a larger vertical disk that displayed a radial pattern of alternating black and white sectors. Individual bees were trained to one of several such radial gratings, each of which had a different spatial period. In tests, bees were given a choice between the training pattern (top row of radial patterns below the graph) and the same pattern rotated by half a period (bottom row of radial patterns). The heavy solid line and dots show the preference score $(n+/\sum n)$ of bees plotted against the period of the gratings that were used during both training and tests. Dashed lines and open symbols show data from control experiments in which bees were trained on the patterns with 11.2-degree and 5.6-degree wavelength and then asked to discriminate them from the radial gratings of lower spatial frequency. In the latter tests, the center of a black sector always pointed downward. (Reproduced from Wehner 1981 by permission of Springer-Verlag.)

rotate the visual pattern in a horizontal plane, allowing the fly to control its horizontal direction of gaze relative to the pattern. But the vertical position of the pattern on the fly's retina was fixed, unless the experimenter moved the pattern up or down with respect to the fly.

For an insect to maintain an appropriate viewing direction without such local props, it needs to set its direction with respect to compass cues. Frier, Edwards, Smith, Neale, and Collett (1996) found that freely flying bees that had no horizontal tube to fix their viewing direction continued to face predominantly in one direction when looking at panoramic patterns. Bees were trained to enter a dustbin with a panoramic pattern of diagonal stripes displayed on the inner wall (figure 9.3). A horizontal tube ran from the center of the bin through a hole in the bin wall to end in a feeding box fixed to the outside of the dustbin. To encourage the bees to notice the pattern on the wall, a second, "negative" bin was placed next to the first. Its inner wall displayed the same pattern, but the bin was rotated 90 degrees relative to the positive one, and the feeding box was empty. Bees readily discriminated between the two bins, and if both tubes and boxes were removed, they hovered preferentially in the bin that displayed the pattern in the positive orientation.

In what orientation did the bee prefer to face the positive pattern? To answer this question, the tube was placed vertically instead of horizontally in the bin, so as to give the bees something at which they could aim, but which did not provide them with a directional cue. When bees were trained with the reward box facing west (as in figure 9.3 middle), they tended to face west when approaching the vertical tube. This experiment was repeated three times with the tube in the rewarding bin facing in the other three cardinal compass directions during training. In each case, when bees were tested with a vertical tube in the positive bin, they faced in the usual direction of the tube in that bin, even though there was no horizontal tube to constrain them. Bees thus learn both the pattern and in what orientation they should view it. Other experiments in which the horizontal component of the magnetic field was rotated or in which celestial cues were obscured suggest that the bees' orientation is controlled by several directional cues, including the earth's magnetic field, celestial cues, and the surrounding visual panorama (Collett & Baron 1994; Dickinson 1994; Frier et al. 1996).

A viewing direction that is learned and is specified by compass cues provides bees with a simple mechanism for placing visual memories that are stored in retinocentric coordinates back into an earth-based coordinate frame. The bee need only face in the same viewing direction it adopted during storage for it to be appropriately oriented to recognize a familiar scene.

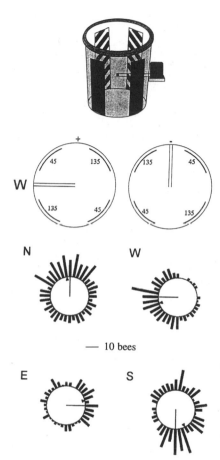

FIGURE 9.3. Viewing directions of freely flying bees inspecting a panoramic pattern. Bees are trained inside a dustbin to enter a horizontal tube that leads to a feeding box with sucrose solution inside. (*Top*) Plastic can displaying a panoramic pattern. The pattern, the feeding tube, and the box are shown. (*Middle*) Plan view of positive and negative bins at the level of the tubes. Numbers give angular orientation of diagonal stripes. Angles are measured counterclockwise from the horizontal. Tube in positive bin faces west. The feeding patterns and boxes are periodically rearranged during training so as to transform the positive bin into the negative bin and vice versa. (*Bottom*) Circular histograms show the viewing directions of different bees trained with tubes in the positive bin facing north, west, east, or south, as indicated by the letter and the horizontal line. During tests, the horizontal tube was removed from the side of the wall and the hole in the wall covered. The tube was placed upright with the bottom end in the center of the floor of the bin. The horizontal orientation of single bees was recorded just before they contacted the vertical tube. Arrows show directions of mean vectors. (Adapted from Frier et al. 1996.)

SCENES AS TWO-DIMENSIONAL VIEWS

A variety of experiments on hoverflies (Collett & Land 1975), ants (Wehner & Räber 1979), and bees (Cartwright & Collett 1983) indicate that remembered views of landmarks resemble other memorized visual patterns and that they are stored as two-dimensional views. One example from honeybees is shown in figure 9.4. Bees were trained to collect sucrose from a small bottle top on the floor of a room that was empty apart from three black cylinders standing on the floor that marked the location of the food. The array of landmarks and feeder was shifted over the floor of the room during training so that only the cylinders could define the feeder's location.

When the feeder was removed, bees spent the most time searching for it at the location defined by the position of the three cylinders. If the

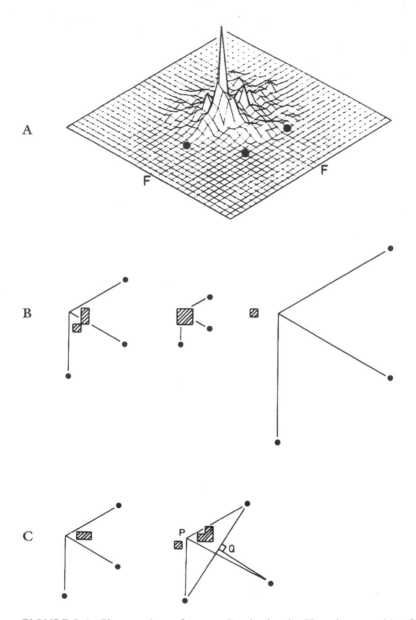

FIGURE 9.4. Places as views of surrounding landmarks. Honeybees searching for a missing source of sucrose at a site defined by three black, upright cylinders. (**A**) Relative time spent by one bee in each cell of an imaginary grid. Position of sucrose during training is marked by Fs on axes. Grid lines are 8.7 cm apart. (**B**) Single bee's search when distance between landmarks is changed from the training situation. (**C**) Bee has choice of searching where bearings P or distances Q are correct. *Leftmost column:* training situation. *Right columns:* distorted arrays. Bee searches where landmarks have same bearings (shown by lines) as those experienced at the feeder during training. Hatched areas show peak search area, which is defined as regions where search density is at least 80 percent of the maximum. (Adapted from Cartwright & Collett 1983.)

cylinders were moved either closer together or further apart, bees searched where the positions of the cylinders on their eye would match those seen from the feeder during training. Bees continued to match retinal positions even though the consequence of doing so was that their distances from the cylinders were grossly abnormal. The bees' search was thus guided primarily by the two-dimensional pattern made by the array. These search patterns suggest that finding a place is in part a process of image matching: the bee moves until the image on its retina is congruent with the two-dimensional scene that it had previously stored when at the feeder.

NAVIGATING WITH STORED VIEWS

From what has been said so far, an insect might be expected to learn a view of a scene when facing in one direction and then to steer by it while facing in the same direction. The flight paths of wasps and bees returning to their nest or to a feeder conform roughly to this expectation (Collett 1995; Collett & Baron 1994; Zeil 1993b). Although a returning wasp (*Vespula vulgaris*) may approach a feeder from a broad range of directions, once it is close, it tends to face in the same preferred orientation on successive visits. In figure 9.5A are shown four flights from the same individual wasp and figures 9.5B and 9.5C give data from numerous flights from 15 wasps. A circular histogram is plotted of the orientation of each wasp when it is near to the feeder. An individual wasp tends to approach the feeder facing in a particular preferred orientation, though with appreciable scatter. Consequently, nearby objects tend to take up relatively constant retinal positions when the wasp has reached its goal. The retinal positions of these objects can thus signal the wasp's successful arrival at the feeder.

With some arrangements of landmarks, the wasp adopts a consistent orientation and approach direction when it is somewhat further away. This is the case for the situation illustrated in figure 9.6. Wasps were trained to collect sucrose from a bottle cap on the ground. The feeder was marked by a nearby cylinder and a slightly larger and more distant cone. At the top of the figure are shown single approach flights from two different wasps with the cylinder shown as a small filled circle and the cone as a larger open one. For about 20 such approaches, we plot the orientation of each wasps' longitudinal body axis and the trajectory of the cylinder and the cone over the wasp's retina as a function of the wasps' distance from the feeder.

There are several notable features of these trajectories. First, the wasp's orientation is constant over the whole approach. Second, it first fixates each

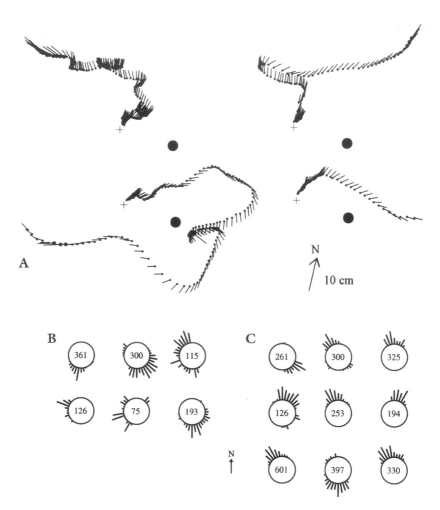

FIGURE 9.5. (**A**) A wasp approaches a feeder facing consistently in the same direction. Four approach flights of the same individual *Vespula vulgaris* to a feeder (+) on the ground with an upright cylinder (●) placed nearby. The arrow points north, and its length represents 10 cm on the ground. The wasp's position and orientation are plotted every 20 ms. The wasp generally aims at the cylinder before approaching the feeder. Once close to the feeder, she tends to adopt a preferred orientation that is constant over many flights. (From Collett 1995.) (**B** and **C**) Circular frequency distributions of the horizontal orientation of the body axis of 15 wasps when they flew within an annulus of 3–6 cm distance from the center of the feeder. Bins are 10 degrees wide. Numbers of frames comprising each distribution are shown in the center of each circle. Number of recorded flights varied from 10 approaches (75 frames) to 51 approaches (601 frames). Feeder marked by a single cylinder (**B**). Feeder marked by cylinder and cone as in figure 9.6 (**C**). (From Collett & Rees 1997.)

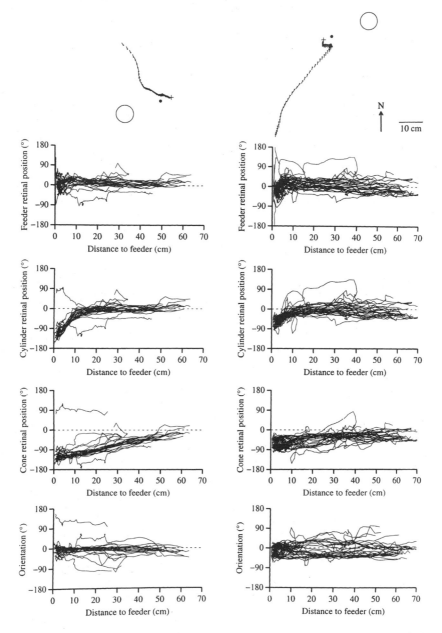

FIGURE 9.6. Superimposed trajectories of 23 approach flights from one wasp (*left*) and of 25 flights from a second wasp (*right*) toward a feeder marked by a cone and a cylinder. (*Top*) Sample flight from each wasp and the arrangement of feeder (+), cylinder (●), and cone (○). (*Bottom*) Trajectories of the retinal positions of the centers of the feeder and of both landmarks plotted against the wasp's distance from the feeder. Positive angles are to the left of the midline and negative angles to the right. Cylinder and cone migrate from the central visual field to the periphery of the right eye as the wasp approaches the feeder. Panels below show the wasp's orientation relative to its preferred orientation. (From Collett & Rees 1997.)

landmark with frontal retina and then moves so that the landmark migrates consistently toward the lateral fields of the same retina. Third, there is a pronounced sideways component to the wasp's flight as it nears the feeder (see also figure 9.5). This last point emphasizes that the wasp does not just fly straight at the feeder, but that it can adjust the direction of its trajectory without changing its orientation.

The sideways component of flight is shown more explicitly in figure 9.7. Here is plotted, for the two wasps of figure 9.6, the difference between the wasp's longitudinal body axis and its flight direction over the course of an approach. When this difference is zero, the wasp is flying straight ahead and when it is 90 degrees the wasp is flying sideways. In both examples, the wasp tends to fly forward at the beginning of the approach and to develop a sideways component of movement toward the end, as it slows down and homes in on the feeder.

Many flying insects have the capacity to fly sideways and so can uncouple their direction of flight from their viewing direction (Collett & Land 1975). This capacity is particularly useful during image matching. One obvious limitation of exclusively forward motion is that objects imaged on the retina can only move from the front of the eye to the periphery. Image motion in the reverse direction, toward the front of the eye, is accomplished by rotation. This asymmetry vanishes with the development of sideways and backward flight. Consequently, an insect returning

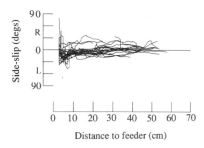

FIGURE 9.7. Sideways component of motion for the approaches illustrated in figure 9.6. The difference between the wasp's viewing direction and its direction of travel is plotted against its distance from the feeder. When this angle is zero the wasp flies along its long axis. When the angle is 90 degrees, the wasp flies sideways. The sideways component increases as the wasp approaches the feeder.

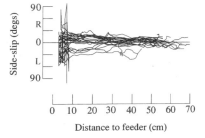

to a goal and moving so that its current image comes to match its stored view can correct any mismatch by translational movements. Its landmark guidance system need not be concerned with rotation.

IMAGE MATCHING AS AN ACTIVE PROCESS

Given that bees and wasps tend to adopt a relatively constant orientation when approaching a goal, it becomes likely that these insects are guided to the goal by a process of retinal image matching that is driven by the discrepancy between the scene viewed from the goal and the insect's current view. There have been a number of models of image matching in bees and robots (Basri & Rivlin 1995; Cartwright & Collett 1983; Hong, Tan, Pinette, Weiss, & Riseman 1991; Röfer 1995; Wittman 1995). But there is scant information about the way in which insects really perform this task.

Junger (1991) has provided a beautiful demonstration of image matching in water striders. These insects use visual landmarks to hold their position on the surface of fast flowing regions of streams where insects trapped in the surface layer are most likely to be caught. The insects will keep station on an artificial stream, when the only visual cue is a single small light bulb seen in its frontal, dorsal visual field (Junger 1991). The bug tends to drift downstream but compensates fully for the drift by discrete jumps against the direction of water flow, so that it holds the light bulb at a constant retinal elevation. If the bulb is suddenly raised, the bug allows itself to drift with the stream until the bulb is returned to its original position on the retina. If the bulb is lowered, the bug jumps forward. The water strider remembers the bulb's desired retinal position, and it moves in the appropriate direction to restore the bulb to that position.

Evidence that wasps (*Vespula vulgaris*) approach a goal by shifting the image of a landmark toward a preferred retinal position has come from analyzing the relationship between the retinal position of the image of a landmark and its retinal velocity (Collett & Rees 1997). If the wasp's search for a goal is controlled by the retinal position of a nearby landmark, then the wasp should move so as to restore the landmark to its preferred retinal location whenever the landmark is incorrectly positioned. A restoring mechanism of this kind would generate a relationship between the landmark's retinal position and the wasp's corrective movements after a delay. Retinal velocity would be zero when the landmark is correctly positioned and any departure from this position would cause the wasp to return the image to the desired location. Retinal velocity and position should thus be negatively correlated, with the data cluster crossing the abscissa at the set point.

Search flights were analyzed for evidence of such a relationship. Wasps were trained to feed at a site defined by a nearby, upright cylinder. In tests with the feeder removed, the wasp first fixated the cylinder, then placed it peripherally on the retina, in this way reaching the expected site of the feeder. After a brief search there, it flew off, sometimes returning for another look.

During search flights, there is a negative correlation between the retinal position of the cylinder and its retinal velocity some time later. The correlation peaks with a delay of about 100 ms. Scatter plots of the retinal velocity of the cylinder against its retinal position at the optimum lag are shown for the foraging visits of one wasp to a feeder (figure 9.8A) and for search flights of the same wasp when the feeder was removed (figure 9.8B). The similarity of the two plots suggests that the same control system operates during both normal foraging visits and searches. Retinal position at zero velocity is -34 degrees for landing flights and -25 degrees for searches, and the slopes of the best linear fits are -5.413 and -4.968 degrees.s^{-1} per degree respectively. Thus, the larger the discrepancy, the greater the restoring force. The wasp notices the landmark when it is away from the stored retinal position and has some measure of the magnitude of the error that needs to be eliminated.

MULTIPLE STRATEGIES OF LANDMARK GUIDANCE

The appearance of a small array of landmarks close to a goal will change substantially as the insect moves away. Consequently, a single stored view taken at the goal can only guide the insect to the goal from a circumscribed area around it. Outside this catchment area, the mismatch between current and stored images will be too great for the insect to know in which direction it should move so as to reduce the discrepancy between the two. Other navigational strategies are needed to bring the insect to within the catchment area.

Recent work has identified two such strategies. The first is *beacon aiming*. It has long been known that bees will use prominent objects such as trees that lie along or close to their path as guideposts, and that they will aim at these objects over long distances (see, e.g., Chittka, Geiger, & Kunze 1995; Chittka, Kunze, Shipman, & Buchmann 1995; von Frisch 1967). In addition, when a bee or a wasp is near to its goal and flying low over the ground, small objects close to the goal that are just a few centimeters high can also come to serve as beacons. The insect will face such objects and fly toward them (Collett 1995; Collett & Baron 1994).

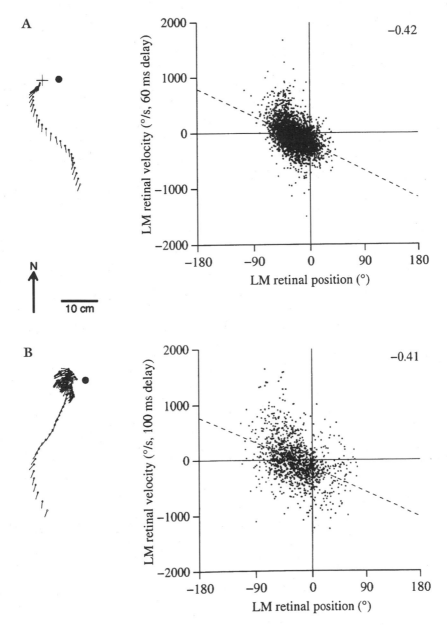

FIGURE 9.8. The control of the retinal position of the landmark during approach and search flights. The angular velocity of the landmark across the retina is plotted against the horizontal retinal position of the landmark after the delay specified in the ordinate. (**A**) Data from approaches to feeder. (**B**) Data from search flights. Dashed lines are fitted regression lines to the data points. Number in upper right-hand corner of panels shows correlation coefficient. Data from 92 approach flights (circa 2600 data points) and 17 search flights (circa 1250 data points). Sample landing and search flights are shown on left. (Adapted from Collett & Rees 1997.)

When there is more than one object close to the goal, as in figure 9.6, the insect may follow a sequence of beacons with the goal itself as the last one. In switching its attention from one beacon so that it can approach another, an insect begins by relinquishing fixation of the first beacon and then must place itself in an appropriate position to gain a standard view of the next. It requires some mechanism for changing between beacons or local views in a controlled manner.

This brings us to the second navigational strategy: *linking trajectories to views of beacons.* One way that bees engineer a controlled transition from beacon to beacon or from beacon to goal is to link a flight trajectory in a given direction to a close-up, frontal view of a beacon. Such an association can be seen in an experiment in which bees were trained to forage at a site midway between two objects: a black cone 15 cm west of the feeder and a blue cylinder 15 cm to the east (Collett & Rees 1997). Over some hours, a bee would approach the array from a constant direction. Sometimes it aimed at the cone, sometimes at the cylinder, and sometimes at both objects in turn. The flight shown in figure 9.9A is from a bee that usually approached from the north. In this approach, it initially fixated the cylinder, secondly the cone, and thirdly the cylinder once more. It then moved toward the feeder so that both landmarks traveled toward the periphery of the retina. Because the bee arrived from the north, the cone moved over the right eye and the cylinder over the left.

Occasional tests were introduced in which the feeder and one of the flanking objects were removed. When just the cone was present, the bee approached it and moved to place the cone to its west and when the cylinder was present the bee moved to place the cylinder to its east. Superimposed trajectories of the path of the object over the bee's retina accumulated over several tests show that after fixation the object traveled from the front toward the periphery of the retina. Movement was mostly across the right eye in tests with the cone and across the left eye in tests with the cylinder (figure 9.9B). The insect seems to have learned distinguishing features of the two objects and to have associated a different trajectory with each object. Such an association can help provide for an orderly transition between different segments of a route.

Beacon aiming and linking trajectories to beacons can provide ways of coping with the complex image transformations that occur when an insect moves in the close vicinity of an array of local landmarks. Beacon aiming would allow the insect to ignore much of the swiftly changing scene on its retina. The insect need only be familiar with the beacon itself. In a similar vein, an open-loop trajectory linked to a beacon enables the insect to

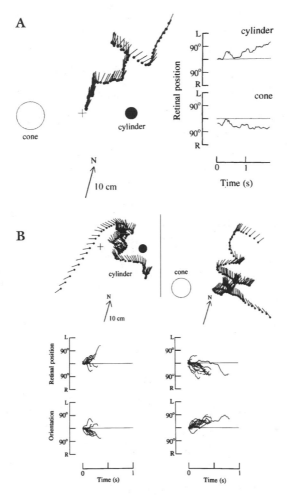

FIGURE 9.9. Linking trajectories to views of beacons. A single bee is trained to forage at a feeder (+) with a cone 15 cm to the east and a cylinder 15 cm to the west. (**A**) Approach flight viewed from above. Circles and tails show the position of the bee's head and the orientation of its body axis every 20 ms. Graphs show the retinal position of the cylinder and cone during the course of the flight. The bee fixates first on the cylinder and then on the cone. Finally, both objects travel toward the lateral visual fields of the left and right eyes, as the bee moves toward the feeder. Arrows point north and their length represents 10 cm on the ground. (**B**) Tests with one object and no feeder. The bee fixates on the object and then flies so that the object moves laterally over the retina. In tests with the cylinder, the bee moves so that the cylinder usually travels across the left eye, whereas in tests with the cone the bee moves so that the conte travels across the right eye. Bees thus link a different movement to each object. Bottom panels show the retinal trajectories of cylinder and cone obtained from several tests together with the accompanying change in body orientation. Trajectories start with first fixation of landmark. (From Collett & Rees 1997.)

disregard the image motion of close landmarks during the trajectory. It is not yet known whether bees do in fact economize on visual learning in these ways.

THE STORAGE OF MULTIPLE VIEWS

The pattern of search flights close to a feeder suggests that bees and wasps store at least one view of the surrounding panorama from a vantage point at or near the feeder, while the results of the previous section imply that bees also acquire at least one frontal view of individual landmarks. How many stored views are needed to guide an insect to its goal? Take an ant walking across a flat surface toward a cone to collect sucrose at its base. Although the appearance of the cone is invariant with viewing direction, its size and position on the retina will change as the ant approaches. These transformations make it questionable that a single stored view will be sufficient to guide the ant along the whole route toward the cone. Detailed analysis of the ant's approach to cones and edges reveals that the insect stores a number of views of the cone from different positions along the approach route (Judd & Collett 1998).

Wood ants (*Formica rufa*) will readily learn the appearance of a cone that they approach. In the experiment that is illustrated in figure 9.10, ants from a laboratory colony were trained to approach an upright, solid black cone with sucrose at its base. In tests with no food present, ants distinguished the cone from an inverted cone of the same size and shape and walked preferentially toward it (figures 9.10A, B). Conversely, ants from a second colony that were trained to approach an inverted cone chose that in preference to an upright cone (figure 9.10C).

Evidence for the storage of multiple views comes from reconstructing how the image of the cone or an edge moves over the ant's retina during its approach. On the assumption that the ant's head is mostly in line with its body, the cone's retinal position can be estimated from the relative positions of ant and cone and the horizontal orientation of the ant's long axis.

The path of an ant toward the cone is shown in figure 9.11A, and the central panel (figure 9.11B) plots the trajectory of one edge of the cone over the ant's retina as a function of the insect's distance from the cone. The edge does not move smoothly over the retina. It is held at one position for a period and it then moves rapidly to another position, where it is again kept steady. An analysis of these plateau regions reveals that they tend to occur when the edge of the cone reaches one of a number of preferred reti-

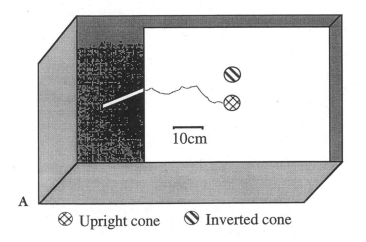

⊗ Upright cone   ⊘ Inverted cone

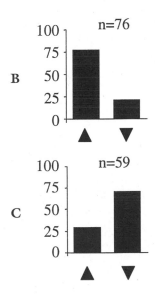

FIGURE 9.10. (**A**) View of ant nest and testing apparatus. A shelf on which wood ants forage for 1 M sucrose is fixed above the nest. It is reached by way of a removable paper ladder. Foragers were trained to collect sucrose at the base of either an upright or an inverted cone (12-cm-high, 7-cm-diameter base). During training one cone was rewarded, but both cone types were present and their positions swapped frequently. After training, ants were periodically exposed to both cones and no sucrose. Olfactory cues were eliminated in tests by covering the shelf with fresh paper and by using different cones. Ants, in these tests, tended to approach the cone at which they had been rewarded. The line going from ladder to cone is a typical example of an ant's path. (**B** and **C**) Histograms showing percentage choice in tests between the rewarded and unrewarded cones. Ants chose preferentially the rewarded cone type. Statistical significance was assessed with the binomial test (**B**, upright cone rewarded, $p < 0.001$; **C**, inverted cone rewarded, $p < 0.005$); n is number of tests. (From Judd & Collett 1998.)

nal positions. The right panel (figure 9.11C) plots the dwell time of the edge of the cone as a function of retinal position during the plateau periods. The distribution is clearly multimodal.

Ants will also approach a single, long black-white edge that has the same orientation as one of the edges of the cone. The ant's path and the trajectory of the edge over the retina are comparable to those generated in response to the cone (figure 9.11D, E, G, H). As with the cone, the single edge is held steady in preferred positions on the retina. This can be seen in the multimodal distributions that emerge when the dwell time of the edge during the plateau period is plotted against retinal position (figure 9.11F, I). The two edge configurations (figure 9.11D, G) correspond to the right and left sides of the cone. The peaks in the two plots are distributed symmetrically about the midline, implying that the ant takes snapshots with the cone fixated frontally.

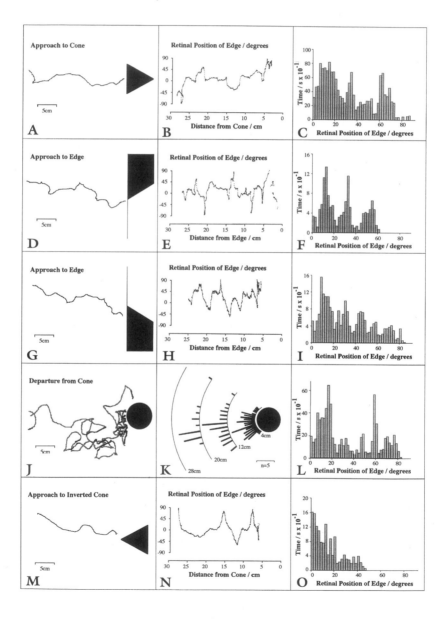

The underlying similarity of the multimodal distributions produced by edges and cones was assessed by cross correlating the two distributions over a range of angular separations. For the approaches in figure 9.11, the peak correlation occurred with an angular separation of 2 degrees. Given that the interommatidial separation is at least 4 degrees, we conclude that the peaks occur in essentially the same retinal positions, whether the ant approaches a cone or an edge.

The multimodal peaks are best explained in terms of the matching of edges to multiple stored views. The argument is simplest with a single edge. Suppose that the ant tries to keep the edge positioned where it best matches a stored view. If there were a single stored view, the edge would be ideally kept in the same position throughout the trajectory (figure 9.12C). With several stored views (see figure 9.13), the edge would migrate from one preferred retinal position to another and so would generate the multimodal peaks found in figure 9.11.

The matching of cones to their stored views is somewhat more complex, but the outcome will be multimodal peaks corresponding approximately to those generated by edges. Consider a single stored view. A perfect match can only occur if the ant is looking from where the view was taken. When the ant is further away, the best partial match requires no more than that the cone is contained within the view. Unless the stored view

FIGURE 9.11. A wood ant's approach to a cone and a black-white edge, and its return path from the cone. Data in **A–L** come from one individual: (**A**) Trajectory toward a cone. (**B**) Estimated position of the right edge of the cone on the retina is plotted against the ant's distance from the cone. Position of the edge is measured relative to the bottom of the edge. Notice the flat or plateau regions of the trajectory during which the ant walks toward the cone while holding the edge steady on its retina. (**C**) Dwell time of right edge in different positions on the right retina during the plateau phases of 14 approaches to the cone. Bins are 2 degrees wide. (**D**) Trajectory of ant toward black-white edge, parallel to the right edge of the cone. (**E**) Estimated position of edge on the right retina during approach to edge. (**F**) Dwell time of edge in different positions on the right retina during the plateau phases of four approaches to the edge. (**G**) Approach of the same ant to a black-white edge parallel to the left edge of the cone. (**H**) Estimated position of the edge on the left retina during the approach. (**I**) Dwell time of the edge in different positions on the left retina during the plateau phases of five approaches to the edge. (**J**) Trajectory of the same ant during its first departure from the cone. Trace is thickened when ant turns and walks back toward the cone. (**K**) Circular distribution of inspections of the cone accumulated over 14 departures. Inspections are placed in bins 8 cm apart measured from the cone and occur less frequently with increasing distance. (**L**) Dwell time of right edge in different positions on the right retina during the inspection from 14 departures. (**M–O**) Approaches of ant to an inverted cone. (**N**) Trajectory of the edge of the cone over the retina. (**O**) Dwell time of edge in different retinal positions during the plateau phases of five approaches to the cone. Similar data were recorded from three more ants trained to inverted cones. (From Judd & Collett 1998.)

emphasizes edges, in which case the cone will have two preferred positions, one at each side of the template, the cone will move freely within the template (figure 9.12A, B).

A good check of this line of reasoning is to examine where ants position edges and cones on their retina when they are trained to approach an inverted cone. The pattern of results should be quite different. Because the ant's horizon is only just above the ground plane, the apex of the inverted cone will be imaged at the same vertical position at all distances so that

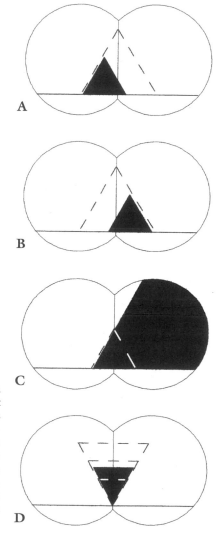

FIGURE 9.12. Postulated template on ant's retina when trained to an upright cone or an inverted cone. When the ant is more distant from the cone than the vantage point of a template, the image of the cone can fit in several positions within the template (**D** and **B**), but the black-white edge has only one optimal position (**C**). Apex of inverted cone falls on the same retinal position whichever template is engaged (**D**).

the bottom part of all the stored views will overlie each other (figure 9.12D). Consequently, there will be only one preferred position for an edge. The ant's behavior nicely matches this prediction. Instead of a multimodal distribution of dwell times there is one peak centered along the midline (figure 9.11O).

LEARNING VIEWS

When a wasp or a bee first leaves its nest or a newly discovered foraging site, it performs an elaborate flight during which it acquires information that is later used to guide its journey back to the same place (see Wehner 1981; Zeil, Kelber, & Voss 1996). One important characteristic of these learning flights is that the insect pivots around the site to which it will return, moving in a sequence of arcs of increasing radius (Zeil 1993a). This pattern of motion allows the insect to pick out objects close to the site that will best define the site's location. A less conspicuous feature is that, toward the end of the arcs, the insect faces the site with frontal retina; it may be at these moments that the insect stores views (Collett 1995; Collett & Lehrer 1993).

On their first few visits, wood ants leaving the cone perform a complex maneuver in which they also periodically turn and walk back toward the cone. The spatial distribution of these inspections of the cone suggests that this maneuver is concerned with obtaining views of the cone for guiding later approaches. One example of a departure is shown in figure 9.11J. The trace is thickened when the insect turns round and moves toward the cone. The middle panel (figure 9.11K) plots the radial positions and distances from the cone of the inspections made by the ant over many departures. Inspections occur relatively frequently over a wide range of viewing angles when the ant is close to the cone, and they become rarer and are more narrowly distributed as the ant retreats.

The strongest indication that the ant learns views when inspecting the cone on departure comes from the distribution of retinal positions. The right-hand panel shows a distribution of the dwell time of one edge of the cone during the inspections that occurred over 14 departures. This distribution was cross-correlated with that of the comparable distribution from the same ant obtained during its approaches to an edge on the same day. The correlation peaked with a lag of 2 degrees.

THE DISTRIBUTION OF STORED VIEWS

The exact positions of the peaks in these multimodal distributions vary from insect to insect and may change for the same individual if it is tested after an interval of several days. Despite this variability, the distribution of views is interestingly lawful. The peaks in figure 11F occur at roughly 20-degree intervals (52, 32, and 12 degrees) corresponding to views taken at 4.5, 6.6, and 16.8 cm from the center of the cone. Ants have adopted a strategy of sampling more densely close to the beacon where image size changes rapidly and less frequently farther away where change in image size is less dependent on range (figure 9.11K).

Sampling the cone at equiangular separations gives the ant a series of views that covers evenly the transformation of the shape from its first sighting to the ant's arrival at its base (figure 9.13). An excess of stored views brings the danger of "overloading" its memory (Willshaw, Buneman, & Longuet-Higgins 1969), while an insufficiency would lead to regions of the approach in which there might be mistakes in matching.

Sampling over a wider range of viewing directions when close to the cone may also have functional value. The ant approaches the cone in roughly the same direction on each trip, as will also happen when it approaches a landmark in more natural situations. Far from the object, any minor sinuosity in the ant's path will have little effect on its viewing direction so that it is enough to learn the cone's appearance over a narrow range.

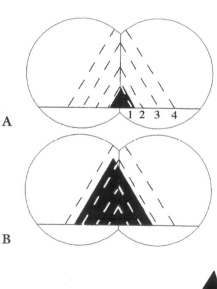

FIGURE 9.13. Views of the cone taken at equiangular steps correspond to the ant storing more views when it is close to the cone than when it is far from it.

But, as the ant comes closer, deviations from a straight path have an increasing impact on viewing direction, making it advisable to learn views over a larger range.

MODELING NAVIGATION
WITHOUT TOO MANY PRECONCEPTIONS

Navigation as illustrated by a wasp's or by an ant's path to a goal is the outcome of many mechanisms, the designs and workings of which are determined in many unknown ways. Constraints on how the nervous system is configured may arise from the evolutionary history and development of the organism, its habitat, the large variety of tasks the organism performs, and so on. This complexity makes it difficult to be at all sure that interesting characteristics of a wasp's flight, such as its tendency to fly sideways or its use of a consistent compass direction, are really advantageous to the insect in the way we have supposed. It is always possible that they are present for some unrelated and unidentified reason. A new way to ask whether such properties are useful in navigating to a learned goal is to evolve simulated creatures (SCs) that are selected for their efficiency in performing just a single navigational task (Harvey, Husbands, & Cliff 1994; Holland 1975). It is then possible to discover whether an SC that is subject to more circumscribed and identifiable constraints will make use of sideways flight or a compass if given the opportunity to do so. One of us (K.D.) is exploring the behavior and "neural networks" of artificially evolved creatures of this kind.

THE TASK AND THE ENVIRONMENT
The SC operated in a flat, two-dimensional environment. It was placed randomly within a 30-degree sector north of a single landmark (figure 9.14), consisting of an octagon with a Lambertian surface lit from all sides. This object was the only reflecting surface in the SC's environment. The light sources themselves were not visible. The SC's task was to reach as rapidly as possible an invisible target that was located a little to the west of the landmark.

THE BUILDING BLOCKS
Selection operated on the "synaptic" connections between a number of irreducible components. Information about the outside world came from a one-dimensional eye with a ring of 20 facets that covered 360 degrees horizontally. Each facet gave an output proportional to the total light falling within its 36-degree field of view. Information also came from a compass

composed of four channels, each of which was maximally sensitive to a different cardinal compass direction and gave an output that dropped as the SC turned away from the channel's most sensitive direction.

Forward speed was governed by the activities of left and right motors. Each motor was fed by two motorneurons that could either be on or off, allowing the SC to travel at full or half speed. When both motors operated at the same speed, the SC moved forward; differences in speed added a turning component. In the simulations considered here, the SC was provided with two additional motors that could drive it left or right in a direction orthogonal to the main motors. The maximum speed of the sideways motors was half that of the main motors.

Information was passed from input neurons to motorneurons either directly or by way of four intermediate or hidden neurons. Potentially any hidden neuron or motorneuron could choose to receive input from any neuron in any layer with synaptic weights that could vary between -1 and $+1$. Selection acted to adjust the synaptic weights between neurons and to set the threshold of the whole neuron somewhere between 0 and 1.

Noise was mixed with the sensory input to mimic a noisy environment. To do this, the input signal was multiplied by a random number that lay anywhere between zero and 30 to 40 percent of the maximum sensory input. Noise was not added to the motor output.

THE GENETIC MATERIAL, ITS VARIATION AND SELECTION

The weights and synapses are encoded as a single string of 190 eight-bit characters. Each hidden neuron and motorneuron potentially has six inputs that can be assigned either randomly or through selection to any neuron in the net. Genetic material is exchanged between individuals by random crossovers that occurred in all pairings. Variability within an individual's genome is introduced by random mutation at a rate of five bits per individual per breeding cycle.

The bit string can be in any of 2^{1520} states and the job of artificial evolution is to scour the space of possibilities for regions that will generate SCs with networks that can perform the navigation task. In order to guide the search through this genetic state space, there has to be a measure of the fitness of each individual at the task so that selection can operate and the search can be centered around the region of genetic state space occupied by the fittest individuals in the current population. The fitness measure must thus enable individuals to be ranked at all levels of competence, from those that are very bad to those that are very good at the task. To explore hopeful regions, the fittest individuals are bred and the chromosomes of the progeny are made to differ slightly from their parents and each other through recombination and mutation.

Artificial evolution can only be an effective search tool if there is at least some local mapping between proximity in genetic state space and proximity in fitness value. This is the case, by and large, in the present model where small genetic changes correspond to small changes in synaptic connectivity. Because the network may be richly interconnected, however, small genetic changes can occasionally lead to large jumps in fitness, if, for example, a change in a single synapse or a threshold level has widespread effects.

To initiate an evolutionary run, the bit strings of 50 SCs were set at random. Each SC was placed within the starting area and allowed to move according to the dictates of its randomly connected nervous system for 150 time units. Fitness at each time unit of a trial was defined as $1/d^2$, where d is the distance of the SC from the target. Fitness for the whole trial was the sum of the fitness values over all time units. The SC's fitness was evaluated over six such trials and the lowest recorded fitness allocated to it. Each individual was then given a probability of breeding according to its fitness rank within the population. The outcome was a new population of 50 SCs in which the offspring of the fittest tended to replace the least fit.

Initially, the population is randomly distributed over genetic state space and it explores several regions for possible solutions. The breeding scheme will gradually lead the population to converge upon a single solution that may vary from experiment to experiment.

In the context of this chapter, the interesting question is whether the SCs evolved navigational strategies that resembled those of wasps. Do the SCs, if given the opportunity, make use of a sideways component of motion and follow a single compass direction? An affirmative answer would indicate that the strategies found in wasps have wide utility because the detailed inner workings of wasps and SCs are obviously very different.

EVOLVED BEHAVIOR USING SIDEWAYS MOTION AND COMPASS

Figure 9.14B shows two typical trajectories generated by a single, evolved SC. Four phases of the left-hand trajectory have been labeled. In the first phase, the SC turns on the spot until it assumes its preferred orientation. It then moves directly toward the landmark. At the start of the second phase it changes orientation a little and starts to move sideways. Sideways motion continues until the beginning of the fourth phase. The sideways motor is

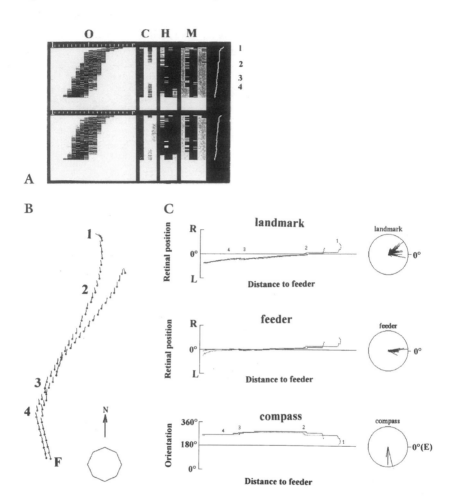

FIGURE 9.14. Trajectory of a simulated creature moving toward an invisible feeder (F) placed west of an octagon. (**A**) Behavior of visual neurons (O), compass neurons (C), hidden or interneurons (H), and motorneurons (M) during two trajectories. The left two motorneurons control the sideways motors, the right four the forward motors. Level of activity is indicated by the density of the black bars. Time runs from top to bottom. (**B**) Position and orientation of the SC during two trajectories. (**C**) Retinal positions of the center of the octagon and the feeder and the SC's orientation are plotted against the distance of the SC from the feeder. Circular histograms to the right show accumulated retinal positions and orientations during these trajectories.

switched off when the edge of the octagon is imaged by a particular ommatidium. The SC then heads directly for the feeder, facing in the same direction as it did in phase one.

The activities of the neurons during the two trajectories are plotted in figure 9.14A, with time running from top to bottom. Shown from left to right are the visual input neurons, the directional sensors, the left and right sideways motorneurons, and the four motorneurons controlling the forward motors. The direction of the trajectory seems to be governed by the right sideways motor which is turned on during phase two and turned off abruptly at the beginning of phase four. Throughout the approach SC travels at half its maximum speed, suggesting that the control system is unable to cope with full speed. Time and again, the outcome of relatively unconstrained evolution was a system that relied on sideways motors to control the trajectory. When sideways motors were not available to the evolving SCs, they were normally much less accurate in reaching their goal.

There are intriguing parallels between figure 9.14C and the trajectories of real wasps shown in figure 9.6. In both cases, the same compass orientation is maintained over many approaches and there is a pronounced sideways component of motion. The feeder, which is invisible in the case of the SC, is "fixated" throughout the approach. The landmark is first imaged frontally and then moves to the periphery of the eye. The major difference between the SC and real wasps is in the very last phase: The sideways component found in wasps (figure 9.7) is absent in the SC.

CONCLUSIONS

This chapter is limited to a very small fragment of a hymenopteran's navigational repertoire. Nonetheless, something of the insect's cognitive style has emerged from the discussion. It seems to be driven by the demands of view-based guidance. Because familiar places are recognized by the idiosyncratic details of the surrounding landscape, a navigating insect must rely heavily on a memory for arbitrary visual patterns. Because visual memories of scenes are retinotopically organized, motor strategies are carefully tailored to make best use of these memories. Insects try to view scenes from a constant direction and so voluntarily limit their rotational freedom. Because a close view of a scene that is stored retinotopically can only be recognized from within a small area, ants, and probably other insects as well, have evolved schemes to store enough but not too many views. Even when navigating over a small area, insects use stored views in several different ways to guide their paths, and this multiplicity of navigational strategies may also help limit the number of views that insects need store.

Something that has not been stressed but which is obvious from scrutinizing almost all the graphs is that the system seems to operate sloppily. Where does the noise or slop reside? There are several reasons to suppose that views are recorded at well-defined positions and that the errors occur in movements and their control. Data from Wehner (figure 9. 2) and from Junger's (1991) experiments suggest that small movements of the image away from a stored template are detectable. The peaky search pattern in figure 9.4 argues the same way. If this is the case, there remains the interesting question of how precise views are acquired from a noisy platform.

ROSWITHA WILTSCHKO
WOLFGANG WILTSCHKO

10

Compass Orientation as a Basic Element in Avian Orientation and Navigation

Birds are famous for their navigational abilities. Every year, many birds temporarily leave their breeding grounds to avoid adverse conditions. In the tropics, this involves migrations in accordance with the wet and the dry seasons; at higher latitudes, it is the approaching winter with short days, cold temperatures, and lack of adequate food that forces birds to head for distant regions of the world. The distances traveled are impressive. The record is held by sea birds like the Arctic Tern, *Sterna paradisea,* which breeds in the Arctic tundra and migrates around the globe to spend its winters near the shores of Antarctica. Yet migration is not the only behavior that involves spatial tasks exceeding those of most other animals. The home range of birds usually covers several kilometers, involving distances that are considerably larger than in mammals of comparative size. Furthermore, birds are known to return after displacement from totally unfamiliar areas. This is particularly true for pigeons, whose capability of homing has been famous since antiquity. Man domesticated the wild Rock Dove, *Columba l. livia,* and used it regularly for carrying messages.

   This chapter is devoted to the question of how birds cope with spatial problems. The discussion focuses on two behaviors: homing and migration. The orientation tasks involved are fundamentally different, and so are the birds' strategies. These strategies, however, have one important feature in common: in homing as well as in migration, birds make use of an external directional reference that is a *compass*.

ORIENTATION BASED ON AN EXTERNAL REFERENCE SYSTEM

The most striking feature of bird navigation is the long distance involved. Birds rarely have direct contact with their destination. This means that in most cases, they cannot be guided by cues emerging from the goal itself, but must establish contact to the goal indirectly. When a bird wants to fly from a site A to a distant site B, the position of B with respect to A must be defined in a way that the bird can derive information leading it from A to B. This would require a specification equivalent to our human use of compass directions, when one says: "B lies *south* of A."

This example illustrates the role of an external reference system in defining the position of places relative to each other. Cues that are independent of any specific location, being accessible from everywhere within the home range and beyond, are best suited for this purpose. Pardi and Ercolini (1986) called such cues "global cues," in contrast to "local cues," which emerge from the local situation and indicate the direction to a goal area directly. Two types of environmental factors possess the required characteristics: the *geomagnetic field* and *celestial cues*. Birds make use of both, and thus have more than one compass mechanism at their disposal.

COMPASS MECHANISMS

Birds can locate directions with help of the geomagnetic field. Their ability to perceive the direction of the magnetic field lines provides them with compass information. Directions are no longer indistinguishable—the field lines define a prominent reference direction. *North* and *south*, or, considering the functional mode of the birds' magnetic compass as an "inclination compass" (see R. Wiltschko & Wiltschko 1995), *poleward* and *equatorward*, can be distinguished. With respect to these directions, any other direction can be established. The magnetic compass is innate in the sense that, in contrast to the celestial compass mechanisms, it does not require learning and is available to birds right from the beginning. By testing birds in magnetic fields with altered north direction, it has been demonstrated in 18 species of migratory birds and in homing pigeons. It seems to be rather widespread not only among birds, but also among other animals (see R. Wiltschko & Wiltschko 1995 for review).

Birds also make use of celestial cues for determining directions. Here, however, they face the problem that these cues move across the sky in the course of the day. When using the sun as a compass, birds must compensate for these movements by estimating the sun's current azimuth with the help of their internal clock. This important role of time in sun compass orientation provides an easy means to demonstrate the use of the sun compass by the classic clock-shift experiments. Birds are subjected for a few days to an artificial light regime, with the light-dark period beginning and ending 6 hours before (fast shift) or after (slow shift) sunrise and sunset (see Schmidt-Koenig 1958 for details). This treatment resets their internal clock, causing them to misjudge the time of the day, and as a result, the azimuth of the sun. When such pigeons are released away from their home, they show a characteristic deviation from the direction taken by untreated controls; this deviation indicates that the sun compass is used (figure 10.1 left). At the same time, clock-shift experiments illustrate the general importance of the sun compass in the orientation of birds. Manipulating the internal clock interferes only with the sun compass; the fact that pigeons are misled by their manipulated sun compass, although their magnetic compass could have given them correct information, clearly demonstrates the dominant role of the sun compass as long as the sun is visible

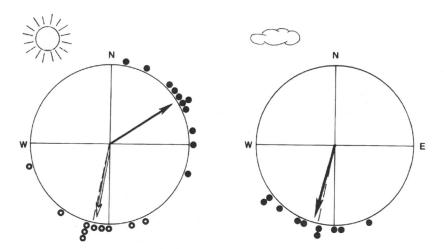

FIGURE 10.1. Compass use in pigeons. The home direction is indicated by a dashed radius; the symbols at the periphery of the circle indicate vanishing bearings of individual birds; the arrows represent the mean vectors. (*Left*) Orientation under sun; 6-hour clock-shifted birds (solid symbols) show a characteristic deviation from untreated controls (open symbols), indicating sun compass use. (*Right*) Orientation at the same site under overcast conditions; the birds presumably use the magnetic compass.

(figure 10.1 left). Under overcast sky, however, the orientation of pigeons is equally good (figure 10.1 right; Keeton 1969; R. Wiltschko 1992a), indicating that, when information from the sun is not available, the sun compass can be replaced by the magnetic compass without apparent loss.

Sun compass use is possibly restricted to orientation within the home range and homing, because during migration, the dependency of the sun's arc on geographic latitude would require continuous adjustments (see Munro & Wiltschko 1993 for discussion). For migrants, the sun and sun-related factors play an important role only at the time of sunset (see Able & Bingman 1987). Nocturnal migrants, however, use the stars for direction finding. A star compass has been demonstrated only in night-migrating birds and might be a special adaptation to guide their nocturnal flights. The mechanism is based on the configuration of stellar patterns relative to the celestial pole, this way being largely unaffected by the latitudinal shifts that birds experience in the course of migration (see Emlen 1975 for details).

THE ORIGIN OF THE SET DIRECTION— WHAT COURSE TO FLY?

A compass alone is not sufficient for successful navigation—it helps only if the course to the goal is known. In other words: the ability to locate north, east, south, and west is only helpful if the bird knows that its destination lies, for example, to the south. The crucial question is, how do birds determine their respective course? It is this aspect of navigation where the mechanisms employed in homing and migration differ widely, as a function of the fundamentally different nature of the navigational tasks.

HOMING

Homing involves returning to a very specific site, the bird's home. This site and the area around it are very familiar to the bird. Normally, it is the starting point of the travel, which, under natural circumstances, might be an active foraging flight; under experimental conditions, it is a passive displacement. The home course is variable; it depends on the bird's present location when it decides to return home and may thus lie in every possible direction. The bird has first to determine the correct course.

GENERAL STRATEGIES

Several fundamentally different types of navigational strategies have been suggested for birds: dead reckoning, orientation by landmarks, and several types of bicoordinate navigation. Experimental evidence allows us to limit our considerations to certain strategies and focus on those that involve a compass.

The Map and Compass Model

G. Kramer was the first who realized that for navigation, birds generally rely on an external reference provided by a compass. He described avian navigation as a two-step process: in the first step, the bird determines the compass course leading to the goal, then, in the second step, it uses a compass to locate this course (Kramer 1959). An example may illustrate the crucial point: A bird, displaced to the north, determines its home direction in terms equivalent to "180 degrees" or "south." Then it uses a compass to find out where south lies, obtaining a specification like "this way" or "go there," thus transforming the compass course into an actual direction of flight (figure 10.2).

Kramer's (1959) *map and compass* concept thus regards an external reference as an essential component of the navigation process. His model represents the basic theoretical concept in avian navigation. In the testing of it, the birds' preference of the sun compass over the magnetic compass provides an important methodological tool, because sun compass orientation can easily be demonstrated in free-flying birds (cf. figure 10.1 left). By indicating sun compass use, clock-shift experiments demonstrate a general involvement of compass orientation. This, in turn, is crucial for deciding whether or not the orientation strategy used in a given case follows the map and compass model.

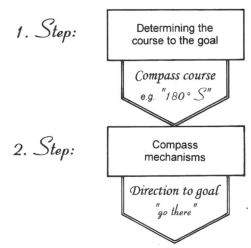

FIGURE 10.2. The *map and compass* model (Kramer 1959) regards avian navigation as a two-step process.

Clock-shift experiments with homing pigeons have produced the typical deflection in all directions from the loft and at distances ranging from less than 1.5 km (e.g., Graue 1963; Keeton 1974) to 167 km (Schmidt-Koenig 1965). Thus navigation follows the map and compass model within the entire range studied so far. This is true for familiar and unfamiliar sites (Luschi & Dall'Antonia 1993). Even at extremely familiar sites, the birds responded to clock-shifting in the predicted way (Füller, Kowalski, & Wiltschko 1983; R. Wiltschko 1991). One may conclude that the relation to home is generally established via an external reference, a compass.

Strategies Used by Pigeons

These positive results of clock-shift experiments exclude navigational strategies like dead reckoning in the sense of path integration based entirely on the integration of internal signals generated en route. This had been suggested by Barlow (1964) and was, for example, demonstrated in small mammals moving within the limited area of an arena (Etienne, Teroni, Maurer, Portenier, & Saucy 1985). Likewise, a response to clock-shifting excludes homing based solely on familiar landmarks, as was proposed by Griffin (1955) and has recently been reconsidered as a strategy at familiar sites under certain conditions (e.g., Papi 1986).

What strategies, then, may birds apply? Navigational strategies can generally be divided into two categories, namely those based on route-specific information collected during the outward journey, and those based on site-specific information obtained at the starting point of the return flight. In view of the involvement of compass orientation, two types of strategies appear possible, namely (1) the use of route-specific information based on an external reference, and (2) navigation by site-specific factors whose directional relationship to the goal is familiar to the birds (W. Wiltschko & Wiltschko 1982, 1987).

Available evidence indicates that birds make use of both strategies, but to different extents. During a transient phase early in life, young pigeons appear to rely solely on outward journey information (R. Wiltschko & Wiltschko 1978a, 1985; W. Wiltschko & Wiltschko 1982). This phase is rather short, however, lasting perhaps only two weeks. Pigeons soon begin to change strategy and turn to local, site-specific information obtained at the site where they begin the return flight (R. Wiltschko & Wiltschko 1985). From this stage onward, they are largely unaffected by manipulations that interfere with their access to the various types of navigational information during the outward journey (e.g., Keeton 1974; Kiepenheuer 1978; Matthews 1951; Walcott & Schmidt-Koenig 1973;

Wallraff 1980; R. Wiltschko & Wiltschko 1985). This indicates that experienced birds can derive their home course from site-specific information alone, and leads to the concept of the navigational "map."

THE NAVIGATIONAL MAP

The map and compass model, assuming compass orientation to be an integrated component of navigation, has an important implication: it implies that spatial information is organized in a directionally oriented way. Models on the navigational "map" take this into account.

The Model

The concept of a navigational map was inspired by the commonly used grid map with its worldwide grid of coordinates. It dates back to the last century (Viguier 1882). The present model of the map has been described in detail by Wallraff (1974, 1991) and by W. Wiltschko and Wiltschko (1982, 1987, in press). Two types of maps supplementing each other have been proposed.

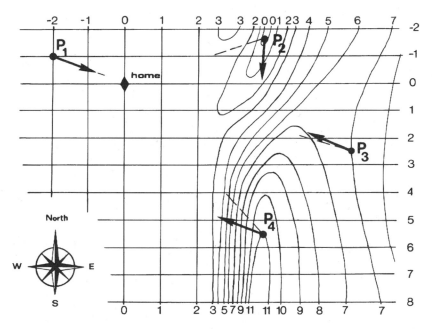

FIGURE 10.3. Model of the navigational map, a directionally oriented mental representation of the spatial distribution of navigational factors. Birds use gradients to determine their position relative to home. The values of the gradients are given relative to the home values. Irregularities in the course of these gradients cause deviations from the true home course known as "release site bias." (After W. Wiltschko & Wiltschko 1982.)

The navigational map or grid map is assumed to be a *directionally oriented representation* of the spatial distribution of navigational factors. These factors—there must be at least two—have the nature of *gradients,* that is, their values vary continuously in space. They must not intersect at angles which are too acute (figure 10.3). Birds know in what direction the respective values increase and decrease and can thus derive their home course by comparing the local scalar values at their present location with the ones remembered from home. In the example in figure 10.3, the birds know that one gradient increases to the east and another to the south. If they find themselves at a location where the local values of both are lower than the home values (P_1 in figure 10.3), they know that they are northwest of home and have to head toward the southeast.

In the vicinity of home, this grid map is supplemented by a *mosaic map* of landmarks (Graue 1963; Wallraff 1974, 1991), which is assumed to be a *directionally oriented representation* of the lay of the land with the positions of salient local features (figure 10.4). The mosaic map corresponds

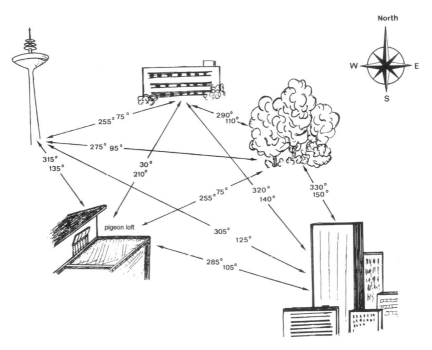

FIGURE 10.4. Model of the *mosaic map,* a directionally oriented mental representation of the spatial distribution of familiar landmarks in the vicinity of home. (After W. Wiltschko & Wiltschko 1982.)

to the grid map in that it gives the direction to the goal as a compass course. Yet instead of a few continuous gradients, positions are indicated by numerous separate entities, like landmarks, whose directional relationship is familiar to the birds. The landmarks involved are mostly visual, but may also be of other sensory qualities, like local sources of infrasound, magnetic anomalies, and so on. As a consequence of being based on local features, the mosaic map covers only terrain that birds know from direct experience and, maybe, a small area beyond (see Baker's 1982 concept of the familiar area map). For the following considerations, the main focus is on the navigational or grid map, which allows birds to return from distant, unfamiliar sites.

Findings on Pigeon Homing

A model of the navigational map should be able to account for the following findings:

1. Birds are able to determine their home direction at totally unfamiliar, distant sites far beyond the range of immediate experience, even if they are deprived of all known navigational information during the outward journey.
2. The directions in which pigeons depart often do not coincide with the true home direction; however, the deviations are usually small, rarely exceeding 30 to 45 degrees (see R. Wiltschko 1993).
3. Pigeons head immediately, that is, in less than 20 seconds (e.g., Pratt & Thouless 1955), into the direction in which they finally depart.

The first two observations—orientation at unfamiliar sites and headings deviating from the home course—are explained by the assumption that the map is based on environmental gradients. Birds use site-specific factors indicating positions they can read because they know them from other locations and can extrapolate them beyond the range of direct experience. At distant sites, birds have to base extrapolation on their "map," that is, on the gradient directions as they are represented according to the experience within the home area. If the navigational factors are not completely regularly distributed, this might cause birds to misjudge their position and, as a consequence, depart in directions deviating from the home course (see figure 10.3, sites P_2, P_3, and P_4). The model of the navigational map thus attributes the frequently observed deviations from the true home course to unanticipated irregularities in the distribution of navigational factors (cf. Keeton 1973; W. Wiltschko & Wiltschko 1982).[*]

---

[*]A totally different explanation of release site biases, based on the assumption of a preferred compass direction, has been forwarded by Wallraff (1978).

From this interpretation, it follows that deviations should be similar in all releases from a given location. This is confirmed by data from numerous sites (e.g., Keeton 1973; Schmidt-Koenig 1963; Wallraff 1959, 1970; R. Wiltschko 1993). There may be considerable variability, but at certain sites, the pigeons depart almost always to the right of home, while at others, they tend to depart to the left. The size of the deviations is likewise characteristic (figure 10.5; for more examples, see R. Wiltschko 1992a, 1993). The phenomenon is referred to as *release site bias* (Keeton 1973, as a translation of the German terms *Ortsmißweisung* by Schmidt-Koenig 1958 and *Ortseffekt* by Wallraff 1959).

Release site biases are not restricted to pigeons, but are observed in other species of birds as well (for summary, see R. Wiltschko 1992a). In the case of Bank Swallows, *Riparia riparia,* the similarity of biases to those of pigeons could be directly demonstrated: Birds from a colony near the pigeon loft showed the same deviations from the home course as pigeons when they were released at the same sites (Keeton 1973, 1974). This points out that the factors causing these deviations act on the various bird species in a similar way. Apparently, at many locations, the navigational system regularly indicates courses that deviate from the true home direction. Consequently, at those sites, the bias direction and not the home direction represents the baseline for clock-shift responses (Keeton 1973; Schmidt-Koenig 1958; R. Wiltschko, Kumpfmüller, Muth, & Wiltschko 1994).

The observation that birds head immediately into the direction in which they will later vanish from sight means that the navigational process does not require extended searching flights. This excludes the possibility that pigeons need to scan local gradients to learn gradient directions and is in accordance with the idea of the map operating on a comparison of scalar

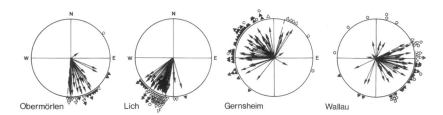

FIGURE 10.5. Examples of the behavior of pigeons at frequently used release sites. The arrows represent the vectors of single releases. The symbols at the periphery indicate the respective mean directions: open triangles, homeward oriented; solid triangles, release site biases; open circles, nonsignificant behavior. (After R. Wiltschko 1992a.)

values. Pigeons have at least a rough idea about their home direction before they even start flying, as could be demonstrated by releases from a cage with several openings (e.g., Chelazzi & Pardi 1972). The model of the navigational map explains these findings by assuming that birds know in which direction the gradient values increase and decrease and have this information incorporated in their map. This leads to the question of how pigeons acquire the necessary knowledge.

THE PROCESSES INVOLVED IN
ESTABLISHING THE NAVIGATIONAL MAP

The so-called maps are based on experience. In the case of the mosaic map, it is obvious that the positions of landmarks and their relationship to each other and to the loft have to be learned. Yet learning processes also establish the navigational map or grid map. None of the known environmental gradients are completely regular, so that the birds have to familiarize themselves with the distribution of these factors within their home region. Learning processes, on the other hand, enable birds to include all suitable factors into their map, including those of only regional importance. Little is known about the respective learning process, because it largely escapes experimental studies. Some details may, however, be inferred from the behavior of young pigeons.

A First Navigational Strategy

Experiments with young pigeons indicate different phases in the development of the navigational system, involving a change in navigational strategy. During an early phase, young birds show excellent initial orientation, with long vectors close to home. When such birds were displaced in a distorted magnetic field, they were no longer oriented. The same magnetic treatment after arrival at the release site had no effect (figure 10.6; R. Wiltschko & Wiltschko 1978a). The fact that only treatment during displacement was effective suggests that young pigeons rely for navigation on magnetic information obtained during the outward journey. They appear to derive their home course by recording the direction of the outward journey with their magnetic compass, integrating turns and detours, if necessary; reversing this direction produces the homeward course. This strategy may be classified as *route reversal* (Schmidt-Koenig 1975) or path integration based on an external reference.

The effect shown in figure 10.6 is restricted to very young, untrained pigeons. It decreases rapidly as the birds grow older than 12 weeks. Trained pigeons are generally unaffected (R. Wiltschko & Wiltschko

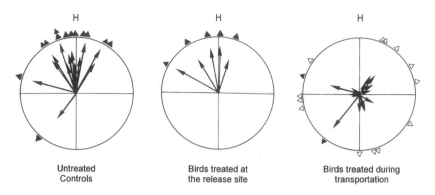

Untreated Controls **Birds treated at the release site** **Birds treated during transportation**

FIGURE 10.6. The effect of a distorted magnetic field on the orientation of very young, inexperienced homing pigeons. The arrows represent the mean vectors of single releases with respect to home (H). The triangles at the periphery indicate the respective mean directions: solid, significant; open, nonsignificant. (Data from R. Wiltschko & Wiltschko 1985.)

1985). This suggests a change in navigational strategy at about the pigeons' third month of life: They no longer rely exclusively on route-specific information, but gradually turn to information obtained at the release site— they begin to use the navigational map. At the same time, as the young pigeons grow older, deviations from the home course become more frequent. The occurrence of release site biases also indicates the onset of a gradual transition from the use of outward journey information to the use of the navigational map. The variability among groups is enormous, which is not surprising considering that spontaneous learning processes are involved.

Forming the Map
The above-mentioned findings suggest that the navigational map normally becomes functional sometime around the third month of life. At the same time, they point out possible mechanisms involved in its formation. A component of the early navigational strategy based on the net direction of the outward journey may be part of the learning processes, as it means that the young birds continuously keep track of the direction of their movements. They may combine this information with registered changes in gradient values encountered en route, associating changes in gradients with specific directions. Together, this information is incorporated into the map, leading gradually to an internal representation of the navigational factors which reflects the distribution of gradients within the home region.

Forming the map takes place during spontaneous flights. When young Rock Doves, *Columba l. livia* (the species from which the carrier pigeon was bred), start to fly, they soon join their parents on extended foraging flights and thus have an opportunity to establish their map under their parents' guidance. Young carrier pigeons, being fed at the loft, do not undertake such foraging flights. After reaching a certain level of flying skill, however, they spontaneously venture farther and farther away from their loft, thus gradually becoming familiar with the vicinity of their home. During these flights, they probably begin to establish a mosaic map by storing in memory the position of salient features of the terrain and their directional relationship to home. Later, on more extended spontaneous flights and on training flights, when they are released at increasing distances from their loft, they familiarize themselves with a growing area of their home region and thus establish their grid map by recording in what direction the gradient values increase and decrease.

The development of the navigational mechanisms in birds as described previously shows some striking parallels to the acquisition of spatial knowledge in humans as outlined by Siegel and White (1975) and Golledge, Smith, Pellegrino, Doherty, and Marshall (1985), in particular with regard to the role of locomotion and the effects of increasing experience. An important difference stems from the fact that birds (and many other animals), in contrast to humans, are provided with compass information: birds store the locations of landmarks and the gradient values in their directional relationship with respect to an external reference. This gives the mosaic map and the grid map the characteristics of *survey maps*, which is important for animals that are not restricted to specific routes, but move freely from starting points to goals in three-dimensional space. Another obvious difference is the time scale: the navigational system of birds develops within a few weeks—in accordance with the fact that young pigeons, for example, have to take care of themselves at the age of about three months.

Forming the grid map during long-term exposition at the home loft also has to be considered (e.g., Papi 1982; Wallraff 1974). This would, however, appear possible only with certain types of factors, such as airborne cues. The regional distribution of gradients can only be recognized during extended flights where a change in gradient values is directly experienced. The important role of flying for the development of the navigational system is amply documented in pigeons (see R. Wiltschko 1991) as well as in wild migratory birds (Berndt & Winkel 1980; Löhrl 1959; Sokolov et al. 1984). Extended flights appear to be necessary to acquire the crucial information that enables migratory birds to return to a specific home area.

Why Prefer Site-Specific Information?

At first glance, the indicated change in navigational strategy from the use of route-specific to that of site-specific information seems surprising, as the first strategy appears to produce good homeward orientation. The reasons become clear when the advantages and the disadvantages of the two strategies are compared. Route reversal has one crucial advantage: it does not require any foreknowledge. The relevant information is obtained during the outward journey; all that is needed is a functioning compass and the ability to record and process directional information, including the integration of detours. The basic computations may be complex, but they, too, may represent an innate ability, thus being carried out automatically without requiring any learning.

Yet any strategy based entirely on information obtained during the outward journey has one crucial disadvantage: there is no possibility to correct for errors. The accuracy of navigation depends on how accurately birds can measure and process directional information. Any initial mistake bears the danger that the birds miss their home at a certain distance—this might become crucial when the young pigeons begin to venture farther away. The mosaic map, and at greater distances the grid map, allow birds to determine the home course from any given site; birds can redetermine their home course as often as they feel necessary.

Yet the use of the maps requires detailed knowledge on the spatial distribution of the factors involved—knowledge that is not ad hoc available, but must be acquired by learning processes that need a certain amount of time. When the young birds begin to fly, they have only one option: they must rely on information obtained during the outward journey with the help of their magnetic compass, the only innate mechanism available to them. This is sufficient for first short flights, but when the birds extend their flights, the inability to correct for errors becomes a growing danger. The increased safety resulting from the use of map information must be assumed to be the primary reason for changing navigational strategy as soon as a map becomes functional (see R. Wiltschko 1991 and W. Wiltschko & Wiltschko 1982, 1987, in press for a more detailed discussion).

Updating the Map

The map remains flexible, and later experiences are regularly included. Old, experienced pigeons showed a marked improvement in homeward orientation when released a second time from distant sites 120 to 200 km from the loft (Grüter & Wiltschko 1990). Interestingly, even birds that were only familiar with an area covering 60 to 80 percent of the route from the loft to the distant site, but not with the site itself, showed better orientation than birds that had their entire experience in a different region. This suggests that the extended knowledge on the distribution of navigational factors in the region halfway to the distant release site provided a better basis for interpreting the navigational factors at the distant site (Grüter & Wiltschko 1990).

These findings show that even pigeons several years old extend their map and include new information when they find themselves in a region where they have never been before. Details on the processes updating the navigational map are not known; one might assume that they are similar to those establishing the map during the birds' first year. One important difference is indicated, however: Early experience seems to set the general rules for the use of the navigational map, specifying how navigational factors are to be extrapolated, whereas the processes updating the map appear to modify the map only locally, without affecting the pigeons' interpretation in other regions (see W. Wiltschko, Wiltschko, & Keeton 1984 for a detailed presentation of the relevant data and a detailed discussion). This procedure guarantees the formation of a highly effective map for an extended area, which allows birds to deal with some irregularities in the distribution of environmental gradients. It is helpful to adapt the regional part of the map to the true distribution of the navigational factors in the respective region, at the same time leaving unchanged the rules for extrapolation in other regions, where the specific distribution of navigational factors may be different and where slightly different rules may be required.

THE NAVIGATIONAL MAP IN USE

When the maps become functional, they offer an effective means for orientation within the home region and beyond. Figure 10.7 summarizes the current ideas of how the maps are established and used, indicating the important role of compass orientation in the navigational system.

So far, the navigational map is described as an open system, without naming the factors included. These may be of a world-wide distribution, like parameters of the geomagnetic field, but also, because the home range of birds seldom exceeds 10 to 20 km, factors of only regional distribution, provided they are available within an area of sufficient range. The number of factors is not limited; any suitable gradient extending over a large enough area might be incorporated. Their nature is still unclear; the ones in discussion are given in figure 10.7. Here, however, the specific nature of navigational factors will not be a major concern; focus will be on how such information may be represented and processed.

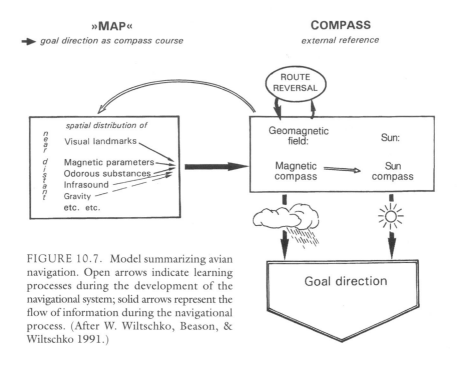

FIGURE 10.7. Model summarizing avian navigation. Open arrows indicate learning processes during the development of the navigational system; solid arrows represent the flow of information during the navigational process. (After W. Wiltschko, Beason, & Wiltschko 1991.)

Extrapolation of Map-Factors at Unfamiliar Sites

There is no direct evidence that birds extrapolate gradients; however, the behavior at the edges of the familiar area and beyond indirectly reflects some features about the use of the grid map. In their first year of life, pigeons of the Frankfurt loft are subjected to a series of standard training flights up to 40 km in the cardinal compass directions, which familiarizes them with the area around the loft. Figure 10.8 shows the deviations from home and the vector lengths recorded from trained pigeons at more than 120 sites around the Frankfurt loft, and their variation with distance. Up to 40 km and just beyond, the orientation was usually good, with mostly long vectors and small deviations from home. Beyond this range, the orientation first became slightly worse, but improved again beyond 80 to 100 km (see figure 10.8; R. Wiltschko 1992a).*

In interpreting these findings, two aspects of extrapolation must be considered. First, as the gradients cannot be assumed to be completely reg-

*A pattern with superficially similar characteristics—a marked deterioration in orientation (much stronger than the one illustrated in figure 10.8) from 20 km onward, followed by an improvement beyond 80 km—has been described for some lofts, whereas no such patterns were observed at others (for summary, see Wallraff 1974). The reasons might lie in different training procedures (for discussion, see R. Wiltschko 1992a).

ular, birds tend to make certain mistakes that would increase with increasing distance from the familiar range. Second, the effect of any mistake of a given absolute size decreases as the difference between the gradient values at the release site and at home increases. Both phenomena overlap, resulting in the pattern documented in figure 10.8. The observed decrease of initial orientation at a certain distance beyond the edges of the familiar area may be attributed to the first, the improvement observed farther out to the second. Vector lengths indicate agreement among pigeons of the same group; the shorter vectors beyond 150 km might reflect increasing individual differences in extrapolating when the birds are far away from their familiar range (R. Wiltschko 1992a).

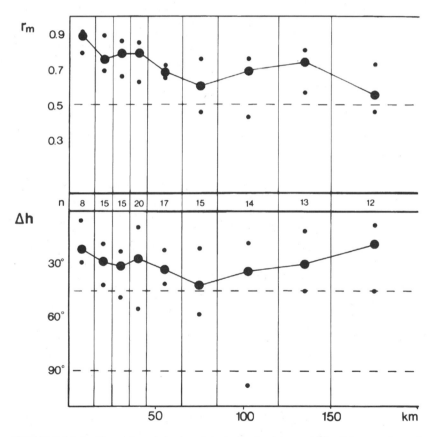

FIGURE 10.8. Vector length (r_m) and deviation from home (Δh) of the experienced pigeons as a function of distance. The line n gives the number of sites included in the analysis; the large symbols mark the median values, the small symbols, the quartiles. (From R. Wiltschko 1992a.)

Free Movements between Various Goals?

The learning processes establish the map center around the bird's home, yet neither the mosaic map nor the grid map is assumed to be home-centric in the sense that movements are restricted to returning home. The map should allow direct movements between arbitrary sites. The home or loft may represent a very prominent site in the life of a bird, but functionally it should be just one site among others. Birds should be able to determine the course from any site to any other site, provided the coordinates of the goal are known.

Within the home range, such free movements between multiple goal sites are usually implied, even if they escape experimental analysis. Within the range of the grid map, one would also expect that birds are able to determine the compass course from any location to any goal, provided they know the local gradient values of the goal site. An interesting experiment by Baldaccini, Benvenuti, and Fiaschi (1976) indicates that birds can indeed head for more than just one goal. Old, experienced pigeons were moved into a new loft about 90 km from the old one and kept there in an aviary for about six months. When these birds were afterwards released from sites between their old and new home, they departed heading toward, and returned to, the loft that was *closer*, irrespective of whether it was the old or the new loft (figure 10.9). These findings show that the pigeons were able to choose between the two goals and determine the respective compass courses, indicating that the map indeed allows free movements between arbitrary goals.

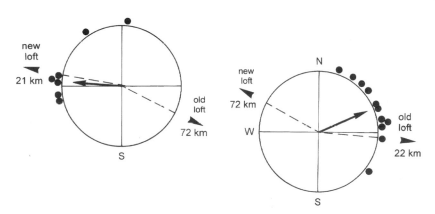

FIGURE 10.9. Orientation of homing pigeons that had been confined to a new loft for about six months, at releases between the old and the new loft. Symbols as in figure 10.1. (After Baldaccini et al. 1976.)

Information on Distance?

Another interesting point concerns the question whether the navigational map is a *metric map* in the true sense that it includes information on distance. Do birds know how far they are from their goal? Distance would not seem to be a necessary component, as long as the birds can reliably determine the direction to the goal and recognize it when they have arrived. Several findings indirectly suggest, however, that birds might have rather precise ideas about distance.

One indication results from the homing performance of wild birds that were displaced over various distances in the range of several hundred kilometers. An analysis of their homing speeds revealed that the highest speeds were usually recorded when the birds homed over the longest distances or when they were back on the day after release (see R. Wiltschko 1992a). Both observations imply that the birds knew how far from home they were released: In the first case, they seemed to be aware that long distances had to be covered and made haste; in the second, they seemed to know that they could reach home with one day's flight if they hurried up. The latter observation even suggests a quantitative estimate of distance.

The best evidence that birds are aware of the distance to a goal is provided by the two-loft experiment of Baldaccini et al. (1976) mentioned above (cf. figure 10.9), an experiment whose important theoretical implications have not received due attention so far. The observation that pigeons headed toward, and homed to, the loft that was *closer* indicates that the birds were able to compare the distances from the release site to the respective goals. This suggests quantitative information on distance and implies that the navigational map represents the distribution of gradients roughly true to scale. How accurate this information might be and whether it may change with increasing distance from the home range is unknown, however.

The Size of the Map

A last question concerns the range of the navigational map: What area does it cover? The estimates of both boundaries vary considerably. For the inner boundary, that is, for the distance where the transition from the mosaic map to the grid map occurs, estimates range from 5 to 10 km (e.g., Michener & Walcott 1967) to about 20 km (e.g., Wallraff 1974, 1991). Release site biases have been observed less than 10 km from the home loft, which indicates that within this radius, pigeons still rely on gradients. Experiments with pigeons deprived of object vision by frosted lenses also indicate that nonvisual navigational factors can be used up to a rather short

278 ROSWITHA WILTSCHKO and WOLFGANG WILTSCHKO

distance from home (Schlichte 1973; Schmidt-Koenig & Walcott 1973). There may be a wide area of overlap between the mosaic map and the grid map, and they must be expected to jointly guide the final part of the flight approaching home. Only the very last step in homing seems to depend on the view of landmarks, if at all.

Concerning the outer boundary of the navigational map, the findings on pigeons are controversial, leading to estimates ranging from world-wide in earlier studies (Wallraff & Graue 1973) to less than 700 km (Ioalè, Wallraff, Papi, & Foa 1983) and to less than 120 km (Benvenuti, Ioalè, & Nacci 1994). The latter appears too small by far considering the distances from which pigeons deprived of outward journey information departed non-randomly (e.g., Wallraff 1980 and unpublished data). Displacement experiments with wild birds also indicate a map of a much larger size. Migrating Starlings (*Sturnus vulgaris*) were caught as transmigrants in the Netherlands and displaced perpendicular to their normal migration route (see figure 10.10). Later, ringing recoveries indicated that birds displaced about 600 km to Basel, Zurich, and Geneva in Switzerland mostly returned to their normal breeding area in spring (Perdeck 1958, 1983). In contrast, another group, displaced about twice as far to Barcelona, Spain, appeared largely unable to return (Perdeck 1967). This suggests a functional navigational map that extends over large parts of Europe. Displacements of more than 1000 km into unknown territory, however, seemed to exceed the capacity of the navigation system in most Starlings.

A COGNITIVE MAP?

In the general discussion on mental representations of space, findings on birds have received little attention so far. When cases are discussed, they predominantly involve mammals (e.g., Ellen 1987) or, more recently, insects such as honey bees and ants (e.g., Wehner 1992; Wehner & Menzel 1990). If birds are mentioned at all, the emphasis is usually on small-scale orientation within the limited space of a room or an arena (e.g., Gallistel 1990; Vauclair 1987; Wehner & Menzel 1990). Homing of displaced birds is seldom considered in this context, which is surprising insofar as the general strategies have been discussed ever since the end of the last century (Viguier 1882), and theoretical models of the map are well developed (e.g., Wallraff 1974; W. Wiltschko & Wiltschko 1978b, 1982, in press).

In a previous paper (W. Wiltschko & Wiltschko 1987), we called the navigational map of birds a *cognitive map*, because it seemed to fit the definition by Tolman (1948) and O'Keefe and Nadel (1978): It is a representation of the environment, and it allows the birds to make decisions about where to fly. The crucial element, namely the ability to design novel routes, is also found in birds. Displaced birds can home from distant, unfamiliar sites and thus use novel routes, even if in this case it is a completely novel route from a novel starting point, rather than the novel shortcut that is usually emphasized in the discussion on cognitive maps. The map is established by experience, and from all that is known about the learning processes, they appear to be analogous to the procedures suggested by Gallistel (1990), because information about the route traveled is combined with information from the current position (see R. Wiltschko 1992b), resulting in an "allocentrically organized representation of environmental features" as defined by Thinus-Blanc (1988).

There are, however, certain differences between our model of avian maps and the currently discussed concepts of a cognitive map. A most striking aspect is distance. The discussion about whether or not animals have cognitive maps usually refers to orientation tasks in the range of a few meters up to a few hundred meters at the most, whereas avian navigation involves distances up to several hundred kilometers. This is to be considered when landmark use, routes, and so on are discussed. The essential difference, however, is that the cognitive maps discussed for other animals are taken to consist of separate entities like landmarks and the relationships between these landmarks; in this way, they show parallels to the mosaic map postulated for birds. The grid map of birds, in contrast, is assumed to be based on environmental gradients, that is, continuous factors whose distribution varies in space. This permits the use of the navigational map at distant locations beyond the range of immediate experience and gives the avian map an entirely different dimension that maps based on landmarks can never reach.

Another crucial difference involves the role of compass information. The discussions on cognitive maps emphasize that the relationship between landmarks is represented, but it is not always clear in what terms this relationship is specified. It might be in relation to other landmarks, or the walls of the enclosure. Specifications in terms of compass direction have also been discussed (e.g., Gallistel 1990), but have not received very much attention so far. In the mosaic map as well as in the grid map of birds, directional information is an essential component. Both maps indicate compass courses and require a compass to be transferred into an actual heading for the flight. Thus, in birds, landmarks and compass orientation are not alternatives like in insects that can reach their goals using either (e.g., Dyer & Gould 1981; von Frisch & Lindauer 1954). Analysis of the spatial representation of bees and ants has only just begun, however, and the difference in the role of compass orientation might turn out smaller than it appears at the moment.

MIGRATION

In contrast to homing, *migration* means moving to a distant region of the world. During their first migration, young migrants face the task of reaching a particular geographic region that is yet unknown to them—the species-specific wintering quarters. This means that they must head toward a goal without being familiar with the local conditions, regional distribution of potential navigational factors, and so on. On the other hand, the goal is not a specific location, but a more or less extended area offering suitable conditions. Also, the directional relationship between the starting point and this goal is constant; it is given by the geographic positions of breeding and wintering areas.

INNATE INFORMATION ON THE MIGRATORY COURSE

First indications about the mechanisms enabling first-year migrants to head for the still-unknown wintering area come from large-scale displacement experiments. The most prominent one, performed by the Dutch Bird Banding Station, involved more than 10,000 Starlings of presumably Baltic origin. These birds, heading for their wintering area in northern France and southern Britain, were displaced perpendicular to their normal migration route from Den Haag to Switzerland (figure 10.10). Subsequent banding recoveries revealed an interesting difference between juvenile and adult birds. The adults that had already spent a winter in their traditional winter quarters changed their course and moved northwestward, whereas the juveniles continued in their normal migratory direction, which led them parallel to their normal route to areas in southwestern France and Spain (Perdeck 1958). This clearly showed that the young migrants have information about their migratory course, but not about their goal.

Innate information on the migratory course is also indicated in numerous experiments with hand-reared migrants. These birds, raised in isolation from experienced conspecifics, in autumn spontaneously prefer the species-specific migratory direction (e.g., Able & Able 1990; Beck & Wiltschko 1982; Bingman 1983; Emlen 1970). In a study cross-breeding blackcaps, *Sylvia atricapilla*, from populations with diverging migratory directions, the offspring showed intermediate directional preferences (Helbig 1991), indicating that the migratory course is based on genetically coded information that is transmitted from one generation to the next.

ESTABLISHING THE MIGRATORY COURSE

Before their first migration, young migrants must transform this genetically coded directional information into the population-specific course of

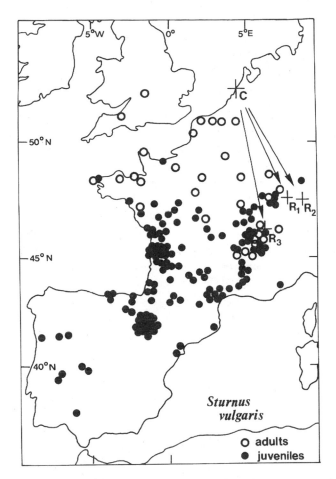

FIGURE 10.10. Displacement experiments with Starlings during autumn migration. C is the site of capture near Den Haag, the Netherlands; R_1, R_2, and R_3 are release sites at Basel, Zurich, and Geneva, Switzerland. Dots mark the recovery sites during autumn and the following winter. (Data from Perdeck 1958.)

migration. These processes appear to take place after fledging during the *premigratory period* so that the necessary information is available to the young birds when migration begins.

Two Reference Systems

The fact that an external reference system is required for transforming innate information into an actual course leads to the question about the nature of possible reference systems. Two such references have been described: *celestial rotation,* indicating geographic north (Emlen 1970), and the *geo-*

magnetic field, indicating magnetic north (e.g., Beck & Wiltschko 1982; Bletz, Weindler, Wiltschko, Wiltschko, & Berthold 1996). Young garden warblers, *Sylvia borin,* nocturnal migrants, were able to orient with the magnetic field as the only cue (W. Wiltschko & Gwinner 1974) and with stars as the only cues (figure 10.11). In the latter case, the stars could be stationary and of arbitrary configuration, as long as the birds had observed these stars rotating around the celestial pole during the premigratory period (W. Wiltschko, Daum, Fergenbauer-Kimmel, & Wiltschko 1987). These findings, showing that birds have no innate knowledge of the stars and their patterns, demonstrate the crucial role of celestial rotation in establishing stellar orientation.

The above-mentioned experiments involved hand-raised young migrants tested with only one set of cues. The results seemed to suggest that each system alone can provide young migrants with information on their southerly course in autumn. In nature, however, geographic and magnetic north do not always coincide, and so the question arose as to how birds cope with such situations. During the premigratory period, hand-raised birds were exposed to celestial cues in a shifted magnetic field so that the center of rotation and magnetic north deviated by 90 to 120 degrees. Later, during autumn migration, when these birds were tested with the magnetic field as their only cue, they had modified their magnetic course according to celestial rotation

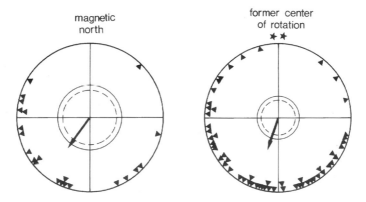

FIGURE 10.11. Orientation of the Garden Warbler, indicating the use of two reference systems for determining the migratory direction. (*Left*) Birds hand-raised and tested with the magnetic field as their only cue. (*Right*) Birds hand-raised under a rotating artificial sky (in the local geomagnetic field) and tested under the same, now-stationary sky in the absence of information. Symbols at the periphery indicate the birds' headings in single tests; the arrows represent the mean vectors. (Data from W. Wiltschko & Gwinner 1974; W. Wiltschko et al. 1987.)

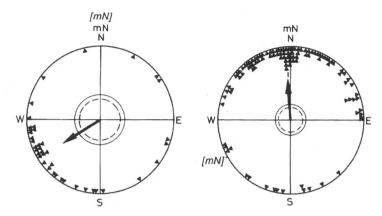

FIGURE 10.12. Orientation in tests with the geomagnetic field as the only cue, documenting the effect of celestial rotation on the magnetic compass course. (*Left*) Pied Flycatchers that, during the premigratory period, had been exposed to the natural sky in the local geomagnetic field preferred the appropriate southwesterly course. (*Right*) Birds that had been exposed to the natural sky in a magnetic field with north shifted to 240 degrees WSW—so that during exposure magnetic north coincided roughly with the migratory direction—preferred a northerly course. Symbols as in figure 10.11; mN is magnetic north during testing; *[mN]* indicates magnetic north during the premigratory period. (Data from Prinz & Wiltschko 1992.)

and now preferred the magnetic course that had been pointing to their geographic migratory direction during exposure (figure 10.12; e.g., Able & Able 1990; Bingman 1983; Prinz & Wiltschko 1992). This clearly shows that, as a reference system, celestial rotation dominates over the magnetic field. This is true for the rotating day sky and night sky alike (Able & Able 1993).

Celestial Rotation—Reference Direction Only?

The above-mentioned findings seem to suggest that the migratory direction is coded twice, with respect to magnetic north *and* to the center of rotation, the latter being dominant. Yet both mechanisms are components of one complex system. Normally, the specific migratory direction is established by interactions between the two. This was indicated by experiments testing the influence of magnetic information on stellar orientation.

Many birds take population-specific routes that avoid ecological barriers like high mountains and minimize sea and desert crossings in the course of migration. This requires population-specific migratory courses, or sequences of courses. For example, the Garden Warblers of Central Europe first head southwestward to Iberia, then switch to a more southerly or southeasterly course to reach the wintering quarters in Africa south of the

Sahara (figure 10.13; Gwinner & Wiltschko 1978). Recent experiments showed that Garden Warblers obviously need simultaneous access to celestial rotation *and* to the magnetic field for establishing their population-specific migratory direction with respect to the stars. When tested with stars as their only cues, control birds that had experienced celestial rotation together with the geomagnetic field preferred their normal southwesterly course, whereas experimental birds that had observed celestial rotation in the absence of magnetic information headed due south (figure 10.14; Weindler, Wiltschko, & Wiltschko 1996).

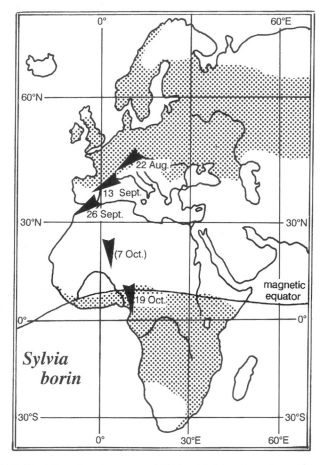

FIGURE 10.13. The migration route of Garden Warblers from Central Europe. The birds leave their breeding grounds on a southwesterly course. In the beginning of October, after leaving Iberia, they change to a southerly or southeasterly course. (After Gwinner & Wiltschko 1978.)

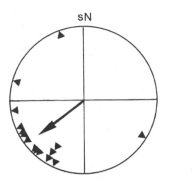
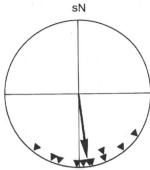

FIGURE 10.14. The orientation of hand-raised Garden Warblers during the first part of autumn migration, tested under an artificial sky without magnetic information, documents the influence of the magnetic field on stellar orientation. (*Left*) Birds that, during the premigratory period, had been exposed to the same, then-rotating sky in the geomagnetic field preferred the appropriate southwesterly course. (*Right*) Birds that had been exposed to the rotating sky without magnetic information preferred southerly courses. Symbols at the periphery of the circle indicate the grand mean vector. sN, stellar north, corresponds to the center of rotation during the premigratory period. (After Weindler et al. 1996.)

Apparently, celestial rotation alone was not sufficient to set the southwesterly starting course with respect to the stars. The behavior of the experimental birds suggests that celestial rotation alone provides only a reference direction away from its center, corresponding to geographic south in the northern hemisphere. As a migration course, this is not totally inappropriate, as it will always lead birds into regions with less severe winters. But the population-specific migratory direction that allows birds to take optimal routes appears to require additional information from the magnetic field, which implies that it is coded with respect to the magnetic field only.

At first glance, this is surprising in view of the finding that celestial rotation dominates over the magnetic field as a reference. Apparently, celestial rotation does not simply override magnetic information. A possible explanation for the interaction of celestial rotation and the magnetic field is that the innate magnetic information does not specify a particular course, such as "230 degrees" southwest, but instead specifies a deviation from a reference, something like "50 degrees to the right" of a basic direction. This means that the migratory course is composed of two components, as indicated by figure 10.15. Birds would establish a basic *reference direction* provided predominantly by celestial rotation indicating geographic south; if celestial rotation is not available, the magnetic field provides magnetic south.

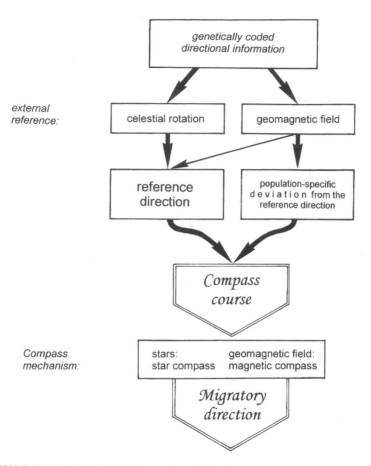

FIGURE 10.15. Model of the interaction of celestial rotation and magnetic information in the processes establishing the migratory course: celestial rotation provides a reference direction, which can also be provided by magnetic north if celestial cues are not available. A population-specific deviation is specified in terms of the magnetic field. Both pieces of information are combined to form the population-specific migration course, which can then be located with the magnetic and stellar compasses.

In a second step, the innate magnetic information would define a specific deviation from this reference direction, resulting in the population-specific migration course. The course established this way would then be memorized with respect to the magnetic field as well as with respect to the stars so that later, when the birds begin to migrate, it can be located with either compass (Weindler et al. 1996).

Why Combine Celestial and Magnetic Information?

One can only speculate on the reason why migrants combine information from two systems to establish their migratory course. Obviously, because of the weather dependency of celestial cues, they need the magnetic field; yet theoretically, they should be able to rely on the magnetic field alone. The experiments with hand-raised birds show that this is indeed possible (cf. figure 10.11 left). Yet birds normally include information from celestial rotation, which even proved dominant (cf. figure 10.12). The reasons may lie in the advantages and disadvantages of the two cue systems.

When innate information is to be transformed into an actual course, it is crucial that this result in a particular geographic direction. This is guaranteed by using as directional reference celestial rotation, which provides a simple, stable reference direction. Yet the direction "away from the center of rotation" is a mental construct based on the integration of visual stimuli; the stars themselves rotate. Because of this, celestial cues might be ill-suited for defining specific directions deviating from the rotational axis. The magnetic field, in contrast, is a stimulus that can be directly perceived. At higher latitudes, birds face the problem of large magnetic declination (the difference between local magnetic north and geographic north) and, even more important, fairly rapid changes of declination due to secular variation. As a result, any magnetic course would have to be subject to continuous modifications. This could be avoided by using the magnetic field to code not a specific course, but a deviation from a reference direction provided by celestial rotation.

An additional problem may arise from long-term changes in the ecological situation along the migration route and in the winter quarters, caused by climatic conditions. Selection will give rise to the required adaptations of the migratory course or sequence of courses, gradually modifying the innate information accordingly. Here, it might be advantageous to code these slightly varying courses with respect to one system only. The magnetic field appears to be the most suitable one for this purpose.

ORIENTATION DURING MIGRATION

The considerations in the previous section refer to the establishment of the starting course during the premigratory period and perhaps to the very first part of migration, as long as the migrants are still near their home region (see Able & Able 1995). In the further course of migration, the birds leave this region behind and follow their population-specific routes until they reach their winter quarters thousands of kilometers away. During this main part of their travels, the control of the migratory course appears to follow different rules.

Ranking of Magnetic and Celestial Cues

Magnetic compass information and compass information based on celestial cues continue to be used together; however, during migration, magnetic information becomes dominant. Numerous migrants have been tested under conflicting cues, under the natural sky in altered magnetic fields or in the local geomagnetic field with celestial cues manipulated in various ways (for review, see R. Wiltschko, Wiltschko, & Munro 1997). At first glance, the results seem rather diverse: in some cases the birds seem to follow celestial cues; in others, they seem to follow magnetic cues; and in still others, they appear confused and are no longer oriented. On closer inspection, however, a clear picture emerges. When the same birds were repeatedly tested under conflicting cues, they eventually oriented as indicated by the magnetic field (cf. figure 10.16, upper diagrams). Unlike the situation that occurs during the premigratory period and at the beginning of migration, magnetic cues dominate over celestial cues. There are differences among bird species, though: some birds immediately follow magnetic cues, whereas others first follow celestial cues and switch to magnetic cues only after a certain delay. The latter species appear to prefer using celestial cues and to check these against their magnetic compass only every few days (R.Wiltschko et al. 1997).

Subsequent experiments with only one set of cues reveal how the birds solve the cue conflict for themselves. As an example, figure 10.16 shows the orientation behavior of birds first tested under conflicting celestial and magnetic cues—they were exposed to the natural evening sky in a magnetic field with magnetic north deflected to 240 degrees WSW—and then tested under each set of cues alone. Under conflicting cues, the experimental birds preferred ENE, which is their southwesterly migratory direction as indicated by the altered magnetic field (figure 10.16, upper diagrams). In later indoor tests with the local geomagnetic field as the only cue, they preferred southerly directions like their controls (figure 10.16, lower diagrams), while under the natural sky in a compensated magnetic field, they continued to head toward ENE (figure 10.16, center). That is, they maintained with the help of celestial cues alone their migratory direction as defined by the altered magnetic field. These findings clearly show that in case of conflict, birds do not simply ignore celestial cues, but instead recalibrate them according to the ambient magnetic field (R. Wiltschko, Munro, Ford, & Wiltschko in press). In the example in figure 10.16, the celestial cues

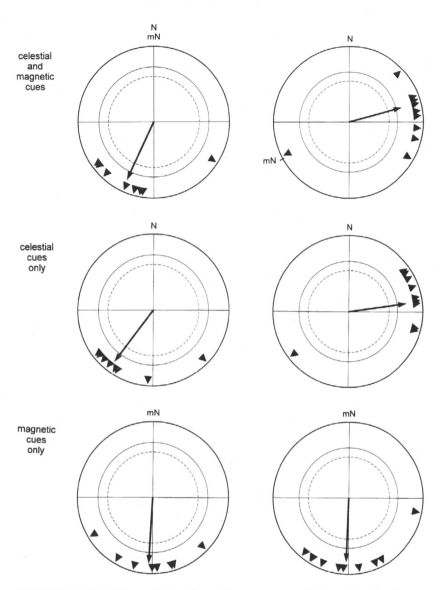

FIGURE 10.16. Effect and aftereffects of exposure to conflicting celestial and magnetic cues on migratory orientation. Upper diagrams: Magnetic and celestial cues available. (*Left*) Control birds, tested under the natural sky in the local geomagnetic field.
(*Right*) Experimental birds, tested under the natural sky with magnetic north deflected
to 240 degrees WSW. Center diagrams: Same groups subsequently tested under the natural sky in a compensated magnetic field, meaning that celestial cues were the only cues.
Lower diagrams: Same birds subsequently tested indoors with the local geomagnetic field
as the only cues. N represents true north as indicated by celestial cues; mN represents
magnetic north. Symbols at the periphery indicate the mean directions of individual birds;
the arrows represent the mean vectors. (Data from R. Wiltschko et al. in press.)

were the pattern of polarized light at sunset; other studies show that stars are recalibrated in a similar way (for summary, see R. Wiltschko et al. 1997 for summary). Readjusting the directional significance of celestial cues seems to be a common response to conflicting cues during migration. As a result, the celestial compass mechanisms are again in harmony with the magnetic compass, and both systems provide the same type of directional information.

The increasing importance of magnetic factors during migration reflects the usefulness and reliability of the various orientation cues as the birds move to lower latitudes: the view of the sky changes rapidly; the familiar stars lose altitude and finally disappear below the northern horizon, while new ones become visible. The geomagnetic field, on the other hand, becomes rather regular at lower latitudes. Declination also decreases with increasing distance from the poles. Magnetic information thus can easily be used to calibrate the new stars appearing in the southern sky.

Reaching the Winter Quarters

To reach the species-specific wintering area, migrants also need information on distance. This information is provided by an *endogenous time program* that determines distance by controlling the duration of migratory activity. Recording the cage activity of captive night migrants revealed that these birds show nocturnal activity for approximately the same time period that their free-living conspecifics need to reach their winter quarters. The total amount of activity differed between species, reflecting the distance of migration (Gwinner 1968). A migration program consisting of pre-programmed responses to celestial rotation and the magnetic field and a time program thus provides young, inexperienced migrants with the direction and distance of the route to reach their still-unknown goal (for review, see Berthold 1988).

When the young birds approach the region of their winter quarters, migration does not end abruptly. The birds search around until they have found a suitable place to settle down for the winter. Here, innate habitat preferences may play a crucial role: inexperienced birds have certain pre-programmed expectations that must be met. This helps them to establish themselves at a place that will probably fulfill their ecological requirements and thus increase their chance of survival.

Return Migration and Later Migrations

In spring, when migrants return to their breeding ground, they must roughly reverse their autumn direction. Yet they do not simply retrace the route that brought them to their winter quarters. Species that take non-straight routes in autumn (cf. figure 10.13) frequently head home on a more direct route. The control of this return migration is poorly known. Tests with caged migrants show that birds have available innate information on their return direction (e.g., Gwinner & Wiltschko 1980), and the time course of migratory activity suggests a migration program similar to that controlling autumn migration.

Yet birds need not rely on this innate information alone, because spring migration is always a return to a familiar goal. Even young birds that complete their first migration head toward an area they know from the year before. While flying around during their postfledgling period, they had established a navigational map of their home area. This means that for spring migration birds can use mechanisms of navigation and homing. This is even more true for adult birds on their second and later migrations. Banding recoveries revealed that many birds return to the same wintering site and the same breeding site year after year. For them the migration route becomes increasingly familiar. This means that adult birds can fully profit from previous experiences. Experienced migrants must be expected to have navigational maps that include not only the breeding area and the winter home, but also large parts of the migration route. During migration, they can use the mechanism of navigation described in the first part of this chapter. Everything that was said about the navigational map of homing pigeons also applies to that of migrants; however, the size of the map will differ according to the distances of migration. For Starlings of Baltic origin migrating to northern France and southern Britain, data from displacement experiments (Perdeck 1967, 1983) suggest that the map covers an area of between 600 and 1000 km; long-distance migrants crossing the equator might be expected to have more extensive maps.

By using their navigational map, birds can head directly toward their respective goals, revisit good stopover sites en route, and carefully avoid areas that proved unfavorable the year before. In view of these advantages, it is not surprising that navigation is the preferred strategy of experienced birds. The adult birds displaced from the Netherlands to Switzerland (cf. figure 10.10) *changed* their normal course and headed for the traditional wintering areas. Young birds continued on their migratory course in autumn, but in spring they returned to their traditional breeding grounds—a trip that involved a return route that differed in direction and length from the normal route between the traditional wintering area and the breeding grounds.

In migration, as in homing, the preferred strategy changes with increasing experience. The first migration to the still-unknown goal is controlled by an innate program, but once the birds have established a navigational map, they make use of their experience, and the map is preferentially used.

DIFFERENT STRATEGIES BASED
ON EXTERNAL REFERENCES

There is a fascinating variety in the way birds utilize potential orientation cues in their navigational strategies. Comparison of the processes involved in homing and in migratory orientation reveals striking parallels and differences. In homing, the goal is familiar, but the birds must determine their home direction from any direction; in migration, they must reach a still-unfamiliar goal in a distant region of the world, but there is a constant directional relationship between the starting point and the goal. In both cases, the young birds are provided with their first means of navigation based on innate mechanisms. The strategies, however, differ greatly according to the respective tasks. In homing, young pigeons use their magnetic compass to record the direction of their flights or of passive displacement, thus deriving the home course from information obtained during the outward journey. As long as they stay in the vicinity of their home, this strategy allows navigation with sufficient accuracy, but navigation becomes increasingly inaccurate over longer distances. In migration, birds rely on genetically coded information on the course and the distance of migration. This information reflects the past experiences of their population; it is transmitted from one generation to the next and leads young migrants into the region of their wintering area. The magnetic field and celestial rotation provide predictable reference systems, which interact to mediate the transformation of the innate information into an actual course for the migration flight.

As the birds grow more experienced, these first mechanisms are replaced by learned mechanisms that permit determination of the course to the goal from any site and offer greater flexibility and safety. These advanced mechanisms make use of the regional and local distribution of various factors whose existence is reliable, but whose specific manifestations cannot be anticipated. For example, it is absolutely certain that birds will encounter landmarks and suitable factors to be incorporated into the mosaic map and the grid map; however, neither the specific nature and position of the landmarks, nor the regional distribution of gradients, can be predicted. Thus, in order to make use of these factors, animals must acquire the relevant information through individual learning processes.

Because they are based on experience, the resulting mechanisms are always perfectly adapted to the local situation. This process, together with flexible compass mechanisms, guarantees fast and efficient movements within the home range between arbitrary goals.

In all these very different navigational strategies, external references provided by the magnetic field and celestial cues are essential components, indicating directions as compass courses. This is true for navigation based on outward journey information, for the innate migration program, and for the grid map and the mosaic map (the latter two being directionally oriented representations of the environment). Compass orientation is thus the backbone of the birds' representation of space. This organization of space ensures that birds—facing spatial tasks that by far exceed those confronting most other animals—have a highly efficient navigational system at their disposal.

CATHERINE THINUS-BLANC
FLORENCE GAUNET

11

Spatial Processing in Animals and Humans

THE ORGANIZING FUNCTION OF REPRESENTATIONS FOR INFORMATION GATHERING

Most studies of spatial cognition in animals and man refer to the notion of a cognitive map, first coined by Tolman in 1948. Cognitive maps are classically considered internal representations of the environment where places and the spatial relationships (angles and distances) among them would be charted. Though widely used, the notion has been extensively criticized (see Thinus-Blanc 1996). Indeed, the expression is antinomic and can casily lead to misunderstandings. The term *cognitive* refers to a dynamic process (process of knowing), whereas the word *map* refers to a product of this process, that is, to a static picture of the real world.

The aim of this chapter is not to enter into this long-standing debate, but rather to argue that although spatial representations are useful for orienting in a given environment, they also contribute to the organization of gathering new incoming spatial information. This second function implies that spatial representations are endowed with a level of abstraction that allows

Some of the studies reported here were supported by grants from the Cogniscience Program (National Center for Scientific Research—CNRS) and from the Ministry of the Army (Direction des Recherches et Etudes Techniques). Parts of this chapter have been drawn from Thinus-Blanc (1996).

for the generation of organizing rules or schemata sufficiently general to be implemented in a wide range of physical environments.

Reciprocal and functional links between human representations in general and knowledge organization have been developed by several authors (e.g., Partridge 1996), but this issue has never been specifically considered in the domain of spatial processing in animals and humans. Such ideas can be applied to spatial representations, viewed not only as maps of the environment but also as cognitive or active information-seeking structures. Data drawn from animal and human studies and theoretical developments that support this hypothesis are discussed in the following sections. The first section deals with the reaction-to-change tests based on dishabituation of exploratory activity in rodents. In the second section, we develop arguments that support animals' capabilities to construct abstract representations. The third section is devoted to data from human studies that suggest that representations control the organization of spatial information collection while exploring a new environment. Finally, the reciprocal links between representation and action are discussed as a working hypothesis for further research.

ANIMAL STUDIES

THE DISHABITUATION PARADIGM

All mammals react in some way to novelty. Exploratory behaviors are easily observable in rodents. After a short phase of "freezing," a rat or a hamster placed in a new situation displays a feverish investigatory activity, which finds expression in multiple contacts with objects in the environment, sniffing, stopping, rearing up on the hind paws, and so on. One feature of this activity is its decrease over time, a phenomenon called habituation. Depending on the complexity of the situation, its degree of novelty, and the species concerned, the process of habituation varies. It is now widely acknowledged that these behaviors are part of the general process of knowledge acquisition (Toates 1994). Information from previous situations is stored and serves as a reference for detecting novelty in a form of modification to settings previously encountered or exposure to a new setting (Berlyne 1960, 1966). In such cases, dishabituation (re-exploration) is observed. Novelty exists only through reference to familiarity. Consequently, if novelty is detected and triggers a renewal of exploration, that means that some representation of the initial situation has been constructed and stored.

Of course, such reasoning applies to knowledge acquisition in general and to spatial knowledge in particular. According to O'Keefe and Nadel (1978), one of the functions of exploratory activity is to update cognitive maps when the animal is confronted with spatial novelty. The less familiar a situation, the higher the degree of exploration. When placed in a partially or totally novel situation, there is a systematic screening of stored memories of familiar environments. The detection of a discrepancy or mismatch between what is known or expected and what is currently perceived triggers exploratory activity, which is aimed either at constructing a spatial representation, or cognitive map, of a totally new situation or at updating an already internalized model to incorporate modifications. Then, habituation corresponds to the increasing degree of matching between the two situations (the stored and the currently perceived) that are being compared.

In this respect, the dishabituation paradigm or "reaction-to-change test" represents an invaluable means of investigating spatial cognitive abilities in animals. Whatever the procedural details and species under study, the principle of these experiments is the following: first, subjects are repeatedly exposed to a given spatial configuration (usually one or several objects contained in an open field); then after habituation of exploration has occurred, some spatial relationships between the elements of the situation are modified (for instance by displacing one object). What is new is not the characteristics of the individual objects, as these have been previously investigated, but rather their spatial arrangement. If the animals react to the change by displaying a renewal of exploratory activity, then it can be assumed that the detection of spatial novelty relies on a comparison between the currently perceived arrangement and a representation or stored "internal model" of the initial situation.

The dishabituation paradigm allows us to infer spatial features of a situation that are spontaneously encoded during exploration, because the various types of rearrangement differ in the reactions they evoke. In a series of experiments (Poucet, Chapuis, Durup, & Thinus-Blanc 1986; Thinus-Blanc 1988; Thinus-Blanc, Durup, & Poucet 1992; Thinus-Blanc et al. 1987), hamsters were given three 15-minute sessions of exploration several hours apart. The apparatus was a circular open field containing four different objects. Surrounding the field was a curtain that provided a visually homogeneous background except for a striped pattern affixed to the curtain. This set-up was designed to induce the animals to focus their attention on the objects contained in the experimental field by preventing them from being distracted by irrelevant distal cues. The number and duration of their contacts with the objects were recorded. After habituation had occurred from the first to the second session, a spatial change was made in the initial situation by displacing one or more of the objects. For the control groups, the arrangement remained unchanged during the third session.

In most cases, re-exploration was selectively directed toward the object that had been displaced. This occurred when one object was set apart from the others. The same pattern of results was obtained when the objects were identical, and one object was set apart. When one object was moved closer to the others, re-exploration was directed toward all of the objects whether they had been displaced or not. Hamsters reacted as if the situation was completely new. Finally, when the size of the object configuration was increased or decreased by moving all of the objects along the same distance, no reaction was observed (figure 11.1).

Such data demonstrate that hamsters appear spontaneously to encode geometrical relationships because they always react, selectively or not, to a change bearing on this parameter. In contrast, modifying the size but

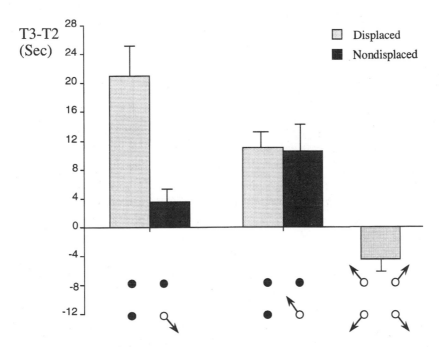

FIGURE 11.1. Differences between sessions 3 (test) and 2 (last session of habituation) in mean time spent investigating either displaced (empty circles) or nondisplaced (filled circles) objects; standard errors are represented by vertical bars. (Adapted from Thinus-Blanc 1996; Thinus-Blanc et al. 1987.)

not the shape of the object configuration did not elicit any reaction (Thinus-Blanc et al. 1987). The latter result suggests that generalization has taken place and that similar geometrical patterns are put into the same category regardless of their size. The fact that selective reactions to the displaced object were often observed supports the idea that initial spatial representations exert some control on the collection of spatial information; the situation is only partly new following the displacement of one object. Thus, the pre-existing representation triggers re-exploration so as to update itself by integrating the modification. If a situation is totally new and consequently does not correspond to any already stored representation, full exploration of the environment is observed until a new representation is constructed. The same process apparently occurred when one object was put closer to the other ones. In contrast to the results observed when a single object was moved away from the set (that "isolated" it from the others), a single displacement toward the set may have led to an updating of spatial relationships among all the objects comprising the set.

LEVELS OF ABSTRACTION IN ANIMALS

Many studies have dealt with animals' general abilities of abstraction and category formation (e.g., learning-sets and reversals), but only a few have specifically addressed the question about spatial processing and learning. For instance, shortcut and detour abilities are considered as reflecting high-level spatial representations that allow reorganization of acquired information so as to bring a new solution to a spatial problem (e.g., when a familiar path becomes impracticable and requires a detour to be made). Shortcuts and detours have been extensively studied in animals (see Thinus-Blanc 1996 for review), but there is no study about the level of abstraction of such an operation. Are animals capable of immediately transferring a solution to a problem similar in principle but presented in an environment of different size containing different landmarks? In a previous study, Thinus-Blanc (1978) showed that learning to discriminate between an open and a closed circular enclosure (i.e., between topological features) could be generalized by hamsters to various sizes of enclosures. An argument in the same vein is provided by dishabituation studies showing that no re-exploration was observed when the four objects in the field were displaced equally so as to form a square larger than the initial one. It is unlikely that such a massive change was not detected by the animals that had been able to detect far more subtle modifications in other tests. Therefore, a process of generalization may have occurred, preventing the animals from reacting to a change that modified neither the shape of the arrangement nor the topological relationships between the objects. Lastly, preliminary (unpublished) results obtained in rats suggest that exploratory activity may be controlled by whether or not animals have the possibility of generalizing spatial knowledge to different situations that share basic geometrical features.

Rate of habituation was examined in two different situations over 12 four-minute sessions of exposure. In the first situation, different objects placed in a circular open field defined an irregular geometrical shape that remained the same across all sessions, but the position of each of the four objects changed from one session to the next so that the shape of the arrangement was permanent but the combination of the objects defining it was variable. In the second situation, the same four objects were used but they defined a different geometrical arrangement on each session. The aim of the experiment was to examine whether rats are able to extract the geometrical structure common to different situations. In such a case, the rate of habituation should have been more pronounced when the configuration remained the same over sessions than when it was modified. As expected, repeated exposures to the same geometrical configuration resulted in a higher rate of habituation in terms of number and duration of contacts (figure 11.2). This finding is convergent with other data (e.g., Cheng 1986, 1987; Cheng & Gallistel 1984) showing that animals spontaneously encode and use geometrical properties of the environment, which may predominate over other types of information. Such encoding is relatively abstract, because it does not appear to be bound to local characteristics (e.g., the objects that define the geometry) but rather may reflect overall generalizable knowledge of the geometrical features of the situation.

There are some arguments that support the notion that rodents form category and relatively abstract representations. In addition, selective re-exploration following a change suggests that exploratory behavior is, to some extent, controlled by a representation of the situation before the change occurred. Is it possible to go further in the demonstration that relatively abstract spatial representations control and organize the gathering of environmental information? That should be possible by examining whether some behavioral regularities can be found in exploratory patterns. Because exploratory activity can be observed, the analysis of behavior patterns should reveal some properties of the unobservable structure that directs and organizes them. In a first attempt made in hamsters (in yet unpublished studies), we sought to identify regularities from recorded sequences of contacts made by animals with objects scattered in an open field. In fact, we failed to demonstrate any emerging patterns. Instead, randomness appeared to be the main "organizing" factor for exploration, at least,

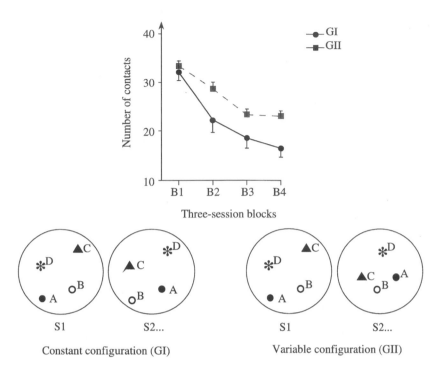

FIGURE 11.2. Evolution of habituation (numbers of contacts) over the three-session blocks of the experiment with a constant (*left panel*) or variable (*right panel*) object arrangement. $N = 8$ in each group. (Adapted from Thinus-Blanc, Save, & Poucet, unpublished study.)

as we quantified exploration, that is, in terms of contacts with the objects. We concluded that the multiple back-and-forth displacements might have an auxiliary function in preventing the repetition of the same patterns of stimulation likely to result in the learning of sequences of stimuli. In a different approach, dealing with the study of spatial performance in blind persons (see below), we examined whether regularities could be found in spontaneously exploring an environment while lacking vision.

HUMAN STUDIES

Data from a number of studies suggest that lack of early visual experience affects the representation of both small and large spatial layouts. Early blind participants have been found to be less successful than late blind and blindfolded, sighted subjects in performing tasks that require a mental manipulation or reorganization of spatial information, such as making shortcuts or mental rotations. However, no differences among the three groups of subjects have been found by some authors in situations that are quite similar to those in which different performances have been reliably observed by others (see Thinus-Blanc & Gaunet 1997 for review). One possible explanation for the divergent data may be related to the selection of early blind subjects (orientation and mobility training methods, individual features, past history, and so on). Such factors are difficult, if not impossible, to control. Therefore, instead of attempting to control factors that cannot be controlled, behavior occurring "upstream" from task achievement, that is, at the moment when spatial information is acquired, has been analyzed. The purpose was to examine whether differentiated performance levels could be correlated with differentiated means of collecting spatial information.

Two reaction-to-change experiments, based on the paradigms used in animal studies, have been adapted for use with human subjects. One of these involved a procedure in "locomotion space," in which participants explored a configuration of easily identifiable objects by walking among them (Gaunet & Thinus-Blanc 1996). The other involved a procedure in "manipulation space," in which seated participants explored a set of objects on a tray in front of them (Gaunet, Martinez, & Thinus-Blanc 1997).

Following a fixed period of exploration, a modification was made to the initial arrangement and participants were asked to explore again until they detected whether a change had been made. During the pretest exploration phases, the sequences of contacts that subjects made with the objects were carefully recorded and analyzed. Two types of patterns were found to stem from exploratory sequences in both locomotion and manipulation spaces: One of them, termed the *cyclic pattern,* consists of visiting all the objects successively, beginning and finishing at the same one. The other pattern—the *back-and-forth* pattern—involves making repeated contacts between two objects (figure 11.3). Both types of exploratory patterns have been found to be implemented by samples from all three populations of participants (early blind, late blind, and blindfolded sighted subjects). But how did strategy use relate to subjects' performance levels? Results showed that early blind subjects were generally impaired at detecting change. Furthermore, early blind subjects made many more cyclic than back-and-forth exploratory patterns, whereas the opposite was observed in the other two groups. Statistical analyses revealed performance level to be correlated with the proportion of contacts involved in each type of exploratory pattern; the higher the performance level, the smaller the proportion of contacts that were of the cyclic pattern type, and vice versa.

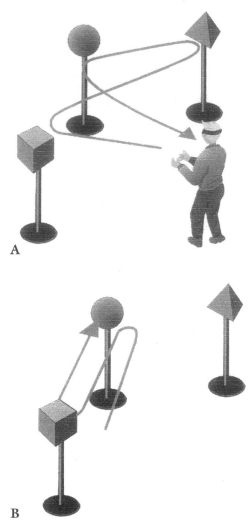

FIGURE 11.3. Example of cycle (**A**) and back-and-forth (**B**) patterns of exploration. (Reproduced from Thinus-Blanc, Gaunet, & Péruch 1996.)

Though they raise many questions, these data support the idea that, at least in humans, some organizing strategies underlie spontaneous exploratory behaviors, which result in more or less accurate spatial representations. In addition, they suggest that our failure to identify any regularities in hamsters' exploratory activity may have been due to the fact that, unlike subjects in the human experiments, animals could pick up distant visual information (many stops with scanning head movements were frequently observed). Therefore, sequences of actual contacts with objects may not have fully reflected the organization of information gathering.

Thus, these data indicate that the accuracy of representations, as reflected by performance levels, depends on the means that have been implemented for constructing them. In addition, the fact that participants grouped on the basis of visual experience used different exploratory strategies when confronted with a new environment indicates that their exploration may have been controlled by some general principles acquired pre-experimentally. The idiosyncratic features of exploratory patterns may be due to the educational methods with which early blind children have been trained for acquiring autonomous mobility. In such a case, early blind subjects' abilities (i.e., potentialities) would be as good as those of visually experienced subjects (along Millar's 1994 distinction between ability, competence, and performance) but they would have developed different and less efficient competences (skills for getting acquainted with space). The prediction is that they should be capable of learning to implement optimal strategies, which should result in an improved performance level. The other alternative is that, regardless of educational methods, early blind persons are unable (lack the ability) to construct as accurate a spatial representation as visually experienced persons. If this were the case, they would have difficulties in learning new competence for coping with space or their performance level would remain poor. Experiments aimed at testing these two alternatives would contribute to understanding the nature of early blind persons' spatial deficits; in connection with the issue debated in the present chapter, such data would help to specify the level (abilities or competence) that predominantly controls the organization of new spatial information acquisition.

THEORETICAL DEVELOPMENTS

NEISSER'S "PERCEPTUAL CYCLE"

Neisser (1976) advanced an interesting definition of a cognitive map as an "orienting schema." The main point of Neisser's theory is that spatial (or nonspatial) representations "direct" perceptual exploration, which, in return, modifies initial representations. Thus, exploratory patterns should reflect some pre-existing organizing schemata. Although Neisser does not elaborate on how this control may be conducted, it is implicit that well constructed schemata do not direct exploration at random. It is likely that general rules are implemented so that information is integrated or assimilated in a manner consistent with the existing schema. Consequently, some of the general features of the spatial representations should be deducible from the observation and analysis of investigatory behaviors.

Data generated from dishabituation experiments in animals and from reaction-to-change tests in blind persons can be interpreted in Neisser's terms. Re-exploration of a spatially modified situation appears to correspond to an updating of the initial representation (Thinus-Blanc 1987) that would control the renewal of exploratory activity, on the basis of a detected mismatch between the initial represented and the currently perceived configurations. At the level of acquisition of knowledge, further studies of exploratory strategies should reveal some properties of the unobservable structure that direct and organize them.

ACTION, REPRESENTATION, AND THEIR RECIPROCAL LINKS
Based on the data presented above, and along the lines of the hypothesis initially advanced by Neisser (1976), a dynamic view of spatial information processing can be portrayed as in figure 11.4. Processing involves two distinct levels, action within the physical world and spatial representation. The first level corresponds to observable behaviors (exploration, oriented trajectories) and information (provided by the sensory environment and movement-generated feedback cues) that can be controlled by manipulating some features of the situation. By its very nature, behavior is organized along an egocentric reference frame (e.g., go ahead, turn left) and is sequentially ordered (go ahead, then turn left), even if this sequence and the bearing of the goal have been determined on the basis of allocentric representation). In contrast, a maplike representation (specific or abstract) holds allocentric information, and its constituent elements are charted regardless of the sequence of actions that took place in the course of their acquisition. External (sensory) and internal (movement-related) inputs generated during exploration have to be matched so as to extract spatial invariants. This stage corresponds to the interface between the level of action and that of representations. Spatial invariants extracted from a given environment are the constituent elements of the representation specific to that environment.

This representation has several functions. First, it may serve as reference for comparing current inputs resulting from the detection of a change brought about to the initial situation. The detection of a mismatch between the stored representation and the sampled modified spatial layout will reactivate exploration in order to update, if necessary, the existing representation. Another role of specific representations is their contribution to the formation of more abstract representations (or "structures" according to McClelland, McNaughton, & O'Reilly's 1995 model; see below) holding general spatial properties common to several specific

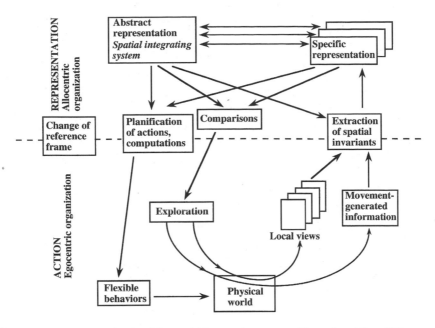

FIGURE 11.4. A sketch of the spatial integrating system. (Reproduced from Thinus-Blanc 1996, by permission of World Scientific Publishing.)

representations. Finally, together with abstract spatial representations, specific representations allow for the computation and planning of oriented trajectories within the physical world. Like the system for extracting spatial invariants the latter process is located at the interface between representation and action, but it operates in the reverse direction. It is assumed to translate an exocentered representation into the form of actions to be executed along a temporally ordered sequence and organized in an egocentric framework. One important feature of this way of representing spatial processing is that it emphasizes, without undue speculation, that exploratory activity is driven by representations.

One may imagine other types of functional loops that do not include high level representations (Thinus-Blanc & Gaunet 1997). In such cases, it is predicted that the dynamic features of exploratory activity will be different. Indeed, abstract representations do not necessarily refer to the highest levels of cognitive processing but to a set of simple functional rules, provided that these can be applied to various situations. The implementation of these functional rules corresponds to an abstract processing because it results in a representation that is not bound to specific sensory features.

Recently, McClelland et al. (1995) proposed a model involving complementary learning systems in the hippocampus and neocortex for learning and memory, which is consistent with the hypothesis described previously. By examining the time course of the respective involvement of the hippocampus and neocortical areas in learning and memory, the authors sketch a picture of their interactions. Though the model mainly focuses on the neural substrates of memory, it proposes that specific memories, initially stored in the hippocampus, become gradually incorporated in the neocortical system in abstract forms, or "structures." The incorporation and consolidation of new information take much more time in the neocortex than in the hippocampus. At the early stage of acquisition the hippocampus exercises a function that precludes interference with memories already stored in the neocortical system. If the incorporation of hippocampal memories in the neocortical system were made rapidly, these would interfere with the system of *structured knowledge* built up from prior experience with other related material. The presentation of new information has to be added gradually, interleaved with ongoing exposure to other domains of knowledge. Thus, the memory trace momentarily stored in the hippocampal system is protected from the risk of interference with the structured knowledge stored in the neocortical system. More importantly, this transient storage allows comparison of that new information with structured knowledge so that it is integrated not only as an episodic event but also as an element that belongs to a "structure" held in the neocortical system or participates in the elaboration of a new structure. The notion of "structure" is central to the model. It is defined as follows by the authors: "What we mean by the term structure is any systematic relationship that exists within or between the events that, if discovered, could then serve as a basis for efficient representation of novel events or for appropriate responses to novel input" (McClelland et al. 1995:436).

This assertion illustrates the possibility of a control system pre-existing with regard to the current experience, providing general rules for coping with any type of novelty. More generally, the model outlines the respective participation of the hippocampal and neocortical systems in the formation of categories and concepts (structures). Concerning the specific domain of spatial representations, the notion of structure can be taken as referring to cognitive maps insofar as one admits that the properties of such high-level spatial representations are not specific to one environment but can be generalized to various situations.

CONCLUSION

Though most studies of spatial representations have focused on their function for orienting, some arguments support the idea that they can also be conceived as active information-organizing structures. In such a dynamic perspective, the term *cognitive map* or *cognitive mapping* is fully justified. As in the case of nonspatial processing, representations constructed over the course of child development and over shorter periods of time in adulthood appear to exert an organizing influence on novel information. In the particular cases when differentiated exploratory strategies are observed, such as in early and late blind persons, the level at which differences originate is difficult to define. Indeed, the whole process is viewed as a *cognitive loop*, with reciprocal feedbacks and interdependent relationships. The nature of spatial representations results from the exploratory strategies implemented to construct them, and these representations, in turn, are supposed to control exploratory activity. Envisioning spatial maps as dynamic loops, serving for both orienting and constructing new maps, may be a useful means to get some new insight in the nature of spatial representations. Besides studies aimed at making more precise spatial representation properties needed for orienting in the environment, more systematic investigations of the relationships between past spatial experience (representation level) and exploratory strategies (action level) should be conducted. For example, animal studies could be conducted in which exploratory strategies exhibited by rats raised from birth in specific environments (e.g., with geometric figures as in the study described earlier or in spatial configurations defining shortcuts or detours) are compared with those exhibited by rats reared in standard or impoverished situations. In the same vein, the observation of children's exploratory strategies after several training sessions with various toys having the same geometrical properties, or games allowing specific spatial problem solving, may contribute to an understanding of the links between representations and information gathering. In addition, such studies would shed some light on animals' abilities to form categories in comparison to those of children at various stages of development. Such studies would strengthen the comparative developmental approach of spatial cognition.

IV

The Neural and Computational Bases of Wayfinding and Cognitive Maps

In this part cognitive neuroscientists Nadel (University of Arizona) and Berthöz and colleagues (Laboratoire de Physiologie de la Perception et de l'Action, Collège de France) examine the neural bases of wayfinding and cognitive maps, and computer scientist Chown (Bowdoin College) discusses their implications for computation and artificial intelligence–based travel. Nadel reviews the neural mechanisms of spatial orientation and wayfinding, Berthöz et al. examine the neural bias of spatial memory during locomotion, and Chown discusses error tolerance and generalization in cognitive maps. Following from the Thinus-Blanc and Gaunet chapter in the previous part, in which the neurocognitive basis of spatial representation is introduced and the likelihood of the hippocampus as a primary focus of spatial representation is examined, the chapters in this section explore recent innovative work such as that in the neurosciences, which focuses on place cells and cell assemblies, and methods for neurological tracing of, or retrieval of, stored information (spatial memory) during locomotor activities.

In chapter 12, Nadel provides an overview of the neural mechanisms of spatial orientation and wayfinding. He suggests that work on the neural bases of wayfinding in mammals has intensified in recent years, building on the discovery of *place* neurons in the hippocampus. The *cognitive map theory* of hippocampal functioning, first put forward by O'Keefe and

Nadel in 1978, suggests that this brain structure is the core of an extensive neural system subserving the representation and use of information about the spatial environment. Nadel argues that evidence supporting this theory comes primarily from brain lesion and neurophysiological recording studies. The former showed that damage in the hippocampus system invariably impairs the ability of animals and humans to learn about, remember, and navigate through environments, while the latter show that neurons in this system code for location, direction, and distance, thereby providing the elements needed for a mapping system. Current work in this area focuses on which stimuli control the activity of these neural elements, and how the system is used in behavior. He cites the fact that the roles of external and internal sources of information are under active investigation. Data that are presently available suggest that both are important. Nadel admits that little is yet known about the use of this system, and research in this area is only just beginning. The advent of new recording techniques in both animals (ensemble recording) and humans (PET and FMRI) provides hope that considerable insights into the neural mechanisms underlying cognitive mapping and wayfinding will emerge in the coming years.

In the next chapter, Berthöz and his associates examine the neural bias of spatial memory during locomotion. This chapter addresses the question of the mechanisms that underlie the capacity to memorize routes and to use this spatial memory for guiding and steering locomotion. A review is presented several paradigms used in their laboratory to study this question. First some previous studies, which have shown that vestibular information about head rotation and translation can be used by the brain to estimate distances, are reviewed. Berthöz et al. claim that such use has been shown by the "vestibular memory contingent saccade task" both in normal subjects and in neurological patients. Second, some recent experiments are described that use the task of walking along a triangular path with or without vision. During this task head position and velocity are measured by videocomputerized techniques. Two main results have been obtained: (1) They have discovered that the head anticipates the body movement during walks around a corner: this anticipation also exists in darkness, suggesting that the orienting system is driven by an internal representation of the trajectory and that the brain uses a strategy of guiding locomotion by gaze (go where you look) even in darkness. (2) When vestibular-deficient patients perform the task they seem to control the total distance but not the direction, suggesting a dissociation between the control of distance and direction in this *locomotor pointing* task.

Berthöz et al. also describe a second paradigm of circular locomotion during which subjects were asked to walk around a circular path with or without vision. Here again the measure of the kinematics parameters of head movement indicates both an anticipation of head direction and a dissociation between the control of distance and direction, and provides clarification of the frequently misinterpreted concepts of course and heading. Finally they review a number of recent results which may lead to an understanding of the neural basis of both anticipation and the role of vestibular cues in the steering of locomotion.

In the final chapter of this part (chapter 14), Chown discusses error tolerance and generalization in cognitive maps. He asserts that it is well known that human cognitive maps are not precise, complete, nor necessarily accurate. Because navigation is so important in everyday life, it is not easy to understand why humans have evolved an internal representation of space that appears to have such basic flaws. The theme of this chapter is that it is exactly the sketchy nature of human cognitive maps that make them such a powerful tool for navigation. There is growing evidence from artificial intelligence and robotics that in real environments, useful representations cannot be achieved without sacrificing completeness and precision. Further, it can be shown that the sketchy nature of cognitive maps more naturally lends itself to error tolerance and generalization than would be the case with alternative structures. Cognitive maps may be sketchy but the information they do store is usually sufficient for human needs. The relationship between human needs and how cognitive maps encode information is discussed in a proposed model called PLAN.

Beyond the basic research need to understand how humans and

nonhuman species interact with, represent, and use the environments in which they live, there is an additional incentive for discovering this knowledge. The chapters in this section explore these matters and lead to speculation about why—today and no doubt in the future—robots and artificial intelligences may be designed to do everything from repetitive and boring tasks in manufacturing and assembling processes to providing in-car guidance systems for travelers in both familiar and unfamiliar environments, from piloting space vehicles to safe landings on unexplored worlds to guiding exploration vehicles in their search for information about those worlds. Chown's final chapter provides an appropriate window on the future with regard to one important consequence of learning about spatial information processing and use by humans and other species. The need is obvious: unmanned vehicles may be the focus of nearby and distant space exploration.

The more we know about how humans or other species can navigate, wayfind, sense, and record and use spatial information, the more effective will be the building of future guidance systems, and the more natural it will be for humans to understand and control those systems. The question of which of the many cognitive mapping, navigational, or wayfinding procedures and behaviors should be taken as the role model for future systems remains unanswered at this stage. Knowing the advantages and disadvantages, the strengths and the shortcomings, the idiosyncrasies and the universals of spatial knowledge acquisition and storage and wayfinding behavior can only lead to the development of systems that are as endemic as path integration, as powerful as cognitive mapping, and as anchored as landmark usage, and that possess the versatility to handle both view-centered and object-centered modes of recording or experiencing new environments.

LYNN NADEL

12

Neural Mechanisms of Spatial Orientation and Wayfinding

AN OVERVIEW

Although discussion of the nature of spatial orientation and wayfinding has a long history in psychology and biology, analysis of the neural bases of these behaviors has begun only recently. For much of this century behavioral neuroscience focused on forms of behavior that were easy to manage in the laboratory but relatively unconnected to the behavior of animals in their natural habitats. Under the assumption that there were general and abstract principles governing all forms of learning, this seemed a reasonable approach. This situation persisted even after the psychologist Tolman (1949) made a very strong case that rats (and presumably other animals) engage in various kinds of learning that have demonstrably different properties. He argued, for example, that rats form internal cognitive maps of experienced environments, thereby laying the basis for the study of spatial cognition in the laboratory. More than 20 years elapsed before this form of learning was subjected to serious study.

In the past 25 years this situation has changed rather dramatically. An important spur was the discovery by O'Keefe and Dostrovsky (1971) of

This research was supported by a grant from the McDonnell Foundation (JSFM 95-14).

neurons in the rat hippocampus whose responses depended on where the animal was located in its environment. Since that discovery considerable research has focused on the role of the hippocampus and its neighbors in spatial learning and memory, and although we now know a great deal about this system, much remains to be determined. In this chapter I consider some of the work on what has come to be called the hippocampal *cognitive mapping* system. I review early results, the initial formulation of the cognitive map theory, and critical findings concerning the role of this system in navigation. I then discuss current quandaries in the field, in particular the still unresolved issue of exactly what controls the activity of hippocampal place cells. Finally, I consider the hippocampus in broader context, as one of many spatial systems.

THE HIPPOCAMPUS AND COGNITIVE MAPPING

Interest in the functions of the hippocampus was spurred by the study of human patients with damage to the medial temporal lobe, including the hippocampus, amygdala, and the entorhinal, rhinal, and parahippocampal cortices. These lesions caused a profound deficit in memory capacity (Scoville & Milner 1957) in the context of apparently spared intellectual ability. This amnesic syndrome involved two components—anterograde amnesia (the loss of the ability to acquire new information) and retrograde amnesia (the loss of information acquired prior to the surgery). Attempts to replicate this syndrome in rats or monkeys with comparable damage to the hippocampus failed. No such general memory defects were observed after experimental lesions. It is now clear that this apparent discrepancy resulted from an incorrect understanding of the nature of memory, as well as the use of tasks in animal studies that were not comparable to those used in the work with amnesic patients (Nadel 1992, 1994). Much as Tolman suggested, there are multiple forms of learning and memory, and different tasks tap different learning-memory systems.

At the start of the 1970s there was considerable confusion about the functions of the hippocampus in nonhumans. Suggestions ranged from motivation, to inhibition, to action. Theories were vague and often vacuous. The discovery of place cells in the hippocampus of the freely moving rat (O'Keefe & Dostrovsky 1971) initiated a new approach to this problem, one that offered a means to understand the selective role the hippocampus might play in learning and memory. Further, it stimulated considerable research emphasis on the neural bases of cognitive maps and navigational behavior that continues to this day.

In the initial study, recordings were made from single neurons in the hippocampus of rats allowed to move about a platform open to the entire experimental room. O'Keefe and Dostrovsky showed that these neurons had a unique property: they were active only when the animal was located in a particular part of the environment, which was called the *place field*. In this situation a rich array of information was available to the animal, any subset of which could, in principle, be responsible for the spatially selective firing patterns observed. This issue of exactly what controls the firing of place cells remains somewhat uncertain.

O'Keefe and I concluded that a collection of these kinds of neurons could provide the basis for something like Tolman's cognitive maps (Nadel & O'Keefe 1974), and we set out to (1) provide additional evidence that the hippocampus was involved in cognitive mapping; (2) develop a theory of what a cognitive mapping system would look like, what it would be used for, and what kinds of information it would have to receive and process in order to play its special spatial role; and (3) consider how this system might function in humans.

EVIDENCE

Evidence was gathered in several domains. First, the single-unit studies continued (O'Keefe 1976, 1979) and were replicated in other laboratories (e.g., Ranck 1973). Second, if the hippocampus was essential for cognitive mapping then lesions in this structure should result in selective deficits in place learning and navigational tasks, with no deficits in other, nonspatial forms of learning. An early lesion study (O'Keefe, Nadel, Keightley, & Kill 1975) confirmed this speculation, as have many others since (e.g., Nadel & MacDonald 1980; Rasmussen, Barnes, & McNaughton 1989; see Nadel 1991 for review). Third, a mapping system would have to receive inputs about distances and directions in the world; a possible correlate of distance information was observed in the theta activity readily recorded in rat (and rabbit) hippocampus when the animal translates itself in space (O'Keefe & Nadel 1978).

Matters have become considerably more complicated in the intervening years, but it suffices here to say that considerable evidence supports the view that the hippocampus is essential for place learning and navigation. What remains to be determined are the details of how these functions are carried out, and how they fit into the broader context of spatial cognition and memory function overall.

COGNITIVE MAP THEORY, 1978

Beginning shortly after the publication of the first paper on place cells, O'Keefe and I began to spell out what a neural cognitive mapping system might look like and how it might work. This required addressing some fundamental questions about knowledge: how it is represented in the brain in general and how spatial knowledge in particular is represented. Philosophers had been discussing spatial cognition for a very long time and had proposed several important notions of how the mind comes by its spatial knowledge of the world.

The major philosophical debate about space concerned whether or not spatial knowledge resulted from experience (the empiricist position), or was instead constructed by the mind (the rationalist or nativist position). A related debate concerned whether our psychological spatial concepts reflected the physical realities of external space, or, in contrast, if there might be differences between psychological and physical space. Empiricists must hold the view that psychological space reflects physical space and our experience with it, but rationalists need not do so.

Two fundamentally opposed views of the nature of physical space exist— the absolute and relative (or relational) views. According to the absolute view, space exists in and of itself, independent of objects. Indeed, absolute space provides the container, or framework, within which objects exist and have location, both with respect to each other and with respect to the framework. According to the relative view, space only exists as a function of extended objects and their relations to each other. The major distinction between the two is their treatment of the notion of empty space. Within an absolute space theory empty space exists; within a relative space theory it is an impossibility. Newton based his thinking on the notion of absolute space; the ascendance of Newtonian physics gave the absolute theory strong support for several centuries, although there were many strong critics, including, most prominently, Leibniz. Not only was absolute space taken as the correct view of physical space, it was assumed that space had a specific metric, namely that it was Euclidean.

Empiricist philosophers of space in the eighteenth and much of the nineteenth centuries necessarily assumed that we acquired a notion of absolute Euclidean space through our early experiences in the world. By contrast, rationalist philosophers such as Kant assumed instead that we had a priori intuitions of absolute space that were correct about the world even though they were not derived by each individual based on experience.

The entire argument was transformed in the nineteenth century by the discovery that Euclidean geometry was not the only geometry consistent with the facts of the physical world. Indeed, the physical world seemed to conform, at certain scales, to quite different non-Euclidean geometries. Shortly thereafter, the development of relativity theory challenged the notion of absolute space. It is in fact difficult at present to render a clear judgment on the nature of physical space and its geometric properties (see Jammer 1993 for a thorough historical review of these issues).

All this has had a profound effect on the debate among philosophers (and more recently among psychologists) about the nature of psychological space. If humans indeed have a psychological notion of absolute, Euclidean space, where does it come from? Experience in a non-Euclidean world should not readily lead to it. This consideration lends weight to the Kantian view that the mind must construct this notion a priori. And this is where the cognitive map theory proposed by O'Keefe and myself starts. It suggests that the brain contains a system generating an absolute spatial framework with a Euclidean metric, and that this system is prestructured to do so without specific spatial experiences during early development. It asserts that this system is but one of many spatial systems in the brain, most of which constitute relative spaces. It further supposes that the absolute spatial system (centered on the hippocampal formation) and various relative spatial systems (scattered throughout the brain) are in constant interaction, and that our behavior in space reflects this interaction.

PREMISES OF O'KEEFE AND NADEL'S
COGNITIVE MAP THEORY

There are a handful of critical premises of cognitive map theory:

1. There are a large number of spatial representational systems that differ with respect to the nature of the information they process and store and with respect to the framework they generate.
2. The hippocampus (and its related neighbors) is the core of a neural system that generates a nonegocentric spatial representation. It does this partly by using inputs from other, egocentric, spatial systems and partly by virtue of its connection structure. It is this latter fact that makes it similar to Kant's a priori intuition of space. That is, the system creates an absolute spatial framework and imposes it upon inputs that by themselves could not constitute absolute space.
3. The cognitive map system allows animals (and humans) to define places, locate themselves in specific places, remember what is located where, and figure out how to get from one place to another depending on current needs and motivations. It is also critically involved in allowing animals to learn about the spatial contexts in which objects and events occur.

4. The hippocampal system depends upon exploration as its means of gathering the information critical to the formation of specific spatial maps. Hence, the hippocampus is responsible for an animal's exploratory reaction to novel environments, novel spatial arrangements, and novel episodes.

5. The hippocampus both forms and stores spatial maps, so it is necessary for learning about and subsequently utilizing information about places and what is in them.

6. The hippocampus in humans is critical for spatial mapping. In addition, it is implicated in those aspects of memory that depend upon information about the environmental context in which some event transpired. Hence it is centrally involved in what is known as episodic memory.

7. The human hippocampus is also involved in those aspects of language that allow us to talk about various features of space, and to use spatial information in the comprehension and production of narratives (Peterson, Nadel, Bloom, & Garrett 1996).

EVIDENCE IN SUPPORT OF CENTRAL ASSERTIONS

In the years since the initial formulation of cognitive map theory its central postulates have received strong empirical support:

— Studies of the activity patterns of hippocampal neurons have confirmed their spatial character (Muller 1996).

— Studies of the impact of lesions in the hippocampus of rats and monkeys have consistently demonstrated drastic impairments in place learning tasks, in navigational tasks, in exploration, and in learning about contexts (Nadel 1991, 1994).

— Studies of humans with damage in the hippocampal formation have confirmed the impact of this damage on spatial tasks, as well as on the ability to acquire, or indeed retrieve, episodic memories (Nadel & Moscovitch 1997; Pigott & Milner 1993).

Although lively debates concerning the details of hippocampal function persist, few would deny the critical role of this structure in spatial mapping and navigation. For present purposes these debates will be ignored, and the chapter will concentrate on what is generally accepted, namely, the role of the hippocampus in representing nonegocentric (or allocentric) space. The questions of interest: How does the hippocampus generate a spatial cognitive map? What forms of information go into the construction of these maps? What does such a map look like in the brain? How are these maps used in behavior?

ELEMENTS OF THE HIPPOCAMPAL COGNITIVE MAP

The fundamental elements of the hippocampal spatial mapping system are the place units first described by O'Keefe and Dostrovsky (1971) and subsequently expanded upon by a number of other research groups. Over the past 20 years some of the basic properties of these place cells have been worked out, although the full story remains to be uncovered.

1. Place cells are the predominant cell type in the hippocampus of the rat, comprising perhaps 90 percent of the cells. In the hippocampus proper they are the pyramidal cells, while in the dentate gyrus they are the granule cells. Cells with place-like receptive fields can be found in other structures in the hippocampal formation as well (O'Keefe 1979; Wilson & McNaughton 1993). Place cells in the rat hippocampus have small, well-localized fields (Recce 1994); this may not be the case for cells with spatial correlates elsewhere in the brain. Place cells have been reported in the monkey (Rolls & O'Mara 1995), but they apparently comprise a much smaller proportion of the population of principal cells. More numerous are *view* cells that are activated whenever the monkey is looking at a particular place in the test environment. These cells have large fields compared to the place cells observed in rats.

2. Place cells have *fields* in multiple environments, indicating that any given hippocampal place cell probably plays a role in many spatial maps (Kubie & Ranck 1983).

3. Place cells acquire their location-specific receptive field in an environment within a relatively short period (at most several minutes) after being placed in that environment, and retain that field as long as there is no serious change to the environment (Hill 1978; Wilson & McNaughton 1993). Exactly what constitutes a "serious" change, of course, is an empirical matter.

4. Place cells are part of a memory system; that is, they continue to show location-appropriate activation even after all external cues have been removed from the environment, assuming the animal has been given the chance to initially determine where it is (O'Keefe & Speakman 1987).

5. Place cells are in most environments *omnidirectional;* that is, they fire regardless of the direction in which the animal is facing. In certain environments, however, they are directionally dependent, firing mostly or only when the animal is facing in a specific direction (McNaughton, Barnes, & O'Keefe 1983). Recent evidence suggests that this omnidirectional character reflects a combination of inputs that are themselves directional (O'Keefe & Burgess 1996).

6. Many place cells participate in coding for any specific location in the environment; that is, the representation of space involves a distributed code rather than a "value" code (Wilson & McNaughton 1993).

7. Place cells can depend critically upon feedback from movement for their activation (Foster, Castro, & McNaugton 1989). Place cells are not driven solely by movement without other information about location, however, because identical movements executed in other places will not cause cell activation.

8. The precise properties of place cells are determined by the various inputs they receive and by the connectivity structure of the hippocampus. Full understanding of the spatial mapping system that the aggregated place cells compose demands an analysis of the informational content of those structures that feed into the hippocampus, as well as an understanding of the transformations wrought by the networks within the hippocampus itself.

Given the distributed nature of place coding, the study of the activity of large numbers of these cells simultaneously helps us to more fully understand how they accomplish their job. Recent technical advances have made this a real possibility, permitting the simultaneous analysis of as many as 50 to 100 hippocampal cells in animals behaving relatively freely in space (Skaggs & McNaughton 1995; Wilson & McNaughton 1993, 1994). In these studies, the ensemble nature of hippocampal place coding is quite easy to see. In any collection of 50 or so cells, a large fraction will have quite localized fields in the environment, a smaller fraction will be inactive, and a handful will comprise the other major class of hippocampal cells, which O'Keefe and I (1978) described as *displacement* cells. These cells fire in phase with the *theta rhythm* in the hippocampus (a 6–12 Hz sinusoidal pattern of neuronal activity), and have been called *theta cells* by most investigators.

The multiple cell recording technique has also made it possible to determine some of the relations between the hippocampal place cells and cells found in other parts of the system. These other components of the spatial mapping system provide essential inputs conveying information about objects, distances, and directions (Mizumori 1994). One of these components comprises the *head-direction* cell population in the postsubiculum discovered by Ranck, Taube, and colleagues (Ranck 1984; Taube 1992; Taube, Muller, & Ranck 1990a, 1990b; see also Chen, Lin, Barnes, & McNaughton 1994a; Chen, Lin, Green, Barnes, & McNaughton, 1994b; Mizumori & Williams 1991, 1993; Wiener 1993). This population of cells fires whenever the animal's head is pointing in a specific direction (relative to an absolute framework), and they can be assumed to provide the directional information that a spatial mapping system requires. Head-direction cells have been found in the postsubiculum, lateral dorsal thalamus, and caudate, that is, areas both afferent and efferent to the hippocampus. That these elements are critical to spatial mapping has been shown by Mizumori, Miya, and Ward (1994); inactivation of the lateral dorsal thalamus disrupted both hippocampal place cell representations and performance on a spatial maze.

Another important input cell type could be the *body-movement* cells found in posterior parietal cortex of the rat (McNaughton, Chen, & Markus 1991). These cells and the head-direction cells contribute movement and directional information that is essential to the use of a spatial map for accurate navigation. Recent work shows that there is a very strong coupling between head-direction cells and place cells. When environmental conditions cause a shift in, for example, a place cell's field, a corresponding shift occurs in the head-direction cells, indicating that both are part of some larger allocentric framework (Knierim, McNaughton, Duffield, & Bliss 1993).

The elements of the hippocampal spatial mapping system thus seem to include

— place cells—convey information about locations
— theta/displace cells—might convey information about distance
— inputs from head-direction cells—convey information about directions
— inputs from polysensory systems—convey information about the landmarks and landscapes that comprise the content of the spatial map
— inputs from vestibular and other systems—convey information about movements through space

How these elements function together to create a working spatial map and how this mapping system is used in navigating through the world to desired goals remains unclear. A number of quite detailed hypotheses have been proposed, however, and computational models based on these hypotheses are currently being tested in several laboratories (see McNaughton 1989; O'Keefe 1989, 1991a; O'Keefe & Burgess 1996; Rolls 1989).

One thing that is clear is that there is not a topographic spatial map in the hippocampus; that is, physically adjacent cells are no more likely to have adjacent place fields in the environment than are any two randomly chosen cells in the hippocampus. In recent work Samsonovich and McNaughton (1997) have sketched a model of the hippocampal cognitive map that considers this structure as composed of a large set of "charts," each of which can be conceived as a plane on which place cells are arrayed near other cells that have place fields adjacent to them in the environment. This "virtual" map can then be viewed as a surface, or state space, through which neural activity moves as the animal moves through the environment. The usefulness of this formalism remains to be determined.

PLACE CELLS AND MAPS: WHAT DRIVES THEM?

In its original formulation, cognitive map theory stressed the fact that spatial information derived from the connection structure of the hippo-campal system and that information about objects in the environment was linked to places in maps as a function of the organism's exploration of the environment, be it through active movement or visual scanning. Most of the work on place cells focused on the role of environmental stimuli in deter-mining the properties of place cells. More recent work has raised two impor-tant questions, hinted at previously: (1) Which environmental stimuli are critical? (2) What role does movement feedback information play?

WHICH STIMULI ARE IMPORTANT

On largely theoretical grounds, and following Hebb's (1938) insight about the crucial importance of distal cues, O'Keefe and I (1978) asserted that distal rather than proximal cues were important in building place rep-resentations and cognitive maps. We argued that only distal cues retained their relative spatial positions as animals moved through the environ-ment, so only distal cues would be useful in building up an allocentric spatial map. Recent evidence supports this argument by showing that proximal cues can be largely irrelevant in defining the place fields of hippo-campal cells (Poucet 1996); supporting evidence comes from analysis of how humans talk about places and what features of environments they emphasize in providing directions to other people. Denis (1996) has shown that the features of proximal objects are most often ignored (although the objects themselves might be used as landmarks) in such place and wayfinding descriptions.

Even though proximal objects seem to play a minor role in defining place cell activity, it does not follow that animals fail to know where these objects are located. Indeed, as Thinus-Blanc, Poucet, and their colleagues have shown in many elegant studies (see Thinus-Blanc 1996 for a review), animals are very aware of the location of objects, they notice when objects shift locations, and the hippocampus is critical to these abilities. Thus, it seems that distal landscape features are important in defining place fields, and that landmarks, though registered in cognitive maps quite accurately, may not be important in driving place cells.

In a recent paper O'Keefe and Burgess (1996) showed that place fields can be controlled by features of distal objects (the walls of experimental boxes), and provided a computational model of how this might work. McNaughton (1996) considered the same data and offered an explanation based on the primacy of movement, or path integration, information instead. It seems most likely that hippocampal place cell activity is multiply determined and that both landscape and movement information are important. One important finding in the O'Keefe and Burgess study is the fact that two place cells do not necessarily represent the same spatial relation in two different environments. This seems to eliminate the possibility that a Kantian absolute space is prewired into the internal connection structure of the hippocampus alone. It leaves us with a quandary, however: how does a particular pairing between two cells (or collections of cells) represent one distance in a particular environment and a different distance in another environment? One possibility is that spatial information about distance and direction is conveyed to the hippocampus and its internal representations via inputs from the parietal lobe (for distance) and the head-direction system noted previously. Distance information would include both sensory and movement-feedback inputs.

In order to capture the a priori nature of this spatial information we would have to speculate that these inputs are fixed in some way, and that the animal could retrieve accurate distance and directional information by "reading" the information in reverse. In order to account for the fact that any particular pair of hippocampal neurons can represent different distances in different environments, we would have to further suppose that there are multiple paths by which inputs can activate the same pair (or ensemble) of cells. This quasi-independence of mappings from one environment to another could mean that there is no overall integrated map, but rather a collection of spatial representations of distinct environments that are not in any way related to each other. This formulation has similarities to the multichart architecture proposed by Samsonovich and McNaughton (though they would give considerably more weight to movement inputs than envisioned here, and would argue for prestructured wiring in the hippocampus itself). It is also a formulation that suggests how the cognitive map system could simultaneously represent the environment at multiple scales. If distance information is conveyed via connections to the hippocampus, rather than in connections within the hippocampus, this multilevel mapping becomes easy to accomplish.

Finally, if the derivation of spatial information is viewed in this way, it becomes formally similar to how best to view the derivation of sensory feature information. That is, information about features is conveyed to the hippocampus, along labeled lines, from the ventral stream in the temporal cortex. This connectivity must be bidirectional, so that the featural aspects of memory for spatial layouts (and episodes) can be readily retrieved when needed.

The role of the hippocampal system in dead reckoning remains to be determined. Path integration can be an important component of spatial wayfinding, as elegant studies in ants, bees, and various mammals have shown (e.g., Etienne, Maurer, & Séguinot 1996; Mittelstaedt & Mittelstaedt 1980; Wehner & Srinivasan 1981). Inputs from path integration certainly play some role in the hippocampal system, but how central they are remains to be determined. Some form of feedback from an animal's actual or intended or imagined movements could readily serve to "move" the animal through its internal map, thereby allowing the animal to anticipate where it will be. How movement information interacts with the sensory landscape information already known to influence the cognitive map is a major question for the future.

HOW MAPS ARE USED IN BEHAVIOR

This obviously central issue has received perhaps the least amount of attention from neuroscientists to date. In our book, O'Keefe and I speculated on various ways in which cognitive maps might be used, including using a map to recognize where one is, where objects and places of interest (food, water, safety) are located, predicting where one will be (and what one will experience) if one makes certain movements, and, perhaps, calculating how to get from where one is to where one wants to be. Although considerable information is available about the neural elements of cognitive maps, relatively little is available about the neural elements of navigation and wayfinding. A recent computational model offers some hope of increased attention to this problem in the future (Touretzky & Redish 1996).

A HIERARCHY OF SPATIAL SYSTEMS

One way to think about the multitude of spatial systems is portrayed in figure 12.1. This scheme starts with egocentered spatial systems that are driven predominantly by sensory inputs and then moves through a set of ever more complex systems, each of which represents information in a different frame of reference, culminating in the hippocampal cognitive map system which represents the world in a nonegocentered reference frame.

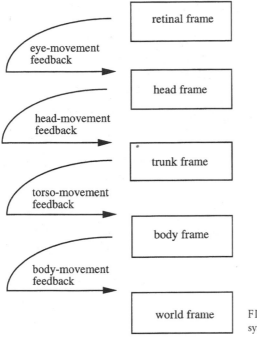

FIGURE 12.1. Multiple spatial systems.

There are two critical features of this scheme that are worth noting. First, the transition from one level to the next always involves a combination of the outputs of the first level with some movement feedback, or efference copy, information that transforms one reference frame into another. Second, the final step, from a parietal lobe system into the hippocampal system, uses movement information from the entire body to transform this most complex egocentric representation to the nonegocentric map found in the hippocampus. Thus, one might say that egocentric coordinates are stripped away from the representation as it is transformed into hippocampal coordinate space. Allocentric spatial coordinates are attached in the hippocampus as a function of the feedback from movement (real or intended), and the consequences this feedback has for determining the locus of activation within the hippocampal system. It is in this sense that O'Keefe and I thought of this movement feedback information as a kind of coding strategy used by the hippocampus to build its spatial maps.

Within the context of this sort of hierarchy, one might ask whether or not there is an interesting evolutionary story to be told about spatial systems. Is there some sort of progression to be noted here, with more "primitive" spatial systems leading to more "complex" ones?

One possible way to think about this is as a progression from the use of cues (or *guidances* as O'Keefe and I called them), through the use of path integration information, to the combination of these two in the use of integrated maps. Two kinds of guidance use can be distinguished: following a continuous cue such as a shoreline, or a river bed, and approaching or avoiding a fixed landmark such as a mountain peak or a tree in a field. These guidance uses differ in the following important way: the latter case involves a cue whose "image" on the retina changes over time. As the organism gets nearer or farther away, image size increases or decreases accordingly. Thus, the animal's behavior cannot be regulated by trying to maintain a fixed retinal image of a given cue, as it can in the case of following a continuous cue such as a shoreline.

The use of path integration information is logically separate from the use of guidances, and one might imagine these having emerged independent of one another in evolution. In its simplest form path integration involves updating a vector pointing back to some starting point as a function of movements, allowing the organism to return to the start point via a "beeline." Somewhat more complex would be the ability to use path integration information to calculate trajectories other than one involving return to a start point.

At some point in evolution the ability to path integrate, and hence get somewhere through dead reckoning, was combined with the ability to use guidances as a corrective. Path integration is subject to cumulative error that only occasional position fixes can correct. It is clear that ants, rats, and humans can use dead reckoning and cues as required.

A distinction has to be made between the ability to use various forms of information and the capacity to create and manipulate internal representations based on these forms of information. Although ants and bees can use much the same information that humans use, there is little reason to think that they create internal cognitive maps permitting allocentric spatial strategies (e.g., detours). Humans not only create such maps and use them in generating novel trajectories, but they can also talk about their spatial knowledge, and indeed even use spatial representational systems as the basis for complex problem-solving behavior.

Another important point that must be stressed in this discussion is the fact that not all of these spatial systems agree with each other. That is, the spatial information present in one system can differ from the spatial information present in another. Depending on how this information is accessed, different behavior results, which is clearly described in the work of Loomis

(1996) and his colleagues. This fact raises the important question of how these contrary representations are reconciled in the overall behavior of the organism.

THE HIPPOCAMPAL FORMATION AND SPACE IN HUMANS

There is considerable evidence in support of the view that human hippocampal formation is also engaged in the form of spatial mapping discussed for rats and monkeys. Damage to the human hippocampus, as observed in amnesia, Alzheimer's disease, and certain other syndromes, invariably results in the loss of wayfinding abilities, often in the context of the preservation of other, egocentric, spatial abilities (e.g., Bohbot et al. 1997; Pigott & Milner 1993). Recent studies of brain activation using functional MRI have demonstrated activation in the hippocampal formation during tasks involving real or imagined traversals of space (Maguire 1997). These studies suggest that there is continuity across species with regard to the neural mechanisms of wayfinding, and encourage the application of knowledge gained in much more detailed and controlled work with rats and monkeys to the human case.

CONCLUSION

Considerable insight into the neural mechanisms of wayfinding has been gained in recent years, but there are still no final answers. A point has been reached where the right questions can be asked, and methods exist that might lead to answers. The neural system representing cognitive maps has been localized, and some notions of what controls its activity have emerged. Less is known about how this map is used in behavior. The next few years should illuminate many of these remaining areas of darkness.

ALAIN BERTHÖZ
MICHEL-ANGE AMORIM
STEPHAN GLASAUER
RENATO GRASSO
YASUIKO TAKEI
ISABELLE VIAUD-DELMON

13

Dissociation between Distance and Direction during Locomotor Navigation

Navigation in a spatial environment requires that the brain update information stored in spatial memory about the orientation and the position of the body. This chapter addresses the question of the mechanisms that underlie this capacity to memorize routes and to use this spatial memory for guiding and steering locomotion.

During blindfolded locomotion, it has been demonstrated that human subjects could reach a previously seen visual target on the floor several meters away with eyes closed (Thomson 1983) even after a number of detours. The results obtained by several groups indicate that information about step length derived from proprioceptive or outflow motor command signals, as well as vestibular signals, could contribute to the updating of the mental representation of a subject's location in space and allow for path integration (Mittelstaedt & Glasauer 1991). Although the possible contribution of proprioceptive or motor outflow signals to spatial updating has been

This research was supported by grants from the Programme Cognisciences of CNRS; fellowships from the Programme Cognisciences to I. Viaud-Delmon; and an EEC Human Capital and Mobility grant to R. Grasso. The authors thank the Centre National d'Etudes Spatiales for providing the equipment for data acquisition.

largely documented (Rieser, Pick, Ashmead, & Garing 1995) the contribution of vestibular cues is under debate.

First, we review some recent results which suggest that the brain can indeed derive from vestibular information an estimation of angular and linear path. We then review recent experiments of our group, which suggest that during locomotion the brain is using a predictive strategy to guide and steer locomotion and that there are separate mechanisms for the control of direction and distance. We suggest that vestibular cues can contribute to the detection and storage of head direction and that proprioceptive and motor outflow information is important for the storage of distance. Finally, in order to document the possible mechanisms that may underlie these processes, we mention recent studies using brain imagery with positron emission tomography (PET) scanning that have explored the cortical and limbic areas involved in spatial memory during mental navigation tasks.

VESTIBULAR CUES

In this section we review results which show that vestibular information can contribute to path integration during small duration passively induced rotations and translations.

MEMORY OF ANGULAR DISPLACEMENTS

Vestibular (inertial) measurement of a locomotor trajectory can be performed by *path integration,* namely the integration of linear and angular head acceleration signals provided by the vestibular organs (the canals measure angular head rotation and the otoliths measure head linear acceleration and head tilt). The contribution of the vestibular system to the orientation and localization of the body in space after a displacement, in animals and humans, has long been suggested (Beritoff 1965; Potegal 1982). Beritoff was the first to report that animals (cats and dogs) can return blindfolded to their starting position after they have been passively transported. Because accurate performance in this task was only achieved in animals with an intact vestibular system, he concluded that the vestibular organs are required for navigation in the dark. He also observed that deaf-mute children with lesioned labyrinths could not, when blindfolded, retrace the route along which they had been taken, while normal children were perfectly able to do so. This result was confirmed in rats by Miller, Potegal, and Abraham (1983), who observed a degradation of the performance after vestibular lesion. These results have been reviewed in Wiener and Berthöz (1993).

Our group has studied the contributions of vestibular cues to memory of whole body passively imposed rotations and translations. In the case of horizontal rotations we have used a new psychophysical test about passive body rotation estimation called the vestibular memory contingent saccade (VMCS) paradigm (Bloomberg, Melvill-Jones, Segal, McFarlane, & Soul 1988). The seated subject is first presented with an earth-fixed visual target, then the subject is rotated by turning the chair about its swivel base (oriented along the earth-vertical axis). During the rotation, the subject has to fixate another target that remains at a fixed position with respect to the head (head-fixed target), in order to suppress the vestibulo-ocular reflex (VOR). After a delay of about 2 seconds, the subject is required to make a saccade toward the "memorized" earth-fixed target, in complete darkness. The results showed that humans can correctly match the amplitude of a preceding head rotation with a voluntary ocular saccade of equal but opposite amplitude. The amplitude of a passive body rotation in darkness can therefore be correctly estimated by the brain, but this information can also be adequately stored and further retrieved and used by the oculomotor system. This is true also for rotations in different planes of three-dimensional space (Israël, Fetter, & Koenig 1993).

In a subsequent experiment (Israël, Rivaud, Pierrot-Deseilligny, & Berthöz 1991), we studied in greater detail the duration of vestibular information storage using delays of less than 20 seconds, as well as 1 and 5 minutes. The results showed that the amplitude of a passive rotation detected by the vestibular system can be accurately memorized for as much as 5 minutes. At such long delays, however, smaller rotation angles (< 20 degrees) are overestimated (i.e., lead to an undershoot). Following these results we inferred the involvement of the hippocampus and cortical structures (frontal and parietal lobes) in the representation of self-rotation detected by the vestibular system (Berthöz 1989).

We have also investigated which areas of the cerebral cortex would be particularly important in the VMCS. We chose those areas of the cortex that are activated during saccade generation or have been suggested to be involved either in visuo-spatial delay tasks and working memory or in the processing of vestibular signals. We applied the vestibular memory contingent saccade task to neurological patients with various cortical lesions and to age-matched control subjects (Israël, Rivaud, Berthöz, & Pierrot-Deseilligny 1992; Israël, Rivaud, Gaymard, Berthöz, & Pierrot-Deseilligny

1995; Pierrot-Deseilligny, Israël, Berthöz, Rivaud, & Gaymard 1993). Anticipation and latency, direction errors and accuracy of the first saccade, stability of eye position into darkness, and final eye position were quantified. Patients were divided into small groups with lesions affecting the following cortical areas: left or right frontal eye field (FEF), left or right prefrontal cortex (area 46 of Brodmann) (PFC), left supplementary eye field (SEF), left or right posterior parietal cortex (PPC), and right parieto-temporal cortex (PTC) (i.e., the vestibular cortex). There were some abnormalities in the results of the right FEF group in regard to anticipation, direction errors, and latency of the first saccade, but no differences from controls in the accuracy of the first saccade or of the final eye position. Results in the left FEF group were normal. Accuracy of the first saccade was impaired in the SEF group bilaterally. Final eye position was also inaccurate in the SEF group. In both PFC groups, significant and, in general, bilateral abnormalities existed for all tested parameters.

MEMORY OF LINEAR DISPLACEMENTS

This ability of the brain to derive angular displacement from semicircular canal information is also valid for the detection of translation by the otoliths. In previous experiments we have shown that subjects can estimate passively induced translational displacements (Israël & Berthöz 1989; Israël et al. 1993). This may be only true for small displacements, however. During locomotion, distance evaluation may be estimated by proprioceptive or motor commands corollary discharge (Mittelstaedt & Glasauer 1991). The exact nature of what is being stored in spatial memory during these passive displacements is not known. Using a mobile robot on which subjects could be transported passively and subsequently asked to reproduce their linear displacement, we have recently shown that linear displacement memory seems to involve a dynamic storage of movement patterns using multisensory cues (Berthöz, Israël, Georges-François, Grasso, & Tsuzuku 1995), an idea which is in accordance with the general view that internally simulated action is a fundamental element of perception and spatial memory (Colby, Duhamel, & Goldberg 1995). These results indicate that vestibular cues contribute to spatial memory during passive transport.

We now describe some experiments that explore the motor and cognitive strategies employed by the brain during active locomotion and the respective roles of the vestibular system, proprioception, and motor commands for the steering of locomotion during simple bidimensional navigation tasks.

ACTIVE LOCOMOTION

TRIANGULAR PATH

In a first set of experiments we asked the following question: How do we locomote along a simple previously seen triangular trajectory and how do we go around corners?

Anticipation of the Trajectory by Head Motion

Repeating a previously seen locomotor trajectory without vision has been examined since Thomson's (1983) experiment on locomotor pointing. Most of the work concentrated on walking toward one target. For two different segments, one straight ahead and the second perpendicular to it, Loomis, Da Silva, Fujita, & Fukusima (1992) showed that subjects are able to reproduce previously seen distances correctly by walking.

A similar task to the one presented here is called *triangle completion*. The subject is guided over two legs, and then he or she attempts to return directly to the point of origin (Loomis et al. 1993; Worchel 1951). Length of the walked segments and their sustaining angle are varied. The measured parameters are the error in a subject's turn toward the origin after walking the first two legs, and the error in the distance the subject subsequently walked to complete the third leg. Both of these errors show a pattern of systematic regression to the mean. Subjects tend to over-respond when the required distance or turn is small and to under-respond when it is large, similarly for both blind and sighted subjects (Fujita, Klatzky, Loomis, & Golledge 1993).

Triangle completion has one major drawback in indicating disturbances in complex spatial understanding in blindfolded individuals. Some errors that are made during the guided walk and the return walk will not be seen in the results. Imagine a subject overestimating his walked distance by a certain factor, but making no other errors: this subject will perfectly perform the triangle completion, but will fail in reaching the first and second corner in our task. Therefore, we have chosen the reproduction of a previously seen path by means of locomotion. The performance of the locomotor pointing allows us to quantify misperception of linear *and* angular self-displacement, although one has to keep in mind that previous experiments (Mittelstaedt & Glasauer 1991) showed that angular as well as linear path integration performance heavily depends on velocity.

In a first series of experiments (Glasauer et al. 1995) we asked normal subjects to perform a triangular path to a series of three targets drawn on the ground. The path consisted of a right triangle with two sides 3 m in length. The three corners were marked on the floor with targets consisting of 7-cm by 7-cm crosses. The subject's task was to walk the triangular path, starting at either corner 1 or corner 3. When the path was completed the subject was requested to turn and face the direction from which he or she started. The verbal instructions given were, "Walk at a comfortable pace, as accurately as possible around the path. The motion should be continuous. The goal is accuracy, with accuracy defined as your ability to 'straddle' the path." For all experiment sessions, two spotters were in the room to prevent any collisions during the eyes-closed tasks.

To control for directional preferences the task was performed in alternating clockwise (cw) and counterclockwise (ccw) directions, but always approaching the right angle (corner 2) of the triangle first. Vision-occluded trials were performed before the eyes-open trials to minimize visual feedback. Also to minimize visual feedback, at the conclusion of each eyes-closed trial, the subject was led in a serpentine path, with eyes still closed, to the next starting point. The subject was instructed to look at the path before starting each eyes-closed trial. The subjects performed 12 trials eyes closed (6 cw and 6 ccw) and 6 trials eyes open (3 cw and 3 ccw).

Each subject wore a helmet with three retro-reflective markers located above the head in approximately the sagittal plane. This helmet was also equipped with headphones that provided white noise to mask out spatial auditory cues and blackened goggles to occlude vision. Head kinematic data were collected with a four camera motion analysis system (from the Motion Analysis Corporation, Santa Rosa, California).

The coordinates necessary to describe head position in all six degrees of freedom were computed from the three-dimensional positions of the head markers. The three translational components were used to identify translational position and compute linear velocity, the three rotational components to express tilt and compute angular velocity of the head. The rotational head position was expressed as quaternions. The corners of the walked trajectory and the maxima of the angular head velocity were determined for each walk. The corner points were used to compute distance errors and mean walking velocity. To evaluate the mean walking direction for each leg of the triangle, lines of minimum least square distance were fitted to the trajectory between the corners. The angle between two lines then gives the amount of turn performed by the subject. The angular deviation from the desired trajectory (i.e., from the triangle's leg) was computed as the difference between the angle turned and the required angle of turn at the respective corner.

Statistical analysis was performed on the mean parameter values of each subject. Two different ways of describing distance errors were applied: (1) the two-dimensional distance error of each corner point to the required corner at the end of a segment (arrival error) and (2) the difference between required length of a segment and actual distance covered (length error). The arrival error gives an absolute estimate of both directional and longitudinal deviations from the required path, while the length error shows purely longitudinal errors in reproducing the segments. Thus, arrival error is cumulative over the walk, and length error is not.

Distance Error

Length error increased from segment 1 to 3 for the eyes-closed condition, and it was largest for segment 2 in the eyes-open condition due to the fact that subjects tended to walk around corners 1 and 2 with open eyes. In the eyes-closed condition, the errors increased more from one segment to the next than with eyes open. Therefore, total distance error was larger in the eyes-closed condition (0.61 ± 0.42 m) as compared to the eyes-open condition (0.22 ± 0.11 m).

Directional Error

The directional error is described as the difference between the mean walking direction during each segment with respect to the previous segment and the required angle of turn from one segment to the next. Therefore, the directional error of the first segment only gives the heading error toward corner 1, while the directional errors during segments 2 and 3 give the errors of angular turn with respect to the preceding segment of the path. Note that directional error as defined here is not cumulative, as it is computed in relative coordinates. The mean absolute directions are evaluated from the lines of minimum least square distance described previously. Directional error was tested only for segments 2 and 3. Mean errors for the eyes-closed conditions (-7 ± 9.8 degrees) showed a trend to underestimate the turns.

Mean Walking Velocity

Mean walking velocity was computed by dividing the walked length by time needed to walk one segment. Subjects walked more slowly in eyes-closed (0.7 ± 0.1 m/s) as compared to eyes-open (0.8 ± 0.1 m/s) conditions.

Corner Parameters

We tried to describe the way subjects walk around a corner using several parameters. The tangential linear velocity turned out to be minimal at the corner point that is preceded by a maximum of angular head velocity. The maximal angular velocity of the trajectory coincides with the minimum of the tangential velocity. This means that prior to walking around the corner the subjects turn their heads in the new direction. The angular velocity vector was computed from the angular position of the head in space.

The maxima of its vertical component, which express the yaw angular velocity, were determined for the head turns at corners 1 and 2. In addition, the time between those maxima and the corner point were evaluated. This parameter is supposed to show the coordination between head and trunk for walking around a corner.

Mean angular head velocity eyes closed was 137 ± 38 degrees per second. Time between maximum head velocity and minimum tangential velocity was always negative, showing that the head turned always before the body went around the corner, as illustrated in figure 13.1.

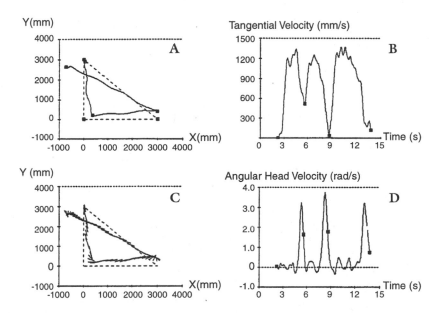

FIGURE 13.1. Locomotor trajectory and head anticipation while walking along a triangular path. Head movement was measured by a video computer system which recorded the movement of two markers placed on the head of the subject in the sagittal plane. (A) Triangular path seen from above. The head trajectory is shown when the subject walks eyes closed from corner 1 (top) to corners 2 and 3 and returns to corner 1. Only the middle point between the two markers is shown here. Dotted lines, ideal trajectory; continuous line, real head trajectory. (B) Head tangential velocity (mm/sec) is plotted against time (sec) during this path. The passage at the corners is indicated by small squares on the curve. (C) Triangular path seen from above. This tracing is from the same sample as in A, but here the line between the two head markers is shown in the form of a little stick that indicates the direction of the head in space. The anticipation of head direction on the trajectory before turning around a corner can be seen clearly. (D) Head angular velocity (rad/sec) plotted against time for the traces in C. The peak value of this parameter occurs before the passage at the corners, which are indicated by the small squares as in B. The lead is 200–300 msec. (From Glasauer et al. 1995.)

Note that for the walks performed with eyes closed, the latency of the anticipation was larger (350 ± 270 ms) than with eyes open (250 ± 100 ms).

DISSOCIATION OF DISTANCE AND DIRECTION ERRORS IN VESTIBULAR DEFICIENT PATIENTS

Glasauer et al. (1994) showed that bilateral labyrinthine-defective (LD) subjects are able to perform linear goal-directed locomotion toward a memorized target, although lack of canal information induced more instability during blind walking (i.e., increased path curvature). In order to further investigate the contribution of semicircular canals input in such a task, the movement of the head in all six degrees of freedom was measured during a blind walk along a previously seen triangular path in both normals and LD patients.

Seven normal subjects and five patients with vestibular deficits (two with unilateral and three with bilateral) were asked to walk a previously seen triangular path first without vision and then with vision. These patients had received a vestibular surgery or acquired a vestibular deficit due to gentamicin at least several months before and had been undergoing rehabilitation procedures. They can be considered as compensated subjects.

The path, marked on the ground by a cross at each corner, consisted of a right triangle with two legs of 3 m in length. The task was performed alternating clockwise and counterclockwise directions, but always approaching the right angle of the triangle first. The subjects were asked to walk the path in both directions three times without vision and then one time with vision. When the path was completed the subject was requested to turn and face the direction in which he or she started. The verbal instructions given were the same as for the experiment described in the previous section.

Before the experiment started, patients were guided eyes closed in the room for a random walk until they felt confident walking without vision, using the classical guiding technique for the blind: holding firmly the arm of the experimenter who walked aside the patient to guide him or her. Although being firmly guided at the beginning of this phase, the patient was progressively given more and more autonomy until he or she could walk alone, being closely followed by the experimenter to avoid falling and bumping into the walls.

Head and trunk displacement and velocities were measured in a similar manner as in the first experiment but with different equipment (ELITE™). Preliminary examples of trajectories are shown on figures 13.2, 13.3, and 13.4.

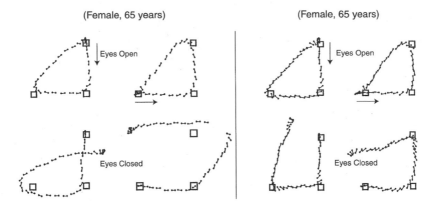

(Female, 65 years)   (Female, 65 years)

Eyes Open

Eyes Closed

FIGURE 13.2. Triangular locomotion in patients with bilateral vestibular deficits. Examples of triangular paths for two patients with bilateral absence of vestibular function. The patients had to walk clockwise along a triangular path drawn on the ground with eyes either open or closed. The initial corner of the triangle was variable. The direction of the walk is indicated by arrows. Note that subjects performed the task very well in the light, indicating that there were no motor deficits. In darkness, however, by contrast with normal subjects who had small errors, these patients made large errors. Analysis of these errors indicates that there was a large variability but only a small error on the *total distance,* but some very large errors concerning the control of *direction.* These results suggest that the vestibular system does contribute to the steering of direction, probably through the head direction cell system.

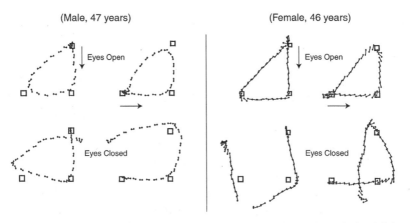

(Male, 47 years)   (Female, 46 years)

Eyes Open

Eyes Closed

FIGURE 13.3. Triangular locomotion in patients with unilateral vestibular deficits. Examples of triangular paths for two patients with unilateral (left) absence of vestibular function. The patients had to walk clockwise or counterclockwise along a triangular path drawn on the ground with eyes either open or closed. The initial corner of the triangle was variable. The direction of the walk is indicated by arrows. Note that subjects performed the task very well in the light, indicating that there were no motor deficits. In darkness, however, by contrast with normal subjects who had small errors, these patients made large errors. Analysis of these errors indicates that there was a large variability but only a small error on the *total distance,* but some very large errors concerning the control of *direction.* These results suggest a dissociation between the control of distance and direction and that the vestibular system does contribute to the steering of direction, probably through the head direction cell system.

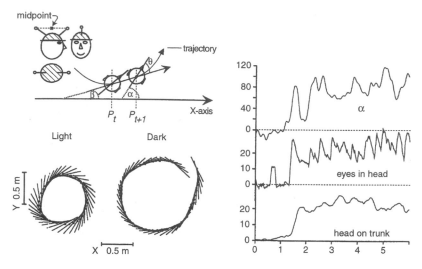

FIGURE 13.4. Head direction anticipation during circular locomotion. (*Upper left*) The subject is asked to walk on a circular path drawn on the ground with eyes either open or closed. Head direction, position, and velocity are measured by a computerized video system that uses markers placed on a helmet. The position of the two infrared reflecting markers on the helmet is indicted; $p_t \ldots p_{t+1}$ denote successive midpoint positions. α, head orientation; β, walking direction; θ, angle between trajectory and head direction. (*Lower left*) The projection of head trajectory onto the XY plane. The sticks (short lines) indicate the direction of the head. (*Right*) Recordings of angular head velocity (p_t, p_{t+1}) in degrees per second, horizontal eye position in the head measured by EOG, and angular displacement of the head on the trunk measured by the relative position of the two markers on the head and the two markers located on the shoulders. The abscissa is the time in seconds. This sample record has been chosen to show that the head lead is accompanied by an eye saccade, which means that it is the mechanisms of gaze orientation that underlie the lead of the head upon the trajectory. This lead is maintained in darkness (blind locomotion), suggesting that the locomotion is driven by an internal representation of the locomotor trajectory that allows this predictive mechanism to operate. (From Grasso, Glasauer, Takei, & Berthöz 1996; see also Takei, Grasso, Amorim, & Berthöz 1997.)

It can be seen that with eyes open, the subjects went along the triangular path with no problem, showing that there was no motor deficits. But when they performed the path eyes closed some striking deficits appeared in the control of the direction of the turns although the total distance was not greatly changed.

Distance Error
Length error is the difference between the required length of a segment and the actual distance covered. Subjects showed a large interindividual variability. No effect of walking direction was found on length error during locomotion without vision. Subjects, whatever their category (normal vs.

patients), significantly overshot the distance of each leg of the path when walking eyes-closed, and undershot it with vision. This result can be due to the fact that subjects performed smoother trajectories (that is, rounded corners rather than sharp triangular turns) when walking with vision.

Directional Error

Directional error is the difference between the mean walking direction during each segment with respect to the previous segment and the required angle of turn. Again, subjects showed a large interindividual variability. Interestingly, two unilateral LD subjects showed opposite directional errors. This might be due to an asymmetric walking behavior present before the vestibular deficit occurrence. Indeed, we observed such an asymmetry for one "atypical" normal subject. We also observed a right spontaneous nystagmus of nonpathological origin in this subject. All our subjects demonstrated an absolute directional error that increased significantly with the amount of directional change to be performed during the walk. Labyrinthine-defective patients made significantly larger errors than normal subjects. Two of the bilateral patients demonstrated an asymmetry in their performance between clockwise and counterclockwise directions. The third performed the same in both directions. It must be noted that the two bilateral patients proved to be areflexiac in both ears with unknown cause and had unilateral complaints at the beginning of their problems.

Mean Walking Velocity

All the subjects walked significantly slower when walking without vision than with eyes open, especially the normal subjects and the unilateral LDs. There was a trend for LD subjects to walk slower than the normal subjects.

Corner Parameters

We analyzed the following corner parameters (for the first and second triangle corners):

1. Angular velocity. With eyes closed, angular velocity decreased. This decrease was larger in bilateral LDs. Regardless of the vision condition, all our subjects showed a significantly larger angular velocity when negotiating the second turn as compared to the first one.
2. Linear velocity. The linear velocity turned out to be minimum near the corner point. Linear velocity was significantly smaller for the second (135 degrees) than the first corner.
3. Head anticipation. Both normal and LD subjects had head anticipation of trajectory. This means that prior to walking around the corner the subjects turned their heads toward the next target.

CIRCULAR LOCOMOTION

In order to further investigate both the anticipation of head direction and the dissociation between the control of distance and direction we asked subject to walk around a circular path either with or without vision (Grasso et al. 1996; Takei et al. 1997). Experiments were carried out in a large room and a two-camera ELITE system was used to estimate head motion in three-dimensions.

Subjects wore a specially designed helmet equipped with two infrared reflecting markers placed in the head sagittal plane and aligned along the naso-occipital axis. The helmet was carefully placed so the midpoint between the two markers (mean value of X,Y,Z coordinates of the two markers) was in the head yaw rotation axis. As illustrated in figure 13.4, instantaneous head orientation (α) was measured from the position of the two markers by applying appropriate trigonometric formulae. The angle formed by the tangent to two successive positions $p_t \ldots p_{t+1}$ of the head midpoint (β) with the X axis describes the instantaneous direction of the trajectory at time t; θ is the instantaneous difference between head and trajectory direction. The mean of the XY coordinates of all successive positions p_t occupied by the head corresponds to the baricenter of the performed trajectory. The mean of the instantaneous distances from the baricenter to all successive positions p_t was taken as a reasonable approximation of the average trajectory radius.

Subjects' stepping rates were computed by means of Fourier analysis of vertical head displacement. Real-time matching between head orientation and trajectory direction was quantitatively assessed in the frequency domain by means of cross-spectral analysis. The averaged cross-spectrum of couples of variables was computed by using a standard fast Fourier transform (FFT). Details of the mathematical procedure can be found elsewhere. The magnitude squared coherence (MSC) function was calculated. It is analogous to the squared correlation coefficient in linear regression: it ranges from 0 to 1 and tests the correlation between synchronous sinusoidal fluctuations (if any) in a given couple of variables. There is one coherence value for each frequency present in the signal. When coherence is significantly high (i.e., >0.5), the phase shift Φ and the transfer function gain G at that particular frequency can be accurately assessed; Φ ($-180° < \Phi < +180°$) quantifies lead-lag relationships and G quantifies the ratio between the amplitudes of synchronous fluctuations. From Φ a corresponding time delay can be calculated by applying the formula: delay = Φ ($2\pi f$), where f is the frequency at which the phase shift Φ is estimated.

Subjects were first trained to walk with eyes open (counterclockwise) at a comfortable speed along a circular trajectory drawn with paint on the ground. Training was stopped when subjects felt confident that they could reproduce the path without watching the painted track (as instructed). The experiment consisted in reproducing the memorized trajectory in three conditions: (1) eyes open (LIGHT), (2) blindfolded (DARK), and (3) while reading a newspaper, held by their own hands, aloud (READ). The experiment was repeated with circles of three different radii (0.5, 0.9, 1.3 m) and at least two complete revolutions were recorded. To avoid any interference from external acoustic cues, subjects wore headphones delivering loud band-limited noise (pink noise). The READ condition presupposes the following rationale: visual pursuit is impeded, a mental work load is executed, and, furthermore, the biomechanical properties of the head-neck system are entirely modified. Fixating the newspaper effectively ties the head to the arms with the consequent effect of increasing the stiffness of the head against rotatory movements.

All participants were able to walk around the three circles in all conditions. The radius of the performed trajectories was highly correlated with and very similar to that of the ideal circle for all subjects ($r > 0.92, 0.96 \leq$ slope $\leq 0.99, p < 0.001$ for all conditions). The trajectory of the performed circles was never completely smooth and often resembled more that of a curved polygon because of the biomechanics of bipedal gait. All participants also displayed fluctuations of head direction in the horizontal plane (which we called *yawing* oscillations).

Such head direction oscillations exhibited at least three typical characteristics: (1) they had half the frequency of the stepping rate, which means that they had the frequency of the whole locomotor cycle; (2) they were always smaller than the synchronous oscillations of walking direction ($G = 0.38 \pm 0.028$ Hz in LIGHT, 0.46 ± 0.014 Hz in DARK, and 0.47 ± 0.030 Hz in READ); and (3) they systematically anticipated (100–200 ms) the correlated fluctuations of the walking direction in all conditions. This phasic anticipation was longer when the radius was shorter ($p < 0.05$).

In the LIGHT the head was tonically oriented toward the inner part of the circle ($\overline{\theta} = 18.3 \pm 305$ degrees mean $\pm$ SEM), although it tended to align with the direction of the trajectory when vision was impeded or distracted by the concurrent task ($\overline{\theta} = 4.5 \pm 4.8$ degrees in DARK and 0.8 ± 3.2 degrees in READ).

These results demonstrate three main phenomena concerning the head orientation during the steering of locomotion:

1. The head oscillates in the horizontal plane in synchronization with the bipedal locomotor cycle (every two steps), although the head oscillates much less than walking direction (gains G always lower than 1). This corresponds to a form of head stabilization similar to what has been previously reported in the sagittal and in the frontal plane (Berthöz & Pozzo 1988; Pozzo, Berthöz, & Lefort 1990; Pozzo, Berthöz, Lefort, & Vitte 1991).

2. There is a consistent deviation of mean head orientation ($\bar{\theta}$) toward the inner part of the curved trajectory, of probable visual origin, because it disappears when visual pursuit of environmental cues is suppressed (in the DARK and READ conditions respectively).

3. Head horizontal oscillations, coupled with the locomotor cycle, are anticipatory with respect to the corresponding walking direction changes. This behavior of head direction strongly suggests the presence of a head nystagmus. An eye-head coordinated nystagmus was reported in the monkey (Solomon & Cohen 1992a, 1992b), and in humans (Bles, de Jong, & de Wit 1984) running in circles both in light and in darkness: its origin has been attributed to a velocity storage mechanism, excited by multisensory (visual, vestibular, and proprioceptive) inputs. In humans, like in the monkey, such a nystagmus continued when the platform was counter-rotated to null angular motion in space, confirming its multisensory origin.

On the basis of these observations we propose that the anticipatory nature of the phasic head orientation with respect to walking direction has to be related to the properties of the vestibular and optokinetic nystagmus that probably occur during the experimental task. Indeed, the rapid phases of eye movements, which are directed opposite to the direction of the displacement of the visual field, are now known to be not only resetting movements but to be genuine anticipatory orienting reactions.

On the other hand, the observed tonic deviation of the head ($\bar{\theta}$) is also reminiscent of the beating field of optokinetic nystagmus (OKN), which is well known to be deviated toward the side of the rapid phase, that is, as if the gaze was anticipating the future heading direction. Because a close coupling between eye and head tonic deviation has been demonstrated in humans (André-Deshays, Berthöz, & Revel 1988; André-Deshays, Revel, & Berthöz 1991), it follows that $\bar{\theta}$ probably reflects the deviation of gaze with respect to the trunk anteroposterior axis.

The anticipation interval of the head direction oscillations (120–200 ms) depends upon the curvature of the walk and seems to be independent of the inertial properties of the head, as shown by the fact that when the subject was reading, and the head was almost immobile relative to the trunk, the pattern was unchanged. Therefore, the phasic component of head orientation might depend directly on the movements of the trunk and thus it would be inherent to the biomechanics of locomotion. Alternatively, the trunk itself might subserve the control of head direction when the head-trunk mechanical degree of freedom is removed. Finally, it cannot be excluded that the tonic and phasic components of anticipation are due to completely distinct mechanisms.

NEURAL MECHANISMS

We now discuss briefly the possible neural mechanisms underlying the potential contribution of vestibular cues to navigation and the dissociation between distance and direction coding.

VESTIBULAR AND VISUAL PROJECTIONS
TO THE PARIETAL CORTEX AND HIPPOCAMPUS

The Parieto-insular Vestibular Cortex

Recent work by Grüsser on monkeys (Bottini et al. 1994; Grüsser, Guldin, Harris, Lefèbre, & Pause 1991; Grüsser, Pause, & Schreiter 1990a, 1990b; Gulyás & Roland 1994) and from PET studies (Vallar, Sterzi, Bottini, Cappa, & Rusconi 1990) in humans have confirmed the existence of an area in the parieto-insular cortex involved in the processing of multisensory (somatic and visual) and particularly vestibular cues about head motion in space. This parieto-insular vestibular cortex (PIVC) is probably the essential station in the transmission of ascending vestibular information from the vestibular nuclei, through the sensory thalamus, to the cortical areas involved in the elaboration of spatial cues, which are now identified as areas 6, 3a, 2v, T3, and 7a in monkeys. The same areas project to the vestibular nuclei in monkeys (Akbarian, Grüsser, & Guldin 1993; Faugier-Grimaud & Ventre 1989; Guldin, Mirring, & Grüsser 1993; Ventre & Faugier-Grimaud 1988).

In order to gain insight into the contribution of the vestibular system to cortical functions we have further studied the vestibular-activated areas using both caloric and galvanic stimulation and recording brain activity by functional MRI of the whole brain. Caloric stimulation was applied to seven subjects by injection of cold water to the left ear and compared with a rest condition. The activated areas were the left temporo-parietal junction (probably the PIVC), the postcentral gyrus, the premotor cortex, the insular cortex, the posterior middle temporal gyrus, the inferior parietal lobule, the frontal cortex, the anterior cingulate gyrus, and the hippocampal gyrus. On the right side these areas were also activated, but more rarely. This list does contain several of the areas already found in other PET studies, but the activation of the hippocampus had not been reported previously.

In order to verify the activation of the hippocampus by vestibular stimulation that had been predicted by previous results obtained in monkeys (O'Mara, Rolls, Berthöz, & Kesner 1997) and rats (Gavrilov, Wiener, & Berthöz 1995; Wiener, Korshunov, Garcia, & Berthöz 1995), we investigated specifically this region in a dedicated campaign (Vitte et al. 1996). Cold water was injected in the right ear during a visual fixation task and compared with visual fixation alone. Ipsilateral hippocampal formation activation (including the subiculum) was found in eight subjects and in three subjects three times. More recently, we have repeated this experiment with bipolar, binaural galvanic stimulation of the labyrinth in six subjects (Wiest et al. 1996). A direct electrophysiological study has also confirmed the existence of additional areas involved in vestibular processing at short latency (Beaudonnière, de Waele, Tran Ba Huy, & Vidal 1997).

The Head Direction Cell System

Vestibular signals may also reach the cerebral cortex and the hippocampus through a second route. A "head direction" cell system has been discovered by Taube, Muller, and Ranck (1990a, 1990b). First discovered in the post subiculum of the rat, these neurons fire whenever the head of the animal is directed toward a particular direction in space independently of where the animal is located in the room. These neurons are influenced by visual cues and their direction is closely related to the organization of the place cells in the hippocampus, which code location of the animal in space. Further studies (Taube 1995; Sharp, Blair, Etkin, & Tzanetos 1995) have revealed that this head direction information from the vestibular nuclei, through the anterior thalamus, the mamillary bodies, and the subiculum could reach the hippocampus after receiving visual environmental information from the parietal cortex, and contribute to the reconstruction at this level of spatial localization and orientation. At the present time only horizontal directions (in the plane of the horizontal semicircular canals) have been found in these cells. Thus the brain has at least two main pathways that seem to carry vestibular information: the PIVC route, which codes angular head rotation in many planes; and the head direction system, which codes static head direction in all horizontal directions. Our hypothesis is that vestibular patients have a specific deficit of the head direction cell system that leads to a deficit in the ability to store direction and lead, in the absence of vision or in the presence of conflicting visual cues, to a deficit in the evaluation of direction.

THE MEMORY OF ROUTES

Our locomotor tasks of blind walking along triangular or circular routes involves some aspects of motor memory combined with a spatial memory task. It is therefore interesting to mention recent work that has tried to reveal the neural mechanisms underlying spatial navigation tasks in humans.

In humans several strategies can be used to remember topographic routes. For instance, if one tries to recall the route from home to office or laboratory, the brain can use a survey strategy and try to imagine a map of the environment and mentally visualize the route on this map. The neural basis for this survey type of strategy has been studied in man (Mellet, Tzourio, Denis, & Mazoyer 1995). On the other hand, subjects can try to remember the sequence of turns and walks in relation to visual landmarks and eventually other cues or actions associated with the route. This type of memory can be subserved by several types of cognitive strategies (Amorim, Glasauer, Corpinot, & Berthöz 1997).

A clear dissociation has been found among these different strategies in patients with brain lesions. Several studies have demonstrated the impairment in topographic memory with a loss of the ability to recognize familiar landmarks (Incisa della Rocchetta, Cipolotti, & Warrington 1996; Whiteley & Warrington 1978), sometimes leaving the ability to describe routes with recognition of landmarks relatively unaffected (Pallis 1955; Paterson 1994). Sometimes patients can also describe routes and recognize landmarks, but they still lose their way because landmarks no longer convey directional information (Hécaen, Tzortzis, & Rondot 1980). The hippocampus is not necessarily the only area involved in these spatial memory deficits. Habib (1987), for instance, has shown that the lesions in patients with topographic memory loss were restricted to the parahippocampus and the subiculum but not extended to the hippocampus. By contrast, patient R. B., who had a severe anterograde amnesia following selective damage to the hippocampus, did not report getting lost in his neighborhood (Zola-Morgan, Squire, & Amaral 1986). The patient studied by Incisa della Rocchetta and colleagues (1996) had a severe deficit in describing familiar routes in her environment but no difficulty in identifying countries from outline maps and in naming cities within a country when they were identified by dots. Both episodic and semantic memory were impaired in this patient. The suggestion is that this patient has a specific deficit for category-specific knowledge of inanimate objects (e.g., hills, buildings). There seems to be a specific coding of topographic objects distinct from other classes of objects. There could also be a separation between the objects that require locomotion to reach them and other objects.

It seemed interesting to explore areas of the brain involving recall or a really executed locomotor route. We therefore investigated brain structures activated during a mental navigation task in which the emphasis was put on memory of self-motion associated with visual landmark recall during route navigation. The details of the experiments are given in Ghaem et al. (1997), so here we summarize the main features of this task. Subjects were driven to a totally unfamiliar urban environment and asked to walk along a previously selected route of about 800 m in the city. The subjects walked three times along the route and were asked to remember the route and in particular seven prominent visual landmarks (e.g., tower, gas station, phone box). The first two times subjects were guided by the experimenter and the third time they walked under supervision. Time of locomotion was recorded.

The day after this learning session and four to six hours before PET acquisition, the subjects executed two tasks that were repeated in the PET.

1. MSR task. In this mental simulation of routes condition the subject was instructed verbally to indicate the name of two landmarks, chosen by the experimenter, between which he or she was supposed to walk mentally. The subject pressed a button upon the mental arrival to the final landmark and the sequence was repeated with another set of two landmarks. This allowed us to measure the mental locomotion time and compare it with the previously recorded actual time in the real environment.

2. VIL task. In this visual recall task the subject was instructed to mentally visualize a landmark and keep it in memory upon hearing its name through the earphones.

The results indicate first that there was a strong correlation between the time to walk mentally between two segments and the real time of locomotion in the city during the training (.99, $p = 0.0003$) and the PET sessions (.92, $p = 0.024$), suggesting that subjects were doing the required task.

The following main areas were activated during these tasks when compared with the rest condition. During the MSR task we found bilateral activation of the dorsolateral cortex, posterior hippocampal areas; posterior cingulate gyrus, supplementary motor area; right middle hippocampal areas; and left precuneus, middle occipital gyrus, fusiform gyrus, and lateral premotor area. During the VIL task we found bilateral activation of the middle hippocampal regions; left inferior temporal gyrus, left posterior hippocampal regions, and precentral gyrus; and right posterior cingulate gyrus. When these two conditions were subtracted (MSR minus VIL) we observed only activation of the left hippocampal regions, precuneus and insula.

IMPACT OF THE FINDINGS

More work is obviously needed to understand the neural basis of anticipation and dissociation between the coding of distance and direction during locomotor navigation. We would, however, like to propose that these findings have some bearing upon the general problem of spatial disorientation. Anxious patients often complain of sensations of floating in space or distortions in their evaluation of distances. Dizziness is one of the key symptoms of panic. For this reason, many studies (Jacobs, Moeller, Turner, & Wall 1985; Swinson et al. 1993) tried to find a vestibular dysfunction in panic and agoraphobic patients, but none of them managed to demonstrate an obvious interrelationship between panic disorder and vestibular function. We propose that their vestibular disorders may not concern low level vestibular functions such as those explored so far (e.g., vestibulo-ocular reflexes), but more generally higher cognitive integration of vestibular self-motion signals to spatial cognition involving structures such as the hippocampus and parietal cortex involvement in spatial memory. The fact that such disorders can be attenuated after active rehabilitation with optokinetic stimulation has been taken to support this interpretation (Tsuzuku, Vitte, Sémont, & Berthöz 1995).

This chapter demonstrates that guiding locomotion during spatial navigation is probably performed not with a continuous intake and treatment of multisensory information but from an internal predictive simulation (Berthöz 1997; Droulez & Berthöz 1988) of the desired trajectory using mental mechanisms such as those described in chapter 6 of this volume (see also Amorim et al. 1997). In addition, if, as suggested by our results, distance and direction are coded separately (a property also thought to be used by the brain for arm reaching), in order to be able to navigate adequately one must be able to bind these two pieces of information and reconstruct a coherent representation of the orientation and position of the body in space from several fragmented pieces of information. This construction of a coherent perception is, in our view, what may be impaired in these patients and more work must be done to fully understand the neural and cognitive basis of these disorders and their potential relation to distress in vertiginous patients and those with agoraphobia (Yardley, Britton, Lear, Bird, & Luxon 1995; Yardley & Putnam 1992).

ERIC CHOWN

14

Error Tolerance
and Generalization
in Cognitive Maps

PERFORMANCE
WITHOUT PRECISION

COMPUTATIONAL LIMITATIONS

OPTIMALITY VERSUS PRACTICALITY

One of the lessons of artificial intelligence (AI) has been that many tasks that seem very simple to humans, such as recognizing one's mother, are actually extraordinarily complex computationally. In the 1960s and 1970s artificial intelligence became known more for what it did not achieve than for what it did. Instead of building robots that were smarter than humans, owing to their superior computational abilities and permanent storage mechanisms, progress in AI was more modest, leading some to conclude that machine intelligence was an impossibility (Dreyfus 1972). During the 1980s there was a growing understanding of the nature of computational intelligence and why it is so difficult to achieve. Results in representational theory (Levesque & Brachman 1985), machine learning (Kearns & Valiant 1994), planning (Chapman 1987), and other areas indicate that many of the basic problems that artificial intelligence researchers would like to "solve" are intractable. From one point of view this could be taken as more evidence that machine intelligence is an impossibility; from another point of view it can be seen as a sign that "intelligence" means something far different than being able to achieve some measure of optimality.

During the same time period psychology was also influenced by the computer revolution. The temptation to think about intelligence in terms of rational models became strong because such models could be implemented and tested on computers. But while computer scientists were discovering that optimal solutions were not possible for many common problems, psychologists were finding that human beings behaved in ways that were far from "rational" (Kahneman, Slovic, & Tversky 1982; Nisbett & Ross 1980). The notion that humans are not rational, or optimal, is not new. Simon coined a term for what humans do—*satisficing*. This term suggests that for humans a good solution is usually enough (Simon 1969). The negative results in various disciplines of artificial intelligence indicate that satisficing may in fact be the *best* solution in many domains. For example, an optimal solution to the traveling salesman problem is not feasible for any computational system given any kind of time constraints. This does not mean, however, that artificial intelligence is a dead end or that problems such as the traveling salesman problem cannot be addressed computationally; it does suggest, however, that brute force computation may not be the answer to many practical problems.

During the 1990s many of the successes of AI research have come as the result of shifting from working on optimal solutions to finding good solutions. For example, Zhang and Dietterich (1995) have worked on job scheduling for the space shuttle, a task for which there is no computationally feasible optimal solution. Their system operates by using heuristics to build reasonable solutions and then improving and refining these solutions as much as reasonably possible. Because optimality is not possible the goal is to find good solutions that satisfy the problem constraints. As the field of robotics has begun to mature it, too, has begun to reflect this insight, and the representations used by robots have grown much more qualitative in nature despite the robotic capacity for precision.

One computational tradeoff that makes optimality impossible for real world problems involves the expressiveness of a representational language versus the tractability of its use (Levesque & Brachman 1985). In short, the more it is possible to encode in a representational system, the more difficult it is to be able to use that system efficiently. For example, first order predicate logic can be used to express any collection of information, but it is this very expressiveness that guarantees that its use is intractable. Predicate logic was abandoned in AI planning systems because it was discovered that any operator that affects the world, such as a robot moving, must be described not only in terms of what changes, but also in terms of everything that does not change. This means that any representational update requires re-evaluating the entire system; this is known as the frame problem. For representations of reasonably large size even a simple change requires more computation than can be efficiently managed.

In trying to understand human cognitive maps, it is useful to keep in mind the lessons of artificial intelligence. The world is sufficiently complex that optimality is not possible. If our internal representations of space were as detailed as the real world their usage would be intractable. Therefore cognitive maps must satisfice by balancing the tension between storing as much useful information as possible against the need to keep the amount of information at a manageable level. Because cognitive maps by their very nature cannot be perfect reproductions of the world, they must have shortcomings. Fortunately many of these shortcomings are limited in scope—indeed humans are remarkably efficient at navigating without the aid of external instruments or tools—and have only come to light with the advent of modern needs.

COMPUTATION AND COGNITIVE MAPS

Representations of space provide an excellent example of the difficulties inherent in trying to create and maintain precise, comprehensive representations. There are several important reasons why qualitative representations of space may ultimately be more useful and efficient than metric representations. Among them: (1) The world is constantly changing, (2) maintaining comprehensive representations is expensive, and (3) all systems are prone to error.

The first reason is simple: The world is constantly changing. Representations that are qualitative require far less maintenance than precise ones. A precise representation must always be fully updated with the most recent information if the precision gain is to be useful. Such maintenance can be costly and inefficient. A representation which only captures details at a schematic level, on the other hand, only needs to be updated when major changes occur. Further, if the world changes often enough the information stored at any given time is unlikely to be accurate.

The second reason is closely related to the first. A system keeping a complete and precise representation of the world must track every part of an environment. Such tracking necessitates studying every aspect of the environment each time it is entered, and comparing the current state of the environment to previous encounters. By contrast, in most environments humans will probably only store the locations of a few objects and only generalized locations at that. Maintaining such representation is simple and only requires gross comparisons for any given experience within the environment. But with extended experience in an environment, humans do have the capacity to learn detail. Even when they do, however, many details are still lacking; simple features like the number of tiles in an office floor or the grain pattern on a wooden door, which could be easily retrieved from a photographic-style representation, are virtually impossible for most people to recall.

Finally, perhaps the most important reason why qualitative representations of space are superior to precise ones has to do with errors. Robots and humans alike are prone to errors; in robots wheels slip, sensors give noisy readings, and so on. Errors cannot be avoided, whether they are due to the agent implementing the plan or to external forces like an unexpected construction zone that forces an alternate route to work. And, because errors cannot be avoided, it is important for any agent operating in the world to minimize the number of errors and the impact of any given error. Consider the difference between an imprecise set of directions and the types of direction that might be given by a robot with a more quantitative representation: (1) "Go straight until you come to a river. Then follow the river to your left until you come to a big tree with red leaves. When you get there look over to your left, you'll see it next to the brick wall." (2) "It is 507.5 meters at a heading 45.78 degrees from where you are." The first set of directions is hopelessly imprecise and does not even give a real feel for how far away the object in question is. The second set is exact. The first is relatively simple to implement and does not require more than the ability to recognize a few simple landmarks along the way. The second set of directions necessitates accurately keeping track of position over arbitrary terrain. Even a small error in initial heading could lead to a large error in final position. Ideally a cognitive map should capture some of the flavor of both sets of directions, neither relying completely upon environmental feedback, nor on the precision of what is internally stored.

The example is instructive with regard to system design. The first set of directions is imprecise and relies on the ability to find and recognize objects in the world. The second set of directions is precise and relies only upon the ability of the agent to successfully implement the directions. To some degree this can be simplified to say that the first set of directions relies on interaction between the agent and the environment, and the second set relies strictly upon the agent. Any internal model of an environment is still only a model, no matter how precise, and is subject to errors as the environment changes. On the other hand, the environment is always a perfect model of itself and therefore, when available as an information source, is generally superior to an internal model. Cognitive maps are essential when planning because only a fraction of the environment may be observable at any given time, but they must necessarily fall short of the real world when it comes to providing accurate, up-to-date information. Ideally, therefore, cognitive maps should contain enough information to make effective plans, but this information does not need to be highly detailed, because the environment can normally be relied upon to provide detail when plans are put into action.

In discussing information processing, Clark (1989) described the "007 principle," whose name reflects the fact that adaptive organisms can only afford to store what they "need to know." Organisms that store too much information are at an adaptive disadvantage because extra information requires extra processing and can also lead to confusion. In a potentially hostile world, survival can depend on quick decisions; too much information can easily slow decision making. Organisms that store only what is necessary can process that information far more efficiently.

In navigation the environment can usually be relied upon to provide any necessary information as plans are implemented so organisms do not usually need detail. Cognitive maps are most useful for planning, at which time the environment might not be available or only partially available. In implementation there is no substitute for the real world; no model can ever hope to match the accuracy or the amount of information provided by the real world. Nevertheless, internal models can be useful at implementation time, especially when conditions are less than ideal, such as at night or during fog.

Arbitrary precision and representational abilities are not practical in any feasible real world system, man or machine. Therefore such agents must adapt their representations to provide the information which is most likely to be relevant to the tasks they are likely to perform (or as Clark puts it, that they need to know). Human cognitive maps are structured on this basis, emphasizing some information at the expense of other data. Further, the organization of human cognitive maps is such that information is readily available in a useable format. All of these computational issues provide useful constraints when considering the underlying structure of cognitive maps. Structures that are exceptionally complex or precise are unlikely because of the information processing costs they necessarily bring. It is far more likely that cognitive maps are relatively simple structures that can be used to acquire useful information quickly.

COMPUTATION AND HUMAN COGNITIVE MAPS

If the computational problem of effectively representing large-scale space is intractable, then it is reasonable to question why humans have success. First, it is worth recalling that the tractability question is highly generalized. It is true that it is hopelessly expensive computationally to store and process every piece of information in an environment, but such extensive actions are not necessary. For tasks such as navigation success is possible using only a fraction of available information. Given this fact a good computational strategy for dealing with new environments is to start by learning what is necessary for some reasonable level of performance, and then with experience gradually increase the amount of knowledge about the environment. Such a strategy allocates resources according to the time spent in individual environments. This is the basis of the theory behind the prototypes, locations, and associative networks (PLAN) model, that cognitive maps are acquired in stages. At each stage of development more information is added to the cognitive map, and with the new information comes additional functionality. In this way cognitive maps are able to balance the computational costs associated with the acquisition of information against the need to have such information available.

The PLAN model is based upon several computational principles: (1) The storage of information is based upon the likelihood that it will be useful in the future, (2) the organization of stored information is analogous to the way in which the information was acquired, and (3) there are recurring spatial patterns in the world that cognitive maps can exploit. All three of these principles have consequences with regard to human performance and will be discussed throughout the chapter. The rest of the chapter is structured according to the manner in which information is built up in cognitive maps. The human visual system, which is the primary source of spatial information, is split up into two halves, the *what* system and the *where* system (Lesperance 1990; Rueckl, Cave, & Kosslyn 1988; Ungerleider & Mishkin 1982). The primary function of the *what* pathway within the visual system is object recognition, whereas the *where* system is mainly involved in determining spatial relationships and characteristics. Essentially the *what* system identifies the individual objects that comprise the world, whereas the *where* system tracks their locations relative to the observer. For cognitive maps to function effectively this information must be synthesized efficiently.

In PLAN, *what* information is integrated into cognitive maps more quickly than *where* information. This is a reasonable supposition because it is necessary to know what the elements of an environment are before putting them into an organized structure. Further, *what* information can be used to quickly build cognitive structures that can be used to effectively navigate in most environments. Although these systems develop in parallel and at different rates, the cognitive system is able to integrate information from both systems into a smoothly functioning whole.

THE "WHAT" SYSTEM

In learning new environments, probably the first information people acquire is about the objects that comprise the environment. Representations of these objects, called landmarks, serve as the fundamental units of cognitive maps. Landmarks are important both because they serve as markers that let one know where one is in an environment, and because they tend to be the places that people need to go to. Because landmarks are so important in navigation and in so many other cognitive tasks, humans have developed a highly efficient object recognition system that comprises one of the visual pathways. Recognizing a familiar landmark is enough to make the difference between being lost and knowing where one is. Further, the ability to recognize an object despite changes in orientation, lighting, and so on, greatly simplifies the computational burden of the rest of the navigation system. A system with a fast, efficient pattern recognition system need not rely upon having the locations of objects specified exactly; "over there" is as good as "30 meters at an orientation of 125 degrees" because the visual system can find and identify objects extremely quickly.

The primary functional use of landmarks in navigation is as a kind of spatial index. Rather than organizing space in terms of some type of coordinate system, in the first developmental stage of cognitive maps humans instead organize space in terms of the objects that make up the world. So instead of specifying location in some absolute coordinate system such as latitude and longitude as a robot might, humans are more likely to think about space in terms of familiar landmarks such as their houses or places of work. When a spatial representation consists of collections of such landmarks a coordinate system is not necessary. With experience, humans can develop internal representations that capture some of the same information that would come from an absolute framework.

A highly developed pattern recognition system enables schematic representations of space that are still highly functional. There are many computational benefits that can be derived from schematic systems. Storage is reduced, so retrieval is simplified. Because stored structures are more compact and contain less information, processing is reduced. Further, the pattern recognition requirements are no greater for a schematic system then for a system that is more precise and detailed. In practice the pattern recognition burden can even be reduced because a detailed system requires extra computation to constantly maintain its structures. It should not be surprising that a schematic system is computationally cheap compared to one rich in detail. What is surprising, however, is that the schematic system loses very little in functionality despite the comparative lack of information. Because human pattern recognition is so efficient there are very few times when detailed information is necessary in navigation; indeed, schematic plans tend to be more robust and error resistant. Although it is true that in the modern world humans often require precision that cognitive maps are not capable of, for example knowing exactly where one's property begins and ends, for the basic navigational tasks of getting from one place to another and occasionally telling someone else how to do it, precision is not generally necessary or even useful. Further, in today's world we have instruments and machines that can provide us with the precision we ourselves are not capable of.

The object recognition system must necessarily be heavily invested in generalization. Just because one is seeing one's mother from a different angle than before does not mean that one should not recognize her. Landmarks must be recognized despite changes in viewing orientation, distance, lighting, and so on. It should come as no surprise that this ability comes with a price; it is often hard to tell the difference between two oak trees, two streets in a residential neighborhood, or two buildings in a modern apartment complex. Cognitive maps directly reflect this tension between the ability to discriminate and the ability to generalize. This has been studied by psychologists since James (1892), who framed the discussion in terms of *synthesis* and *analysis*. In the modern era there is a large literature on classification in both psychology and machine learning. In machine learning, where the distinction between discrimination and generalization can be studied computationally, the dichotomy is between *overfitting* on the one hand and *overgeneralization* on the other. Systems that overfit or discriminate at too fine a grain have a difficult time applying what they have learned previously to new situations, whereas systems that overgeneralize tend to overapply what they have learned. The difference is somewhat like a child learning what a dog is and calling every animal he or she sees a dog, and a child noticing that every dog has different features and not recognizing that there is a general classification covering all of them. Despite extensive research on classification, the mechanics of how humans generalize and discriminate are still not well understood and it remains one of the most important problems in cognitive science. Although the mechanisms of classification are still debated, a great deal is known about behavior, and a number of models have been proposed to account for that behavior. The classification scheme in PLAN combines elements of the prototype construct first proposed by Rosch, Mervis, Gray, Johnson, and Boyes-Braem (1976) with a modernized version of Hebb's cell assembly model (Hebb 1949; Kaplan, Sonntag, & Chown 1991; see also Chown, Kaplan, & Kortenkamp 1995).

Landmarks

One way in which cognitive maps are able to avoid the discrimination-generalization problem is by heavily investing in uniqueness, particularly with regard to landmarks. Appleyard (1969), for example, found that singularity along every dimension he measured (movement, contour, size, shape, etc.) was the critical factor that made buildings identifiable landmarks; for example, a tall building on a block full of smaller buildings is a likely landmark. This is one answer to the question of what one needs to know about an environment. Unique objects are worth learning about because they are not easily confused with other elements of the environment. In terms of representing space, a unique object can serve to efficiently index the representation. In this regard context is an important indicator of whether or not an object is a useful spatial index. One tree in a forest is easily confused with other trees, but a tree in a grass field stands out. In computational models of cognitive maps such as NX (Kuipers & Byun 1991) and the robot implementation of PLAN, R-PLAN (Kortenkamp 1993), landmarks are determined by comparing an object to surrounding objects and determining how different its features are from its neighbors. This is a simplification necessitated by the limitations of computer pattern recognition, but it captures the essence of what happens.

Because landmarks provide the access points to cognitive maps it follows that in environments without many landmarks humans would have a great deal of difficulty, or would fall back on other navigational methods, such as dead reckoning. Probably the best study of such a situation was done by Gladwin (1970) on the navigators of the Puluwat Islands. Gladwin showed that even though the ocean is apparently devoid of landmarks (or seamarks as the case may be) the Puluwat navigators still relied almost exclusively upon landmark-based navigation. The reliance upon landmarks had a cost, however, making navigation a highly specialized skill requiring years of training to be able to recognize the complex signals that made up the seamarks. This is a key point with regard to cognitive maps, because cognitive maps are structured according to a number of assumptions about the world, such as the fact that it is made up of landmarks.

Environments that do not match the assumptions closely will be difficult ones for humans to operate in efficiently. Gladwin's study shows one clear example of this. If human navigation were more heavily invested in dead reckoning then navigating in the Puluwat Islands would not require such specialized skills. The commitment to any representational structure automatically brings with it advantages and disadvantages, but given the nature of evolution, it can be assumed that the disadvantages of cognitive maps are minimal with regard to environments that humans typically interact in.

A further problem with landmarks is that they do not necessarily occur in useful locations. One consequence of this is that people tend mentally to move landmarks in order to make them more useful. A typical example is considering something to be next to a landmark when it is not, or directions such as "turn left at the gas station," when the gas station is not on the corner. Such errors occur because landmarks can compactly represent a great deal of information. "Turn left at the gas station" might not be entirely accurate, but it is simple and one can normally infer the true meaning when actually going past the gas station. In some cases the true meaning is not obvious, however, such as when there are many gas stations or the person receiving the directions is not familiar with what gas stations are; in such situations the schematic nature of the representation is a disadvantage.

In summary, landmarks are one of the most basic elements of cognitive maps. Their primary function is as a kind of spatial index that can be used as a substitute for an absolute frame of reference. Furthermore, it is important to learn about landmarks, because they also tend to be the places in the environment that are important because they are often visited. The question remains as to how landmarks are used in navigation. The scheme in PLAN follows the developmental literature (Piaget & Inhelder 1967; Siegel & White 1975), proposing that objects are organized according to their proximity to each other in structures called route maps.

Route Maps

Route maps draw their name from their apparent organization. For example in describing the neighborhood between her house and her school a young child typically will draw a map by starting at her home and filling in places as though walking from home to school. Her cognitive map is organized along a path in which one landmark follows another. This connectedness is a natural way to organize landmarks because it mirrors the way information is acquired and the way it will be used. In locomoting through an environment one will encounter one landmark after another.

An organization that simply connects landmarks based upon their sequentiality captures significant spatial information without the need for transformations or extraneous cognitive structure. By contrast, building an absolute structure requires determining exactly where the landmarks fit into the absolute framework, a process requiring precise information and significant processing.

As many sequences of landmarks are experienced they will intersect (for example when two different paths contain a common landmark) and the representation will form a kind of network where the nodes are landmarks (figure 14.1). One side effect of the network representation is that even paths that have never been experienced can easily be extracted. In figure 14.1, the path from church to work may never have been taken, but it is easily retrieved from the network structure. Thus behaviors such as path integration and finding shortcuts naturally arise from the structure of the internal representation.

Topological representations of space such as route maps do not contain the full range of spatial relationships, rather they are limited to proximity and order. Fortunately, proximity and order are sufficient for planning a journey. Representing the world as a topologically structured collection of landmarks makes route planning simple: A route consists of going from one landmark to another until the goal is reached. Virtually every computational model of cognitive maps is based upon a topological structure. Examples include TOUR (Kuipers 1978), Navigator (Gopal, Klatzky, & Smith 1989), Traveler (Leiser & Zilbershatz 1989), NAPS (Levenick 1991), and PLAN (Chown et al. 1995). Many of these systems use a version of spreading activation search to find routes between locations. Computer simulations of such systems have been found to be comparable to human navigation (O'Neill 1990, 1991). The spread of activity from one node to another can be thought of as a sort of imaginary journey in the head (except that it is done in parallel over many nodes) and is inspired by the ways neurons and synapses function in the brain. When one of these imaginary journeys

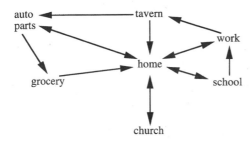

FIGURE 14.1. A network representation of a neighborhood. Any two points in the network can be connected by a path, even if all possible paths have not been traversed.

stretches from start to finish a complete plan can be retrieved. Such a plan also has the benefit of being simple to implement, taking the form: first go to A, then to B, . . . This is exactly the form in which the child imagined her neighborhood: by thinking of going from one place to the next. In terms of implementation of the plan, all that is required is the ability to recognize each landmark along the way. Contrast this with the difficulty of extracting and implementing plans from overhead maps. Such maps are rich with metric detail, but often are lacking in practical information. This is a clear example of how storing less can be beneficial. Having only a few landmarks stored makes path extraction simple; the more information that is stored the more need be processed.

One useful advantage of a topological representation of the style modeled in the NAPS and PLAN systems (Chown et al. 1995; Levenick 1991) is another reflection of the functions of nodes and synapses, that is, that the links between nodes can be variably weighted to capture experience, among other things. One of the oldest learning rules for neural networks, often credited to D. O. Hebb but going as far back as William James, postulates that when objects are experienced together associative links between their representations will be strengthened (Hebb 1949; James 1892). For example, landmarks seen for the first time might only be weakly linked into a cognitive map, whereas landmarks that have been seen many times can be strongly linked. There are a number of advantages to such a scheme. In a spreading activation system well-learned paths will be highly preferred over less well-learned paths because the links between the nodes representing the well-learned path will be stronger, and will spread activity more efficiently. This places an automatic premium on experience and safety because using familiar routes is less likely to result in getting lost or running into unknown (and therefore dangerous) situations. Further, such a scheme can naturally capture the changing nature of real environments. As items come in or drop out of environments links can be added or will naturally disappear, as the case may be. This scheme acquires information incrementally and automatically places a premium on safety and the changing nature of real environments. By contrast a system insensitive to its experience with an environment might not be capable of differentiating permanent objects from objects that are only temporarily in an environment. The alternative to a variably weighted links scheme is to use some sort of bookkeeping system to track the uncertainty inherent in real environments, a process that requires extra computational machinery to store and update all of the factors involved.

THE "WHERE" SYSTEM

Landmarks and route maps represent the first two stages of how spatial information is acquired in cognitive maps. Not coincidentally these stages match the developmental stages identified in children's conceptions of space (Piaget & Inhelder 1967; Siegel & White 1975). The third developmental stage in a child's conception of space also corresponds to a developmental stage in PLAN. In this stage spatial knowledge goes beyond what is captured in a topological network and is expanded to include relative locations. Although topological networks facilitate navigation with relatively little computational cost, they do not account for the full spectrum of spatial relationships. The primary knowledge lacking in a topological network is the relative positions of the objects that make up the network. Having such information available is useful in navigation because it alleviates the need for visual search at each stage in a journey. The developmental term for structures that contain this knowledge is *survey maps* (Siegel & White 1975). Whereas a route map can be represented as a simple linked collection of landmarks, survey maps generally are considered to be more abstract and directly visual in character, taking their name from the fact that a survey representation appears to give one the ability to imagine an environment as though looking at it from a survey viewpoint, above it and off to one side.

The sequential structure of route maps is an analog of one mode of information processing; as humans walk through an environment, or look from one landmark to another, information comes in a natural sequential flow. In contrast, the structure of survey maps reflects a different form of information processing. Typically humans learn about the relative locations of objects by viewing a scene. In the visual system the locations of objects relative to the observer within a scene are processed by the *where* system, located mainly in the posterior parietal and superior colliculus portions of the brain. The information the *where* system processes is much less fine-grained than that processed in the *what* system, which consists of general shape information, texture, and other spatial components. Further, this information is egocentric in nature, because, after all, that is the way the information comes in through the sensory modalities.

In PLAN, where the landmark serves as the basic unit of the route map structure, the scene is the basic unit of the survey map structure. Storing location information as scenes has the advantage of simplicity: no transformations are necessary, information is stored in the format in which it is processed. To know where landmarks are relative to each other in a particular environment, one need merely recall an appropriate scene to re-create relative positions (figure 14.2). This egocentric scheme is attractive because it maintains information in the same format in which it was used. By contrast consider robotic systems that attempt to create absolute models of the world; to succeed they must continually transform the egocentric information they receive into some absolute framework. Further, to put that information to work, they must in turn be able to transform it back into an egocentric format. A key assumption of PLAN is that such transformations rarely need to be made. If one's route from home to work is the same almost every day, then having an absolute representation requires unnecessary processing. The basis of the survey structure found in PLAN is that for environments where one is likely to need information about the relative location of objects, one is likely to need that information in a format analogous to the way in which it was acquired. Planning a journey, for example, typically consists of piecing together pieces of

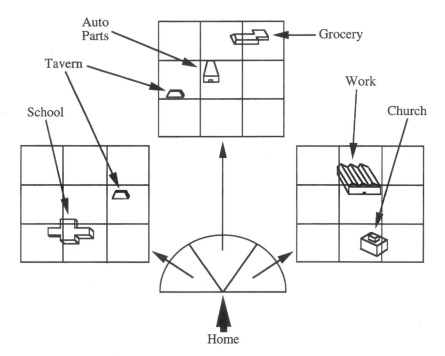

FIGURE 14.2. A local map. A single local map consists of a collection of stored scenes. Each scene corresponds to what can be seen with the head in different positions. In this local map the location of the grocery store relative to the auto parts store is easily determined by recalling the appropriate scene.

journeys previously taken. It is simpler to think of journeys already taken then it is to try and pull a plan out of an absolute representation.

In order to emphasize the egocentric character of the scene construct in PLAN these representations are called *local maps*. Local maps, like landmarks, are connected together in a network that corresponds to the sequential manner in which they are experienced.

Gateways

If the basis of the survey map representation is the scene, there must be a process to determine when pertinent scenes are stored. After all, when locomoting through large environments information comes in a continuous flow. Trying to store scenes at all possible locations is neither practical nor possible. An intermediate solution used by some mobile robot systems has the robot take pictures in fixed intervals, forming a kind of snapshot grid (Asada, Yasuhito, & Tsuji 1988). Such a scheme is still not very efficient, however, requiring snapshots even when there is no new information and not necessarily getting scenes that maximize what can be seen or that occur at useful locations. With the PLAN model it has been proposed that the basis for storing scenes is the usefulness of the location. The major criteria for usefulness in this case are places that are visited often and places where new visual information becomes available. A perfect example of this is a doorway that allows new information to be seen, generally affording a view of an entire room simultaneously, and is visited each time the room is entered. In PLAN these locations are called gateways after Christopher Alexander's design construct of the same name (Alexander, Ishikawa, & Silverstein 1977). Natural gateways include mountain passes, entrances to caves, or the edge of a forest. Manmade environments abound with gateways. Gateways are useful places to structure spatial representations because they break the world up into smaller pieces and they tend to be places that are visited time and time again. For example, it is not possible to enter a room without passing through a gateway—the doorway.

Gateways have a number of computationally advantageous properties. First, gateways are places where people tend to pause to look around because of new information available. This pause is useful when acquiring information because it means extra processing time. Second, in terms of storing visual information efficiently, gateways are ideal locations because they occur where there is hidden visual information that is now in view. This is an excellent heuristic for minimizing the total number of locations that must be stored in order to ensure comprehensive coverage of an environment. Third, gateways are easily and accurately identified, even for robots with limited

pattern recognition abilities. For example, mobile robots are able to identify a gateway and repeatedly stop at the gateway on subsequent visits with
an error of no more than 70 mm in position and 3.5 degrees in orientation (Kortenkamp & Weymouth 1994). Such accuracy is critical for a system that uses scene-based representations because such a system must be
able to resolve what is currently being seen against stored scenes. If the current view is identical to a stored view then the comparison becomes triv-
ial. Furthermore, this precision comes without a price: the ability to accurately
find a gateway is not determined by stored patterns, but comes completely
as a result of information afforded by the environment at any given time.
It is the fact that gateways are repeatedly and precisely visited that allows the
elements of the survey map to be egocentric in nature.

Reliance upon gateways does come with a price. Building a representation out of a collection of egocentric views means that some types of computation will be more efficient than others. One problem of this nature
involves people trying to shift their perspectives to new ones not previously
stored; humans familiar with an environment tend to have a strong preference for their stored points of view. When looking at a map, for example,
some orientations are highly preferred over others (Levine, Jankovic, & Palij
1982). In communicating with others this reliance upon a particular orientation can be detrimental because in conversation it is preferable to take the
perspective of the person one is speaking to (Schober 1993). These sorts of
problems are artificial, however; cognitive maps did not evolve in a world
of cartographic maps, nor are people likely to have to give directions to a
place using orientations they have never experienced. The "price" of a view-
point-based spatial representation is probably a modern problem, that is,
one not directly relevant to the kinds of issues that cognitive maps were
developed to deal with. One real price of gateways, however, is that, as with
landmarks, their usefulness relies upon certain environmental structures existing with some regularity in the world. It is difficult to navigate in areas without them, such as oceans. The example of the Puluwat navigators shows
that even in such cases, people with experience can be surprisingly clever
about recognizing patterns that are not apparent to the casual observer.

Abstracting Scenes
Although scenes represent a significant step beyond the information contained in a topological network, they still do not account for the entire spectrum of human spatial abilities. Fortunately the scene construct is easily
generalized into structures that add yet another level of functionality to
the cognitive map.

The survey maps in PLAN are called *regional maps* and have the same basic structure as the local maps. There are times when distant objects can be seen in relationship to each other, as when observing from a height like a hill; such a view affords uniquely valuable information. By looking from an oblique viewpoint landmarks do not obstruct each other; thus their relationships to each other can be observed simultaneously. Such oblique viewpoints are not always available of course, but the conjecture made in the PLAN architecture is that humans are able to construct representations as though they had experienced them nevertheless.

The network of local maps gives rise to the development of regional maps. As people move through an environment they will activate their corresponding cognitive structures, in this case local maps. Because these local maps are connected in an associative network, when one local map is being processed it will automatically tend to predictively activate the next local map even before the corresponding gateway is reached. For example, "around this corner I would expect to see. . . ." As familiarity with the region grows the predictions become stronger, faster, and more accurate. In time the predictive effects of association will begin to be taken into account in the current local map, particularly if one pauses at a gateway. If one mentally runs through the next part of a journey, a larger structure containing not only the current local map, but parts of neighboring local maps can be created. Such a representation would contain information beyond that which can be seen by taking advantage of the predictive power of a network structure.

At this point it is important to remember that the local map concept was based upon the idea that what should be stored is what can be seen from a particular vantage point. In the regional map case, however, information is stored that cannot be seen from the stored viewpoint. This could potentially present a problem in the use of the structure. For example, if one landmark is directly behind another then they would occupy the same location in the visual field. Thus when the landmark that was behind is added to the local map it would occupy exactly the same portion of the field of view and would appear to be in exactly the same location. Fortunately, there is a simple solution to this problem. If something farther away is thought of as being further "up" or "out" in the visual field, then more distant objects will be placed farther than previous objects on the periphery of the local map structure. Such placement represents a distortion of the true visual field at the location in question. The distortion can be resolved by considering the regional map to occur at a new point directly above the original point, corresponding to an oblique view of the mapped region. Taken further this

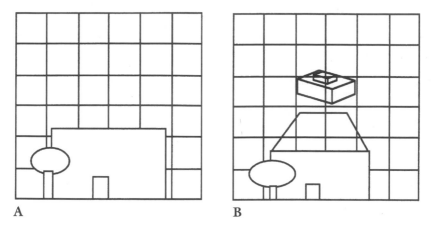

FIGURE 14.3. Two views of the same scene. In **A** the scene is viewed straight on and the second building is obscured by the first. In **B** the view is oblique and the second building can be seen "above" the first in the picture plane.

means that as the area covered by the regional map expands, the perceived "height" of the map will rise (figure 14.3). Thus an oblique viewpoint emerges as one becomes more and more familiar with a large-scale space.

Regional maps are not simply local maps that grow larger and larger as the environment becomes more familiar. The primary function of regional maps is in the planning process. As trips grow long the number of connected elements that must be extracted from the cognitive map grows correspondingly. Managing these long chains can be confusing and would place a high load on short-term memory. On the other hand, a high-level plan of a trip might consist of only a few key landmarks. Each stage of the high-level plan can then be broken down into a smaller plan and the process repeated. Such a scheme is in accord with the well-known study of taxi drivers done by Pailhouse (1969). In this study taxi drivers appeared to have divided the city up into smaller regions. When traveling to a new region they first went to a standard point in that region before proceeding to their ultimate destination. At the high level the city seems to just consist of a few regions and one particular point in each region, but each region in turn has a more detailed representation. Such a framework is also useful in dealing with breakdowns in plan execution. A bridge that is out will not undermine the entire plan, but only one section of it. With such a structure, therefore, a highly reduced representation not only increases efficiency, but robustness as well.

The reduction in information in regional maps comes as the result of the same connection process as is found in route maps. If certain landmarks are used repeatedly in plans then they will become part of the regional map as their connections are strengthened to the high-level structure. Landmarks that are rarely used will fade from the representation just as transient objects fade from route maps. The landmarks that will remain in a regional map will be those that are used over and over again in the context of the large-scale space being represented. Just as with route maps, this process requires no additional overhead or bookkeeping to manage.

Regional maps function within the hierarchical structure of the larger system. Thus, regional maps can be used to generate larger regional maps. The process involved would be virtually the same as in generating regional maps from local maps. There are two major advantages afforded by regional maps. First is the added ability to do hierarchical planning of the type just described. Regional maps can also be useful in performing certain types of spatial reasoning. Because the spatial relationships of distal objects can be "seen" it is possible to some degree to determine whether a certain path is spatially efficient or whether it wanders too far in any direction. These structures remain egocentric in that the information is still organized with respect to a fixed viewpoint, but the collection of them can be apparently allocentric as more information is available than could normally be captured from an actual view, and as one can shift among multiple views of an environment as necessary.

CONCLUDING REMARKS

The PLAN architecture is a description of how cognitive maps can face the computational challenges associated with operating in an information-rich world. The structures of PLAN are simple, reflecting the manner in which information is processed, therefore all but eliminating the need for complex transformations. The information contained in these structures is schematic, reducing the storage demands, while still being extremely functional. Finally, the architecture is incremental; information is added according to experience and the probability of future need. Each new level of the developmental hierarchy of PLAN adds functionality to the previous structure without adding considerable computation. The structure of PLAN is simple enough that is has even successfully served as the architectural basis for an actual mobile robot (Kortenkamp 1993).

Naturally the architecture of a system has consequences, good and bad, on its functioning. As described in PLAN, cognitive maps are schematic and potentially full of distortions. Further, smooth functioning of such a representation requires certain environmental patterns, such as frequent landmarks and gateways. In environments without such patterns, navigation should prove to be difficult at best. There is good evidence that all of these things are true in humans. Human cognitive maps have been called "sketchy" and are known to contain distortions and errors (for reviews, see Byrne 1979; Golledge 1987; Passini 1984b). Human conception of distance, for example, appears to be a product of the number of landmarks between two locations rather than metric distance; as a consequence, people overestimate distances in landmark-dense areas such as cities (Byrne 1979). Such a conclusion supports the topological map underlying PLAN. Humans also tend to represent all directions in increments of 45 degrees, converting 80-degree angles to 90-degree angles, for example (Byrne 1979). Further, Gladwin (1970), in studying the Puluwat navigators, concluded that even when landmarks are few and far between, people find ways to use landmark-based navigation. Such behavioral artifacts do not come as the result of some flaw in cognitive map structure, but rather are part of a natural system designed to cope with the computational demands placed upon them. Issues such as whether humans can accurately judge metric distances are nearly meaningless, because cognitive maps evolved in a context where such judgments were not necessary.

The key issue in terms of efficiency is whether the adaptations are worth the gain in computational efficiency. Trends in robotics would suggest that they are. Robots developed by researchers such as Brooks (1985), Kortenkamp (Kortenkamp & Weymouth 1994), Mataric (1992), and others have sacrificed "advantages" held by robots over humans, such as the ability for precise metric calculations and exact storage of information, in favor of more qualitative representations. Brooks (1985, 1991) has repeatedly and persuasively pointed out that storage of information is often counterproductive when that information can easily be retrieved directly from the environment, even for a perceptually impoverished agent such as a mobile robot. The world is complex enough that any agent, even a computer, cannot thrive by relying upon brute force computation. However, there is enough structure in the world that by intelligent application of computation it is possible to survive and thrive in a wide variety of environments.

Computers and robots are helping to bring a new perspective to the understanding of cognitive maps. Surprisingly this perspective is not in terms of how cognitive maps could be better or why they are flawed, but rather why cognitive maps are structured the way they are in terms of efficiency and performance. One of the advantages of implementing computational models of cognitive maps on real robots is that it can help to sharply differentiate what is only theoretically useful from what is truly practical. As adaptive organisms that operate in a range of often hostile environments, practicality and efficiency are paramount.

References

Able, K. P. (1980). Mechanisms of orientation, navigation, and homing. In S. A. Gauthreaux, Jr. (ed.), *Animal migration, orientation, and navigation* (pp. 282–373). New York: Academic Press.

Able, K. P. (1996). Large scale navigation. *Journal of Experimental Biology,* 199, 1–2.

Able, K. P., & Able, M. A. (1990). Calibration of the magnetic compass of a migratory bird. *Nature,* 347, 378–380.

Able, K. P., & Able, M. A. (1993). Daytime calibration of magnetic orientation in a migratory bird requires a view of skylight polarization. *Nature,* 364, 523–525.

Able, K. P., & Able, M. A. (1995). Interaction in the flexible orientation system of a migratory bird. *Nature,* 375, 230–232.

Able, K. P., & Bingman, V. P. (1987). The development of orientation and navigation behavior in birds. *Quarterly Review of Biology,* 62, 1–29.

Able, K. P., & Gergits, W. F. (1985). Human navigation: Attempts to replicate Baker's displacement experiment. In J. L. Kirschvink, D. S. Jones, & B. J. McFadden (eds.), *Magnetite biomineralization and magnetoreception in organisms* (pp. 569–572). New York: Plenum Press.

Acredolo, L. P. (1981). Small- and large-scale spatial concepts in infancy and childhood. In L. Liben, A. Patterson, & N. Newcombe (eds.), *Spatial representation and behavior across the life span: Theory and application* (pp. 63–81). New York: Academic Press.

Adler, K., & Pelkie, C. R. (1985). Human homing orientation: Critique and alternative hypotheses. In J. L. Kirschvink, D. S. Jones, & B. J. McFadden (eds.), *Magnetite biomineralization and magnetoreception in organisms* (pp. 573–593). New York: Plenum Press.

Aitken, S. C., & Prosser, R. (1990). Residents' spatial knowledge of neighborhood continuity and form. *Geographical Analysis,* 22, 301–325.

Akbarian, S., Grüsser, O.-J., & Guldin, W. O. (1993). Corticofugal projections to the vestibular nuclei in squirrel monkeys: Further evidence of multiple cortical vestibular fields. *Journal of Comparative Neurology*, 332, 89–104.

Alerstam, T. (1996). The geographical scale factor in orientation of migrating birds. *Journal of Experimental Biology*, 199, 9–19.

Alexander, C., Ishikawa, S., & Silverstein, M. (1977). *A Pattern Language*. New York: Oxford University Press.

Allen, G. L. (1981). A developmental perspective on the effects of subdividing macrospatial experience. *Journal of Experimental Psychology: Human Learning and Memory*, 7, 120–132.

Allen, G. L. (1982). The organization of route knowledge. In R. Cohen (ed.), *New directions for child development: Children's conceptions of spatial relationships* (pp. 31–39). San Francisco: Jossey-Bass.

Allen, G. L. (1985). Strengthening weak links in the study of the development of macrospatial cognition. In R. Cohen (ed.), *The development of spatial cognition* (pp. 301–321). Hillsdale, NJ: Lawrence Erlbaum.

Allen, G. L., & Kirasic, K. C. (1985). Effects of the cognitive organization of route knowledge on judgments of macrospatial distance. *Memory and Cognition*, 13, 218–227.

Allen, G. L, Kirasic, K. C., Dobson, S. H., Long, R. G., & Beck, S. (1996). Predicting environmental learning from spatial abilities: An indirect route. *Intelligence*, 22, 327–355.

Allen, G. L., Kirasic, K. C., Siegel, A. W., & Herman, J. F. (1979). Developmental issues in cognitive mapping: The selection and utilization of environmental landmarks. *Child Development*, 56, 1062–1070.

Allen, G. L., & Ondracek, P. J. (1995). Age-sensitive cognitive abilities related to children's acquisition of spatial knowledge. *Developmental Psychology*, 31, 934–945.

Allen, G. L., Siegel, A. W., & Rosinski, R. R. (1978). The role of perceptual context in structuring spatial knowledge. *Journal of Experimental Psychology: Human Learning and Memory*, 4, 617–630.

Alyan, S., & Jander, R. (1994). Short-range homing in the house mouse, *Mus musculus*: Stages in the learning of directions. *Animal Behavior*, 48, 285–298.

Ammeraal, L. (1986). *Programming principles in computer graphics*. New York: John Wiley & Sons.

Amorim, M.-A., Glasauer, S., Corpinot, K., & Berthöz, A. (1997). Updating an object's orientation and location during nonvisual navigation: A comparison between two processing modes. *Perception and Psychophysics*, 59(3), 404–418.

Amorim, M.-A., Loomis, J. M., & Fukusima, S. S. (1998). Reproduction of object shape is more accurate without the continued availability of visual information. *Perception*, 27, 69–86.

Amorim, M.-A., & Stucchi, N. (1997). Viewer- and object-centered mental explorations of an imagined environment are not equivalent. *Cognitive Brain Research*, 5, 229–239.

André-Deshays, C., Berthöz, A., & Revel, M. (1988). Eye-head coupling in humans I: Simultaneous recording of isolated motor units in dorsal neck muscles and horizontal eye movements. *Experimental Brain Research*, 69, 399–406.

André-Deshays, C., Revel, M., & Berthöz, A. (1991). Eye-head coupling in humans II: Phasic components. *Experimental Brain Research*, 84, 1–20.

Anooshian, L. J. (1996). Diversity within spatial cognition: Strategies underlying spatial knowledge. *Environment and Behavior*, 28(4), 471–493.

Appleyard, D. (1969). Why buildings are known. *Environment and Behavior*, 1, 131–156.

Appleyard, D. (1970). Styles and methods of structuring a city. *Environment and Behavior*, 2, 100–117.

Asada, M., Yasuhito, F., & Tsuji, S. (1988). Representing a global map for a mobile robot with relational local maps from sensory data. *Proceedings of the 9th International Conference on Pattern Recognition* (pp. 520–523).

Ashmead, D. H., Davis, D. L., & Northington, A. (1995). Contribution of listeners' approaching motion to auditory distance perception. *Journal of Experimental Psychology: Human Perception and Performance,* 21, 239–256.

Attneave, F., & Farrar, P. (1977). The visual world behind the head. *American Journal of Psychology,* 90, 549–563.

Attneave, F., & Pierce, C. R. (1978). Accuracy of extrapolating a pointer into perceived and imagined space. *American Journal of Psychology,* 91, 371–387.

Axhausen, K. W., & Gärling, T. (1992). Activity-based approaches to travel analysis: Conceptual frameworks, models, and research problems. *Transport Reviews,* 12(4), 323–341.

Axia, G., Peron, E., & Baroni, M. (1991). Environmental assessment across the life span. In T. Gärling and G. W. Evans (eds.), *Environment, cognition and action—An integrated approach* (pp. 221–244). New York: Oxford University Press.

Baddeley, A. (1986). *Working memory.* Oxford: Clarendon Press.

Baird, J. C. (1983). Modeling the creation of cognitive maps. In H. Pick & L. Acredolo (eds.), *Spatial orientation* (pp. 321–344). New York: Plenum Press.

Baird, J. C., Merril, A. A., & Tannenbaum, J. (1979). Studies of cognitive representation of spatial relations: II. A familiar environment. *Journal of Experimental Psychology: General,* 108, 92–98.

Baird, J. C., Wagner, M., & Noma, E. (1982). Impossible cognitive spaces. *Geographical Analysis,* 14, 204–216.

Baker, R. R. (1981). *Human navigation and the sixth sense.* New York: Simon & Schuster.

Baker, R. R. (1982). *Migration—Path through time and space.* London: Hodder and Stoughton.

Baker, R. R. (1985). Magnetoreception by man and other primates. In J. L. Kirschvink, D. S. Jones, & B. J. McFadden (eds.), *Magnetite biomineralization and magnetoreception in organisms* (pp. 537–561). New York: Plenum Press.

Baldaccini, N. E., Benvenuti, S., & Fiaschi, V. (1976). Homing behaviour of pigeons confined to a new loft distant from their home. *Monitore Zoologico Italiano* (n.s.) 10, 461–467.

Barlow, J. S. (1964). Inertial navigation as a basis for animal navigation. *Journal of Theoretical Biology,* 6, 76–117.

Barr, A., & Feigenbaum, E. A. (eds.). (1981). *The handbook of artificial intelligence,* Vols. 1–3. Los Altos, CA: Kaufman.

Barratt, E. S. (1953). An analysis of verbal reports of solving spatial problems as an aid in defining spatial factors. *Journal of Psychology,* 36, 17–25.

Bartlett, F. C. (1961). *Remembering: A study in experimental and social psychology.* Cambridge: Cambridge University Press (originally published 1932).

Basri, R., & Rivlin, E. (1995). Localization and homing using combinations of model views. *Artificial Intelligence,* 78, 327–354.

Beall, A. C., & Loomis, J. M. (1996). Visual control of steering without course information. *Perception,* 25, 481–494.

Beaudonnière, P. M., de Waele, C., Tran Ba Huy, P., & Vidal, P.-P. (1997). Réponses évoquées vestibulaires avant et après neurectomie vestibulaire unilatérale chez l'homme. In M. Collard, M. Jeannerod, & Y. Christen (eds.), *Le cortex vestibulaire* (pp. 95–108). Paris: Editions Irvinn.

Beck, W., & Wiltschko, W. (1982). The magnetic field as reference system for the genetically encoded migratory direction in Pied Flycatchers (*Ficedula hypoleuca* Pallas). *Zeitschrift für Tierpsychologie,* 60, 41–46.

Bell, S. M. (1995). *Cartographic presentation as an aid to spatial knowledge acquisition in unknown environments.* Unpublished master's thesis, Department of Geography, University of California Santa Barbara.

Benhamou, S. (1989). An olfactory orientation model for mammals' movements in their home ranges. *Journal of Theoretical Biology,* 139, 379–388.

Benhamou, S. (1996). No evidence for cognitive mapping in rats. *Animal Behavior,* 52, 201–212.

Benhamou, S., & Poucet, P. (1996). A comparative analysis of spatial memory processes. *Behavioral Processes*, 35, 113–126.

Benhamou, S., Sauvé, J.-P., & Bovet, P. (1990). Spatial memory in large-scale movements: Efficiency and limitations of the egocentric coding process. *Journal of Theoretical Biology*, 145, 1–12.

Benhamou, S., & Séguinot, V. (1995). How to find one's way in the labyrinth of path integration models? *Journal of Theoretical Biology*, 174, 463–466.

Bennett, A.T.D. (1996). Do animals have cognitive maps? *Journal of Experimental Biology*, 199, 219–224.

Benton, A. L. (1969). Disorders of spatial orientation. In P. Vinken & G. Bruyn (eds.), *Handbook of clinical neurology* (Vol. 3, pp. 212–228). Amsterdam: North-Holland.

Benvenuti, S., Ioalè, P., & Nacci, L. (1994). A new experiment to verify the spatial range of pigeons' olfactory map. *Behaviour*, 131, 277–292.

Beritoff, J. S. (1965). *Neural mechanisms of higher vertebrate behavior* (W. T. Liberson, ed. and trans.). Boston: Little, Brown.

Berlyne, D. E. (1960). *Conflict, arousal, and curiosity*. New York: McGraw-Hill.

Berlyne, D. E. (1966). Curiosity and exploration. *Science*, 153, 25–33.

Bernard, J-M. (1994). Introduction to inductive analysis. *Mathématiques, Informatique, et Sciences Humaines*, 126, 71–80.

Berndt, R., & Winkel, W. (1980). Field experiments on problems of imprinting to the birthplace in the Pied Flycatcher, *Ficedula hypoleuca*. In R. Nöhring (ed.), *Acta XII Congressus Internationalis Ornithologici* (pp. 851–854). Berlin: Deutsche Ornithologen-Gesellschaft.

Bernstein, N. (1967). *The coordination and regulation of movements*. New York: Pergamon Press.

Berthold, P. (1988). The control of migration in European warblers. In H. Ouellet (ed.), *Acta XIX Congressus Internationalis Ornithologici* (pp. 215–294). Ottawa: University of Ottawa Press.

Berthöz, A. (1989). Coopération et substitution entre le système saccadique et les réflexes d'origine vestibulaires: Faut-il réviser la notion de réflexe? *Revue Neurologique (Paris)*, 145, 513–526.

Berthöz, A. (1997). *Le sens du mouvement*. Paris: Editions Odile Jacob.

Berthöz, A., Israël, I., Georges-François, P., Grasso, R., & Tsuzuku, T. (1995). Spatial memory of body linear displacement: What is being stored. *Science*, 269, 95–98.

Berthöz, A., & Pozzo, T. (1988). Intermittent head stabilization during complex movements in man. In B. Amblard, A. Berthöz, & F. Clarac (eds.), *Development, adaptation and modulation of posture and gait* (pp. 189–198). Amsterdam: Elsevier.

Bever, T. G. (1992). The logical and extrinsic sources of modularity. In M. Gunnar & M. Maratsos (eds.), *Modularity and constraints in language and cognition* (pp. 179–212). Minnesota symposia on child psychology, Vol. 25. Hillsdale, NJ: Lawrence Erlbaum.

Biederman, I. (1972). Perceiving real-world scenes. *Science*, 177, 77–80.

Biederman, I., Mezzanotte, R. J., & Rabinowitz, J. C. (1982). Scene perception: Detecting and judging object undergoing relational violations. *Cognitive Psychology*, 14, 143–177.

Biegler, R., & Morris, R.G.M. (1993). Landmark stability is a prerequisite for spatial but not discrimination learning. *Nature*, 361, 631–633.

Biegler, R., & Morris, R.G.M. (1996). Landmark stability: Further studies pointing to a role in spatial learning. *Quarterly Journal of Experimental Psychology*, 49B, 307–345.

Bingman, V. P. (1983). Magnetic field orientation of migratory Savannah Sparrows with different first summer experience. *Behaviour*, 87, 43–53.

Bisetzky, A. R. (1957). Die Tänze der Bienen nach einem Fussweg zum Futterplatz unter besonderer Berücksichtigung von Umwegversuchen. *Zeitschrift für Vergleichende Physiologie*, 40, 264–288.

Blades, M. (1991). Wayfinding theory and research: The need for a new approach. In D. Mark & A Frank (eds.), *Cognitive and linguistic aspects of geographic space* (pp. 137–165). Dordrecht: Kluwer Academic.

Blades, M., & Spencer, C. (1986). The implications of psychological theory and methodology for cognitive cartography. *Cartographica*, 3, 1–13.

Blades, M., & Spencer, C. (1987). Young children's strategies when using maps with landmarks. *Journal of Environmental Psychology*, 7, 201–218.

Blaut, J. M., & Stea, D. (1971). Studies of geographic learning. *Annals of the Association of American Geographers*, 61, 387–393.

Bles, W. (1981). Stepping around: Circular vection and Coriolis effects. In J. Long & A. Baddeley (eds.), *Attention and performance* (Vol. 9, pp. 47–61). Hillsdale, NJ: Lawrence Erlbaum.

Bles, W., de Jong, J., & de Wit, G. (1984). Somatosensory compensation for loss of labyrinthine function. *Acta Oto-Laryngologica*, 97, 213–221.

Bletz, H., Weindler, P., Wiltschko, R., Wiltschko, W., & Berthold, P. (1996). The magnetic field as reference for the innate migratory direction in Blackcaps, *Sylvia atricapilla*. *Naturwissenschaften*, 83, 430–432.

Bloomberg, J., Melvill-Jones, G., Segal, B., McFarlane, S., & Soul, J. (1988). Vestibular contingent voluntary saccades based on cognitve estimates of remembered vestibular information. *Advances in Oto-Rhino-Laryngology*, 40, 71–75.

Blouin, J., Gauthier, G. M., & Vercher, J.-L. (1995). Failure to update the egocentric representation of the visual space through labyrinthine signal. *Brain and Cognition*, 29, 1–22.

Bohbot, V., Kalina, M., Stepankova, K., Spackova, N., Petrides, M., & Nadel, L. (1997). Lesions to the right parahippocampal cortex cause navigational memory deficits in humans. Poster at Human Brain Mapping Conference, Copenhagen, Denmark.

Böök, A., & Gärling, T. (1980). Processing of information about location during locomotion: Effects of a concurrent task and locomotion patterns. *Scandinavian Journal of Psychology*, 21, 185–192.

Böök, A., & Gärling, T. (1981a). Maintenance of environmental orientation during body rotation. *Perceptual and Motor Skills*, 53, 583–589.

Böök, A., & Gärling, T. (1981b). Maintenance of orientation during locomotion in unfamiliar environments. *Journal of Experimental Psychology: Human Perception and Performance*, 7, 995–1006.

Bottini, G., Sterzi, R., Paulesu, E., Vallar, G., Cappa, S., Erminio, F., Passingham, R. E., Frith, C. D., & Frackowiack, R. S. J. (1994). Identification of the central vestibular projections in man: A positron emission tomography activation study. *Experimental Brain Research*, 99, 164–169.

Bovy, P.H.L., & Stern, E. (1990). *Route choice: Wayfinding in transport networks*. Dordrecht: Kluwer Academic.

Bremner, J. G. (1982). Object localization in infancy. In M. Potegal (ed.), *Spatial abilities: Development and physiological foundations* (pp. 79–106). New York: Academic Press.

Bremner, J. G., & Bryant, P. E. (1985). Active movement and development of spatial abilities in infancy. In H. Wellman (ed.), *Children's searching: The development search skill and spatial representation* (pp. 53–72). Hillsdale, NJ: Lawrence Erlbaum.

Brooks, R. A. (1985). Visual map making for a mobile robot. *Proceedings of the IEEE Conference on Robotics and Automation* (pp. 824–829). St. Louis: IEEE Computer Society Press.

Brooks, R. A. (1991). Challenges for complete creature architectures. In J.-A. Meyer & S. W. Wilson (eds.), *From animals to animats, Proceedings of the First International Conference on the Simulation of Adaptive Behavior* (pp. 434–443). Cambridge, MA: MIT Press.

Brown, C., Durrant-White, H., Leonard, J., Rao, B., & Steer, B. (1989). *Centralized and decentralized Kalman filter techniques for tracking, navigation, and control* (revised Technical Report 277). New York: University of Rochester, Computer Science Department.

Bryant, D. J., & Tversky, B. (1991). Locating objects from memory or from sight. Paper presented at the 32nd annual meeting of the Psychonomic Society, San Francisco.

Bryant, D. J., Tversky, B., & Franklin, N. (1992). Internal and external spatial frameworks representing described scenes. *Journal of Memory and Language*, 31, 74–98.

Bryant, K. J. (1982). Personality correlates of sense of direction and geographical orientation. *Journal of Personality and Social Psychology*, 43, 1318–1324.

Burda, H., Marhold, T., Westenberger, R., Wiltschko R., & Wiltschko, W. (1990). Magnetic compass orientation in the subterranean rodent *Cryptomys hottentotus* (Bathyergidae). *Experientia*, 46, 528–530.

Burroughs, W. J., & Sadalla, E. K. (1979). Asymmetries in distance cognition. *Geographical Analysis*, 11, 414–421.

Busemeyer, J. R. (1985). Decision making under uncertainty: A comparison of simple scalability, fixed sample, and sequential sampling models. *Journal of Experimental Psychology*, 11, 538–564.

Busemeyer, J. R., & Townsend, J. T. (1993). Decision field theory: A dynamic cognitive approach to decision making in an uncertain environment. *Psychological Review*, 100(3), 432–459.

Buttenfield, B. P. (1986). Comparing distortion on sketch maps and MDS configurations. *Professional Geographer* 38(3), 238–246.

Byrne, R. W. (1979). Memory for urban geography. *Quarterly Journal of Experimental Psychology*, 31(1), 147–154.

Byrne, R. W. (1982). Geographical knowledge and orientation. In A. Ellis (ed.), *Normality and pathology in cognitive functions* (pp. 239–264). London: Academic Press.

Cadwallader, M. T. (1977). Frame dependence in cognitive maps: An analysis using directional statistics. *Geographical Analysis*, 9, 284–291.

Cadwallader, M. T. (1979). Problems in cognitive distance: Implications for cognitive mapping. *Environment and Behavior*, 11, 559–576.

Campos, J. J., Svejda, M. J., Campos, R. G., & Bertenthal, B. (1982). The emergence of self-produced locomotion: Its importance for psychological development in infancy. In D. Bricker (ed.), *Intervention with at-risk and handicapped infants* (pp. 195–216). Baltimore: University Park Press.

Canter, D. V. (1977). *The psychology of place*. London: Architectural Press.

Carlson-Radvansky, L. A., & Irwin, D. E. (1993). Frames of references in vision and language: Where is above? *Cognition*, 46, 223–244.

Carpenter, P. A., & Just, M. A. (1982). Spatial ability: An information processing approach to psychometrics. In R. Sternberg (ed.), *Advances in the psychology of human intelligence* (Vol. 3, pp. 221–253). Hillsdale, NJ: Lawrence Erlbaum.

Cartwright, B. A., & Collett, T. S. (1983). Landmark learning in bees: Experiments and models. *Journal of Comparative Physiology*, 151, 521–543.

Cartwright, B. A., & Collett, T. S. (1987). Landmark maps for honeybees. *Biological Cybernetics*, 57, 85–93.

Casey, S. M. (1978). Cognitive mapping by the blind. *Journal of Vision Impairment and Blindness*, 72(8), 297–301.

Castiello, U., Scarpa, M., & Bennett, K. (1995). A brain-damaged patient with an unusual perceptuomotor deficit. *Nature*, 374, 805–808.

Castner, H. W. (1996). Where am I? Basic concepts in human orientation. *Cartographica*, April, 54–56.

Castner, H. W., & Anderson, J. M. (1996). Orienting ourselves in space: Implications for the school curriculum. In R. C. Colson (ed.), *Cartographica*. Monograph No. 46. Calgary: University of Calgary.

Chalfonte, B. L., & Johnson, M. K. (1996). Feature memory and binding in young and older adults. *Memory and Cognition*, 24, 403–416.

Chalmers, J. D., & Knight, G. R. (1985). The reliability of the familiarity of environmental stimuli. *Environment and Behavior*, 17(2), 223–238.

Chance, S. S., Gaunet, F., Beall, A. C., & Loomis, J. M. (submitted). Locomotion mode affects the updating of objects encountered during travel: The contribution of vestibular and proprioceptive inputs to path integraton.

Chapman, D. (1987). Planning for conjunctive goals, *Artificial Intelligence, 32*, 333–377.

Chase, W. G., & Simon, H. A. (1973). The mind's eye in chess. In W. Chase (ed.), *Visual information processing* (pp. 215–281). New York: Academic Press.

Chelazzi, G., & Pardi, L. (1972). Experiments on the homing behaviour of caged pigeons. *Monitore Zoologico Italiano* (n.s.) 6, 11–18.

Chen, L. L., Lin, L.-H., Barnes, C. A., & McNaughton, B. L. (1994a). Head-direction cells in the rat posterior cortex: Contribution of visual and idiothetic information to the directional firing. *Experimental Brain Research*, 101, 24–34.

Chen, L. L., Lin, L.-H., Green, E. J., Barnes, C. A., & McNaughton, B. L. (1994b). Head-direction cells in the rat posterior cortex. 1. Anatomical distribution and behavioral modulation. *Experimental Brain Research*, 101, 8–23.

Cheng, K. (1986). A purely geometric module in the rat's spatial representation. *Cognition*, 23, 149–178.

Cheng, K. (1987). Rats use the geometry of surfaces for navigation. In P. Ellen & C. Thinus-Blanc (eds.), *Cognitive processes and spatial orientation in animal and man* (Vol. 1, pp. 153–159). Dordrecht: Martinus Nijhoff.

Cheng, K., Collett, T. S., Pichard, A., & Wehner, R. (1987). The use of visual landmarks by honeybees: Bees weight landmarks according to their distance from the goal. *Journal of Comparative Physiology*, A161, 469–474.

Cheng, K., & Gallistel, C. R. (1984). Testing the geometric power of an animal's spatial representation. In H. L. Roitblat, T. G. Bever, & H. S. Terrace (eds.), *Animal cognition* (pp. 409–423). Hillsdale, NJ: Lawrence Erlbaum.

Chittka, L., Geiger, K., & Kunze, J. (1995). The influences of landmark sequences on distance estimation of honeybees. *Animal Behaviour*, 50, 23–31.

Chittka, L., Kunze, J., Shipman, C., & Buchmann, S. L. (1995). The significance of landmarks for path integration of homing honey bee foragers. *Naturwissenschaften*, 82, 341–343.

Chown, E., Kaplan, S., & Kortenkamp, D. (1995). Prototypes, location, and association networks (PLAN): Towards a unified theory of cognitive mapping. *Cognitive Science*, 19, 1–51.

Clark, A. (1989). *Microcognition: Philosophy, cognitive science, and parallel distributed processing*. Cambridge, MA: MIT Press.

Cohen, R., Baldwin, L. M., & Sherman, R. C. (1978). Cognitive maps of a naturalistic setting. *Child Development*, 49, 1216–1218.

Cohen, R., & Schuepfer, T. (1980). The representation of landmarks and routes. *Child Development*, 51, 1065–1071.

Colby, C. L. Duhamel, J. R., & Goldberg, M. E. (1995). Oculocentric spatial representation in parietal cortex. *Cerebral Cortex*, 5, 470–481.

Collett, T. S. (1982). Do toads plan routes? A study of the detour behavior of Bufo viridis. *Journal of Comparative Physiology*, 146, 261–271.

Collett, T. S. (1995). Making learning easy: The acquisition of visual information during orientation flights of social wasps. *Journal of Comparative Physiology*, A177, 737–747.

Collett, T. S. (1996a). Insect navigation *en route* to the goal: Multiple strategies for the use of landmarks. *Journal of Experimental Biology*, 199(1), 227–235.

Collett, T. S. (1996b). Short range navigation: Does it contribute to understanding navigation over longer distances? *Journal of Experimental Biology*, 199, 225–226.

Collett, T. S., & Baron, J. (1994). Biological compasses and the coordinate frame of landmark memories in honeybees. *Nature*, 368, 137–140.

Collett, T. S., Cartwright, B. A., & Smith, B. A. (1986). Landmark learning and visuospatial memories in gerbils. *Journal of Comparative Physiology*, A158, 835–851.

Collett, T. S., Dillmann, E., Giger, A., & Wehner, R. (1992). Visual landmarks and route following in desert ants. *Journal of Comparative Physiology*, A170, 435–442.

Collett, T. S., Fry, S. N., & Wehner, R. (1993). Sequence learning by honeybees. *Journal of Comparative Physiology*, A172, 693–706.

Collett, T. S., & Kelber, A. (1988). The retrieval of visuo-spatial memories by honeybees. *Journal of Comparative Physiology*, A163, 145–150.

Collett, T. S., & Land, M. F. (1975). Visual control of flight behaviour in the hoverfly *Syritta pipiens* L. *Journal of Comparative Physiology*, 99, 1–66.

Collett, T. S., & Lehrer, M. (1993). Looking and learning: A spatial pattern in the orientation flight of the wasp *Vespula vulgaris*. *Proceedings of the Royal Society of London*, B252, 129–134.

Collett, T. S., & Rees, J. A. (1997). View-based navigation in hymenoptera: Multiple strategies of landmark guidance in the approach to a feeder. *Journal of Comparative Physiology*, A181, 47–58.

Collett, T. S., & Zeil, J. (1996). Flights of learning. *Current Directions in Psychological Science*, 5, 149–155.

Conning, A. M., & Byrne, R. W. (1984). Pointing to preschool children's spatial competence: A study in natural settings. *Journal of Environmental Psychology*, 4, 165–175.

Corlett, J. T., Byblow, W., & Taylor, B. (1990). The effect of perceived locomotor constraints on distance estimation. *Journal of Motor Behavior*, 22, 347–360.

Corlett, J. T., Kozub, S., & Quick, H. (1990). Memory for distance to be walked following interfering activities. *Journal of Human Movement Studies*, 18, 117–127.

Corlett, J. T., & Patla, A. E. (1987). Some effects of upward, downward, and level visual scanning and locomotion on distance estimation accuracy. *Journal of Human Movement Studies*, 13, 85–95.

Corlett, J. T., Patla, A. E., & Williams, J. G. (1985). Locomotor estimation of distance after visual scanning by children and adults. *Perception*, 14, 257–263.

Cornell, E. H., & Hay, D. H. (1984). Children's acquisition of a route via different media. *Environment and Behavior*, 16, 627–642.

Cornell, E. H., Heth, C. D., & Alberts, D. M. (1994). Place recognition and way finding by children and adults. *Memory and Cognition*, 22, 633–643.

Cornell, E. H., Heth, C. D., & Broda, L. S. (1989). Children's wayfinding: Response to instructions to use environmental landmarks. *Developmental Psychology*, 25, 755–764.

Cornell, E. H., Heth, C. D., & Rowat, W. L. (1992). Wayfinding by children and adults: Response to instructions to use look-back and retrace strategies. *Developmental Psychology*, 28, 328–336.

Couclelis, H. (1996). Verbal directions for way-finding: Space, cognition, and language. In J. Portugali (ed.), *The construction of cognitive maps* (pp. 133–153). Dordrecht: Kluwer Academic.

Couclelis, H., Golledge, R. G., Gale, N. D., & Tobler, W. R. (1987). Exploring the anchorpoint hypothesis of spatial cognition. *Journal of Environmental Psychology*, 7, 99–122.

Cousins, J. H., Siegel, A. W., & Maxwell, S. E. (1983). Way finding and cognitive mapping in large-scale environments: A test of a developmental model. *Journal of Experimental Psychology*, 35, 1–20.

Craik, F. I. M., & Tulving, E. (1975). Depth of processing and the retention of words in episodic memory. *Journal of Experimental Psychology: General*, 104, 268–294.

Cratty, B. J., Peterson, C., Harris, J., & Schoner, R. (1968). The development of perceptual-motor abilities in blind children and adolescents. *The New Outlook for the Blind*, 62, 111–117.

Cressant, A., Muller, R. U., & Poucet, B. (1997). Failure of centrally placed objects to control the firing fields of hippocampal place cells. *Journal of the Neurological Sciences*, 17, 2531–2542.

Crowell, J. A., & Banks, M. S. (1993). Perceiving heading with different retinal regions and types of optic flow. *Perception and Psychophysics*, 53, 325–337.

Cutting, J. E., Springer, K., Braren, P. A., & Johnson, S. H. (1992). Wayfinding on foot from information in retina, not optical, flow. *Journal of Experimental Psychology: General,* 121, 41–72.

Cutting, J. E., and Vishton, P. M. (1993). Maps, not flowfields, guide human navigation. *Perception,* 22, 11.

Cutting, J. E., Vishton, P. M., & Braren, P. A. (1995). How we avoid collisions with stationary and moving obstacles. *Psychological Review,* 102, 627–651.

Da Silva, J. A., & Fukusima, S. S. (1990). *Scaling egocentric distance in natural indoor and outdoor settings with and without vision.* Paper presented at the 22nd International Congress of Applied Psychology, Kyoto.

Da Silva, J. A., Ruiz, E. R., & Marques, S. L. (1987). Individual differences in magnitude estimates of inferred, remembered, and perceived geographical distance. *Bulletin of the Psychonomic Society,* 25(4), 240–243.

Decety, J. (1991). Motor information may be important for updating the cognitive processes involved in mental imagery of movement. *European Bulletin of Cognitive Psychology,* 11, 415–426.

Decety, J., Jeannerod, M., & Prablanc, C. (1989). The timing of mentally represented actions. *Behavioural Brain Research,* 34, 35–42.

De Loache, J. (1987). Rapid change in the symbolic ability of very young children. *Science,* 238, 1556–1557.

Denis, M. (1996). Language and spatial cognition. Paper presented at ENP School Conference on the Representation of Space, San Feliu, Spain.

Denis, M., & Cocude, M. (1992). Structural properties of visual images constructed from poorly or well structured verbal descriptions. *Memory and Cognition,* 20, 497–506.

Denis, M., & Denhière, G. (1990). Comprehension and recall of spatial descriptions. *European Bulletin of Cognitive Psychology,* 10, 115–143.

De Renzi, E. (1982). *Disorders of space exploration and cognition.* New York: Wiley.

De Vega, M., Intons-Peterson, M. J., Johnson-Laird, P. N., Denis, M., & Marschark, M. (1996). *Models of visuospatial cognition.* New York: Oxford University Press.

Dickinson, J. A. (1994). Bees link local landmarks with celestial compass cues. *Naturwissenschaften,* 81, 465–467.

Diez-Chamizo, V., Sterio, D., & Mackintosh, N. J. (1985). Blocking and overshadowing between intra-maze and extra-maze cues: A test of the independence of locale and guidance learning. *Quarterly Journal of Experimental Psychology,* 37B, 235–253.

Dill, M., Wolf, R., & Heisenberg, M. (1993). Visual pattern recognition in Drosophila involves retinotopic matching. *Nature,* 365, 751–753.

Dodds, A. G., Howarth, C. I., & Carter, D. C. (1982). The mental maps of the blind: The role of previous visual experience. *Journal of Visual Impairment and Blindness,* 76, 5–12.

Donald, M. (1991). *Origins of the modern mind: Three stages in the evolution of culture and cognition.* Cambridge, MA: Harvard University Press.

Downs, R. M. (1981a). Maps and mappings as metaphors for spatial representation. In L. S. Liben, A. H. Patterson, & N. Newcombe (eds.), *Spatial representation and behavior across the life span: Theory and applications* (pp. 143–166). New York: Academic Press.

Downs, R. M. (1981b). Maps and metaphors. *The Professional Geographer,* 87, 287–293.

Downs, R. M., & Liben, L. S. (1985). Children's understanding of maps. In P. Ellen & C. Thinus-Blanc (eds.), *Cognitive processes and spatial orientation in animal and man,* Vol. 2: *Neurophysiology and developmental aspects* (pp. 202–219). Dordrecht: Martinus Nijhoff.

Downs, R. M., & Stea, D. (1973a). Cognitive maps and spatial behavior: Process and products. In R. M. Downs & D. Stea (eds.), *Image and environment: Cognitive mapping and spatial behavior* (pp. 8–26). Chicago: Aldine.

Downs, R. M., & Stea, D. (1973b). *Image and environment: Cognitive mapping and spatial behavior.* Chicago: Aldine.

Dreyfus, H. L. (1972). *What computers can't do.* New York: Harper and Row.

Drose, G. S., & Allen, G. L. (1994). The role of visual imagery in the retention of information from sentences. *Journal of General Psychology*, 117, 277–293.

Droulez, J., & Berthöz, A. (1988). Servo-controlled conservative versus topological (projective) modes of sensory motor control. In W. Bles & T. Brandt (eds.), *Disorders of posture and gait* (pp. 83–97). Amsterdam): Elsevier.

Droulez, J., & Berthöz, A. (1991). The concept of dynamic memory in sensorimotor control. In D. R. Humphrey & H. J. Freund (eds.), *Motor control: Concepts and issues* (pp. 137–161). London: John Wiley & Sons.

Durrant-White, H. F. (1990). Sensor models and multisensor integration. In I. J. Cox & G. T. Wilfong (eds.), *Autonomous robot vehicles* (pp. 73–89). New York: Springer-Verlag.

Dyer, F. C. (1991). Bees acquire route-based memories but not cognitive maps in a familiar landscape. *Animal Behavior*, 41, 239–246.

Dyer, F. C. (1996). Spatial memory and navigation by honeybees on the scale of the foraging range. *Journal of Experimental Biology*, 199(1), 147–154.

Dyer, F. C., & Dickinson, J. A. (1994). Development of sun compensation by honey bees: How partially experienced bees estimate the sun's course. *Proceedings of the National Academy of Sciences USA*, 91, 4471–4474.

Dyer, F. C., & Gould, J. L. (1981). Honey bee orientation: A backup system for cloudy days. *Science*, 214, 1041–1042.

Easton, R. D., & Sholl, M. J. (1995). Object-array structure, frames of reference, and retrieval of spatial knowledge. *Journal of Experimental Psychology: Learning, Memory, and Cognition*, 21, 483–500.

Eby, D., & Loomis, J. (1987). A study of visually directed throwing in the presence of multiple distance cues. *Perception and Psychophysics*, 11, 308–312.

Edelman, G. M. (1992). *Bright air, brilliant fire: On the matter of the mind*. London: Penguin.

Edwards, W. (1962). Subjective probabilities inferred from decisions. *Psychological Review*, 69, 109–135.

Einhorn, H. J., Kleinmuntz, D. N., & Kleinmuntz, B. (1979). Linear regression and process tracing models of judgment. *Psychological Review*, 86, 465–485.

Ekstrom, R. B., French, J. W., & Harman, H. H. (1976). *Kit of factor-referenced cognitive tests*. Princeton, NJ: Educational Testing Service.

Eliot, J., & McFarlane-Smith, I. M. (1983). *An international directory of spatial tests*. Oxford: NFER-Nelson.

Ellen, P. (1987). Cognitive mechanisms in animal problem-solving. In P. Ellen & C. Thinus-Blanc (eds.), *Cognitive processes and spatial orientation in animal and man* (Vol. 1, pp. 20–38). Dordrecht: Martinus Nijhoff.

Elliott, D. (1986). Continuous visual information may be important after all: A failure to replicate Thomson (1983). *Journal of Experimental Psychology: Human Perception and Performance*, 12, 388–391.

Elliott, D. (1987). The influence of walking speed and prior practice on locomotor distance estimation. *Journal of Motor Behavior*, 19, 476–485.

Embretson, S. E. (1987). Improving the measurement of spatial aptitude by dynamic testing. *Intelligence*, 11, 333–358.

Emlen, S. T. (1970). Celestial rotation: Its importance in the development of migratory orientation. *Science*, 170, 1198–1201.

Emlen, S. T. (1975). Migration: Orientation and navigation. In D. S. Farner & J. R. King (eds.), *Avian Biology* (Vol. 5, pp. 129–219). New York: Academic Press.

Ericsson, K. A., & Simon, H. A. (1993). *Protocol analysis: Verbal reports as data* (rev. ed.). London: MIT Press.

Esch, H. E., & Burns, J. E. (1996). Distance estimation by foraging honeybees. *Journal of Experimental Biology*, 199(1), 155–162.

Etienne, A. S. (1980). The orientation of the golden hamster to its nest-site after the elimination of various sensory cues. *Experientia*, 36, 1048–1050.

Etienne, A. S. (1992). Navigation of a small mammal by dead reckoning and local cues. *Current Directions in Psychological Science*, 1, 48–52.

Etienne, A. S., Berlie, J., Georgakopoulos, J., & Maurer, R. (1998a). The role of dead reckoning in navigation. In S. Healy (ed.), *Spatial representation in animals* (pp. 54–68). Oxford: Oxford University Press.

Etienne, A. S., Hurni, C., Maurer, R., & Séguinot, V. (1991). Twofold path integration during hoarding in the golden hamster? *Ethology, Ecology and Evolution*, 3, 1–11.

Etienne, A. S., Joris-Lambert, S., Maurer, R., Reverdin, B., & Sitbon, S. (1995). Optimizing distal landmarks: Horizontal versus vertical structures and relation to background. *Behavioral Brain Research*, 68, 103–116.

Etienne, A. S., Joris-Lambert, S., Reverdin, B., & Teroni, E. (1993). Learning to recalibrate the role of dead reckoning and visual cues in spatial navigation. *Animal Learning and Behavior*, 21, 266–280.

Etienne, A. S., Maurer, R., Berlie, J., Derivaz, J., Georgakopoulos, J., Griffin, A., & Rowe, T. (1998b). Cooperation between dead reckoning (path integration) and external position cues. *Journal of Navigation*, 51, 23–34.

Etienne, A. S., Maurer, R., & Saucy, F. (1988). Limitations in the assessment of path dependent information. *Behaviour*, 106, 81–111.

Etienne, A. S., Maurer, R., Saucy, F., & Teroni E. (1986). Short-distance homing in the golden hamster after a passive outward journey. *Animal Behavior*, 34, 699–715.

Etienne, A. S., Maurer, R., & Séguinot, V. (1996). Path integration in mammals and its interaction with visual landmarks. *Journal of Experimental Biology*, 199(1), 201–209.

Etienne, A. S., Teroni, V., Hurni, C., & Portenier, V. (1990). The effect of a single light cue on homing behavior of the golden hamster. *Animal Behavior*, 39, 17–41.

Etienne, A. S., Teroni, V., Maurer, R., Portenier, V., & Saucy, F. (1985). Short distance homing in a small mammal: The role of exteroreceptive cues and path integration. *Experientia*, 41, 122–125.

Ettema, D., & Timmermans, H. P. J. (eds.). (1997). *Activity-based approaches to travel analysis*. New York: Elsevier.

Evans, G. W., & Gärling, T. (1991). Environment, cognition, and action: The need for integration. In T. Gärling & G. W. Evans (eds.), *Environment, cognition, and action* (pp. 3–13). New York: Oxford University Press.

Evans, G. W., Marrero, G. D., & Butler, A. P. (1981). Environmental learning and cognitive mapping. *Environment and Behavior*, 13, 84–104.

Evans, G. W., & Pezdek, K. (1980). Cognitive mapping: Knowledge of real-world distance and location information. *Journal of Experimental Psychology: Human Learning and Memory*, 1, 13–24.

Eysenck, H. J. (1967). Intellectual assessment: A theoretical and experimental approach. *Journal of Educational Psychology*, 37, 81–98.

Farah, M. J. (1984). The neurological basis of mental imagery: A componential analysis. *Cognition*, 18, 245–272.

Farah, M. J., Hammond, K. M., Levine, D. N., & Calvanio, R. (1988). Visual and spatial mental imagery: Dissociable systems of representation. *Cognitive Psychology*, 20, 439–462.

Farrell, M. J., & Robertson, I. H. (submitted). *Mental rotation and the automatic updating of body-centered spatial relationships*.

Faugier-Grimaud, S., & Ventre, J. (1989). Anatomic connections of inferior parietal cortex (area 7) with subcortical structures related to vestibular-ocular function in a monkey (*Macaca fascicularis*). *Journal of Comparative Neurology*, 280, 1–14.

Ferguson, E. L., & Hegarty, M. (1994). Properties of cognitive maps constructed from texts. *Memory and Cognition*, 22(4), 455–473.

Field, D. J. (1987). Relations between the statistics of natural images and the response properties of cortical cells. *Journal of the Optical Society of America*, A4, 2379–2394.

Fleishman, E. A. (1975). Toward a taxonomy of human performance. *American Psychologist*, 30, 1127–1150.

Ford, J. K., Schmitt, N., Schechtman, S. L., Hults, B. M., & Doherty, M. L. (1989). Process tracing methods: Contributions, problems, and neglected research questions. *Organizational Behavior and Human Decision Processes,* 43, 75–117.

Foreman, N., Orencas, C., Nicholas, E., Morton, P., & Gell, M. (1989). Spatial awareness in seven to 11-year-old physically handicapped children in mainstream schools. *European Journal of Special Needs Education,* 4, 171–179.

Foster, T. C., Castro, C. A., & McNaughton, B. L. (1989). Spatial selectivity of hippocampal neurons: Dependence on preparedness for movement, *Science,* 22(4), 1580–1582.

Frank, A. U., Campari, I., & Formentini, U. (1992). *Theories and methods of spatio-temporal reasoning in geographic space.* Berlin: Springer-Verlag.

Franklin, N., Henkel, L. A., & Zangas, T. (1995). Parsing surrounding space into regions. *Memory and Cognition,* 23, 397–407.

Franz, M. O., Schölkopf, B., Georg, P., Mallot, H. A., & Bülthoff, H. H. (1997). Learning view graphs for robot navigation. In W. L. Johnson (ed.), *Proceedings of the First International Conference on Autonomous Agents* (pp. 138–147). New York: ACM Press.

Freundschuh, S. M. (1991). *Spatial knowledge acquisition of urban environments from maps and navigation experience.* Unpublished Ph.D. thesis, Department of Geography, State University of New York at Buffalo.

Frier, H., Edwards, E., Smith, C., Neale, S., & Collett, T. S. (1996). Magnetic compass cues and visual pattern learning in honeybees. *Journal of Experimental Biology,* 199, 1353–1361.

Frymire, M. (1997). *Dynamic imagery strategies for spatial orientation when walking without vision.* Undergraduate honors thesis submitted to the Faculty in Cognitive Studies, Department of Psychology and Human Development, Vanderbilt University, Nashville, Tennessee.

Fujita, N., Klatzky, R. L., Loomis, J. M., & Golledge, R. G. (1993). The encoding-error model of pathway completion without vision. *Geographical Analysis,* 25(4), 295–314.

Fujita, N., Loomis, J. M., Klatzky, R. L., & Golledge, R. G. (1990). A minimal representation for dead-reckoning navigation: Updating the homing vector. *Geographical Analysis,* 22(4), 326–335.

Fukusima, S. S., Loomis, J. M., & Da Silva, J. A. (1991). Accurate distance perception assessed by two triangulation methods. Paper presented at the meeting of the Psychonomic Society, San Francisco.

Fukusima, S. S., Loomis, J. M., & Da Silva, J. A. (1997). Visual perception of egocentric distance as assessed by triangulation. *Journal of Experimental Psychology: Human Perception and Performance,* 23, 86–100.

Füller, E., Kowalski, U., & Wiltschko, R. (1983). Orientation of homing pigeons: Compass orientation vs. piloting by familiar landmarks. *Journal of Comparative Physiology,* 153, 55–58.

Gale, N. D. (1980). *An analysis of the distortion and fuzziness of cognitive maps by location.* Unpublished master's thesis, Department of Geography, University of California, Santa Barbara.

Gale, N. D. (1982). Exploring location error in cognitive configurations of a city. In R. G. Golledge & W. R. Tobler (eds.), *An examination of the spatial variation in the distortion and fuzziness of cognitive maps.* Final report, December. NSF Grant #SES81-10253.

Gale, N. D. (1985). *Route learning by children in real and simulated environments.* Unpublished Ph.D. dissertation, Department of Geography, University of California, Santa Barbara.

Gale, N. D., Golledge, R. G., Halperin, W. C., & Couclelis, H. (1990). Exploring spatial familiarity. *The Professional Geographer,* 42(3), 299–313.

Gale, N. D., Golledge, R. G., Pellegrino, J. W., & Doherty, S. (1990). The acquisition and integration of neighborhood route knowledge in an unfamiliar neighborhood. *Journal of Environmental Psychology,* 10(1), 3–25.

Gallistel, C. R. (1989). Animal cognition: The representation of space, time and number. *Annual Review of Psychology,* 40, 155–189.

Gallistel, C. R. (1990). *The organization of learning*. Cambridge, MA: MIT Press.

Gallistel, C. R. (1993). *The organization of learning* (2nd ed.). Cambridge, MA: MIT Press.

Gallistel, C. R. (1994). Space and time. In N. J. Mackintosh (ed.), *Animal learning and cognition*, Vol. 9: *Handbook of perception and cognition* (pp. 221–253). London: Academic Press.

Gallistel, C. R., & Cramer, A. E. (1996). Computations on metric maps in mammals: Getting oriented and choosing a multi-destination route. *Journal of Experimental Biology*, 199(1), 211–217.

Garing, A. E. (1998). *Intersensory integration in the perception of self-movement*. Dissertation to be submitted to the Faculty of the Graduate School, Vanderbilt University, Nashville, Tennessee.

Gärling, T. (1989). The role of cognitive maps in spatial decisions. *Journal of Environmental Psychology*, 9, 269–278.

Gärling, T. (1994). Processing of time constraints on sequence decisions in a planning task. *European Journal of Cognitive Psychology*, 6, 399–416.

Gärling, T. (1995a). How do urban residents acquire, mentally represent, and use knowledge of spatial layout? In T. Gärling (ed.), *Readings in environmental psychology: Urban cognition* (pp. 1–12). London: Academic Press.

Gärling, T. (1995b). Tradeoffs of priorities against spatiotemporal constraints in sequencing activities in environments. *Journal of Environmental Psychology*, 15, 155–160.

Gärling, T. (1996). Sequencing actions: An information-search study of tradeoffs of priorities against spatiotemporal constraints. *Scandinavian Journal of Psychology*, 37, 282–293.

Gärling, T., Böök, A., & Ergezen, N. (1982). Memory for the spatial layout of the everyday physical environment: Differential rates of acquisition of different types of information. *Scandinavian Journal of Psychology*, 23, 23–35.

Gärling, T., Böök, A., Ergezen, N., & Lindberg, E. (1981a). Memory for the spatial layout of the everyday physical environment: Empirical findings and their theoretical implications. In A. E. Osterberg, C. P. Teirnan, & R. A. Findlay (eds.), *Design research interactions: Proceedings of the Twelfth International Conference of the Environmental Design Research Association* (pp. 69–76). Ames, IA: EDRA.

Gärling, T., Böök, A., & Lindberg, E. (1979). The acquisition and use of an internal representation of the spatial layout of the environment during locomotion. *Man-Environment Systems*, 9, 200–208.

Gärling, T., Böök, A., & Lindberg, E. (1984). Cognitive mapping of large-scale environments: The interrelationship of action plans, acquisition, and orientation. *Environment and Behavior*, 16, 3–34.

Gärling, T., Böök, A., & Lindberg, E. (1985). Adults' memory representations of the spatial properties of their everyday physical environment. In R. Cohen (ed.), *The development of spatial cognition* (pp. 141–184). Hillsdale, NJ: Lawrence Erlbaum.

Gärling, T., Böök, A., & Lindberg, E. (1986a). Spatial orientation and wayfinding in the designed environment: A conceptual analysis and some suggestions for post-occupancy evaluation. *Journal of Architectural and Planning Research*, 3, 55–64.

Gärling, T., Böök, A., Lindberg, E., & Arce, C. (1991). Evidence of a response-bias explanation of noneuclidean cognitive maps. *Professional Geographer*, 42, 143–149.

Gärling, T., Böök, A., Lindberg, E., & Nilsson, T. (1981b). Memory for the spatial layout of the everyday physical environment: Factors affecting the rate of acquisition. *Journal of Environmental Psychology*, 1, 263–277.

Gärling, T., & Evans, G. W. (eds.) (1991). *Environment, cognition, and action*. New York: Oxford University Press.

Gärling, T., & Gärling, E. (1988). Distance minimization in downtown pedestrian shopping behavior. *Environment and Planning*, A20, 547–554.

Gärling, T., & Golledge, R. G. (1989). Environmental perception and cognition. In E. H. Zube & G. T. Moore (eds.), *Advances in environment, behavior, and design* (Vol. 2, pp. 203–236). New York: Plenum Press.

Gärling, T., & Golledge, R. G. (1993). Understanding behavior and environment: A joint challenge to psychology and geography. In T. Gärling and R. G. Golledge (eds.), *Behavior and environment: Psychological and geographical approaches* (pp. 1–15). Amsterdam: Elsevier Science.

Gärling, T., Karlsson, N., Romanus, J., & Selart, M. (1997a). Influences of the past on choices of the future. In R. Ranyard, R. Crozier, & O. Svenson (eds.), *Decision making: Models and explanations* (pp. 167–188). London: Routledge.

Gärling, T., Lindberg, E., Carreiras, M., & Böök, A. (1986b). Reference systems in cognitive maps. *Journal of Environmental Psychology, 6,* 1–18.

Gärling, T., Lindberg, E., & Mäntylä, T. (1983). Orientation in buildings: Effects of familiarity, visual access, and orientation aids. *Journal of Applied Psychology, 68,* 177–186.

Gärling, T., Säisä, J., Böök, A., & Lindberg, E. (1986c). The spatio-temporal sequencing of everyday activities in the large-scale environment. *Journal of Environmental Psychology, 6,* 261–280.

Gärling, T., Selart, M., & Böök, A. (1997b). Investigating spatial choice and navigation in large-scale environments. In N. Foreman & R. Gillett (eds.), *A handbook of spatial research paradigms and methodologies* (Vol. 1, pp. 153–186). Hillsdale, NJ: Lawrence Erlbaum.

Gaunet, F., Martinez, J.-L., & Thinus-Blanc, C. (1997). Early-blind subjects' spatial abilities in the manipulatory space: A study of exploratory strategies and reaction to change performance. *Perception, 26,* 345–366.

Gaunet, F., & Thinus-Blanc, C. (1996). Early-blind subjects' abilities in the locomotor space: Exploratory strategies and reaction-to-change performance. *Perception, 25,* 967–981.

Gavrilov, V. V., Wiener, S. I., & Berthöz, A. (1995). Enhanced hippocampal theta EEG during whole body rotations in awake restrained rats. *Neuroscience Letters, 197,* 239–241.

Ghaem, O., Mellet, E., Crivello, F., Tzourio, N., Mazoyer, B., Berthöz, A., & Denis, M. (1997). Mental navigation along memorized routes activates the hippocampus, preuneus, and insula. *Neuroreport, 8*(3), 739–744.

Gibson, J. J. (1958). Visually controlled locomotion and visual orientation in animals. *British Journal of Psychology, 49,* 182–194.

Gibson, J. J. (1966). *The sense considered as perceptual systems.* Boston: Houghton Mifflin.

Gibson, J. J. (1979). *The ecological approach to visual perception.* Boston: Houghton Mifflin.

Gilmartin, P., & Patton, J. C. (1984). Comparing the sexes on spatial abilities: Map-use skills. *Annals of the Association of American Geographers, 74,* 605–619.

Giraudo, M.-D., & Pailhous, J. (1994). Distortions and fluctuations in topographic memory. *Memory and Cognition, 22,* 14–26.

Gladwin, T. (1970). *East is a big bird: Navigation and logic on Puluwat Atoll.* Cambridge, MA: Harvard University Press.

Glasauer, S., Amorim, M. A., Bloomberg, J. J., Reschke, M. F., Peters, B. T., Smith, S. L., & Berthöz, A. (1995). Spatial orientation during locomotion following space flight. *Acta Astronautica, 36*(8–12), 423–431.

Glasauer, S., Amorim, M.-A., Vitte, E., & Berthöz, A. (1994). Goal-directed linear locomotion in normal and labyrinthine-defective subjects. *Experimental Brain Research, 98,* 323–335.

Gluck, M. (1991). Making sense of human wayfinding: Review of cognitive and linguistic knowledge for personal navigation with a new research direction. In D. Mark and A. Frank (eds.), *Cognitive and linguistic aspects of geographic space* (pp. 117–135). Dordrecht: Kluwer Academic.

Gobet, F., & Simon, H. A. (1996). Recall of random and distorted chess positions: Implications for the Theory of Expertise. *Memory and Cognition, 24,* 493–503.

Golledge, R. G. (1974). Methods and methodological issues in environmental cognition research. In R. G. Golledge (ed.), *On determining cognitive configurations of a city,* Vol. I: *Problem statement, experimental design and preliminary findings* (pp. 225–242). Final report, NSF Grant GS-37969.

Golledge, R. G. (1976). Methods and methodological issues in environmental cognition research. In G. T. Moore & R. G. Golledge (eds.), *Environmental knowing* (pp. 300–313). Stroudsburg, PA: Dowden, Hutchinson, and Ross.

Golledge, R.G. (1977a). Environmental cues, cognitive mapping, spatial behavior. In D. Burke, et al. (eds.), *Behavior-environment research methods* (pp. 35–46). Madison: University of Wisconsin, Institute for Environmental Studies.

Golledge, R. G. (1977b). Multidimensional analysis in the study of environmental behavior and environmental design. In I. Altman & J. F. Wohlwill (eds.), *Human behavior and environment* (Vol. 1, pp. 2–46). New York: Plenum Press.

Golledge, R. G. (1978a). Learning about urban environments. In T. Carlstein, D. Parkes, & N. Thrift (eds.), *Timing space and spacing time,* Vol. I: *Making sense of time* (pp. 76–98). London: Edward Arnold.

Golledge, R. G. (1978b). Representing, interpreting and using cognized environments. *Papers and Proceedings, Regional Science Association,* 41, 169–204.

Golledge, R. G. (1987). Environmental cognition. In D. Stokols & I. Altman (eds.), *Handbook of environmental psychology* (Vol. 1, pp. 131–174). New York: John Wiley and Sons.

Golledge, R. G. (1990). The conceptual and empirical basis of a general theory of spatial knowledge. In M. M. Fischer, P. Nijkamp, & Y. Y. Papageorgiou (eds.), *Spatial choices and processes* (pp. 147–168). Amsterdam: Elsevier Science.

Golledge, R. G. (1992). Place recognition and wayfinding: Making sense of space. *Geoforum,* 23(2), 199–214.

Golledge, R. G. (1995a). Path selection and route preference in human navigation: A progress report. In A. U. Frank & W. Kuhn (eds.), *Spatial information theory: A theoretical basis for GIS* (pp. 207–222). Proceedings of the International Conference on Spatial Information and Theory (COSIT '95), Semmering, Austria. New York: Springer-Verlag.

Golledge, R. G. (1995b). Primitives of spatial knowledge. In T. L. Nyerges, D. M. Mark, R. Laurini, & M. J. Egenhofer (eds.), *Cognitive aspects of human-computer interaction for geographic information systems* (pp. 29–44). Dordrecht: Kluwer Academic.

Golledge, R. G. (1996). Geographical theories. *International Social Science Journal, Geography: State of the Art I—The Environmental Dimension,* 150, 461–476.

Golledge, R. G., Bell, S., & Dougherty, V. J. (1994). The cognitive map as an internalized GIS. Paper presented at the 90th Annual Meeting of the Association of American Geographers, San Francisco.

Golledge, R. G., Briggs, R., & Demko, D. (1969). The configuration of distances in intraurban space. *Proceedings of the Association of American Geographers,* 1, 60–65.

Golledge, R. G., Costanzo, C. M., & Marston, J. R. (1996a). *Public transit use by nondriving disabled persons: The case of the blind and vision impaired.* California PATH Working Paper UCB-ITS-PWP-96-1. Berkeley: University of California.

Golledge, R. G., Dougherty, V., & Bell, S. (1995). Acquiring spatial knowledge: Survey versus route-based knowledge in unfamiliar environments. *Annals of the Association of American Geographers,* 85(1), 134–158.

Golledge, R. G., Gale, N. D., Pellegrino, J. W., & Doherty, S. (1992). Spatial knowledge acquisition by children: Route learning and relational distances. *Annals of the Association of American Geographers,* 82(2), 223–244.

Golledge, R. G., & Hubert, L. J. (1982). Some comments on non-euclidean cognitive maps. *Environment and Planning,* A14, 107–118.

Golledge, R. G., Klatzky, R. L., & Loomis, J. M. (1996b). Cognitive mapping and wayfinding by adults without vision. In J. Portugali (ed.), *The construction of cognitive maps* (pp. 215–246). Dordrecht: Kluwer Academic.

Golledge, R. G., Parnicky, J. J., & Rayner, J. N. (1980). Procedures for defining and analyzing cognitive maps of the mildly and moderately mentally retarded. In R. G. Golledge, J. J. Parnicky, & J. N. Rayner (eds.), *The spatial competence of selected populations* (pp. 87–114). Ohio State University Research Foundation Project 761142/ 711173. Final Report, November, NSF Grant SOC77-26977-A02.

Golledge, R. G., Rayner, J. N., & Rivizzigno, V. L. (1982). Comparing objective and cognitive representations of environmental cues. In R. Golledge and J. Rayner (eds.), *Proximity and preference: Problems in the mulitdimensional analysis of large data sets* (pp. 233–266). Minneapolis: University of Minnesota Press.

Golledge, R. G., Richardson, G. D., Rayner, J. N., & Parnicky, J. J. (1983). Procedures for defining and analyzing cognitive maps of the mildly and moderately mentally retarded. In H. L. Pick, Jr., & L. P. Acredolo (eds.), *Spatial orientation: Theory, research, and application* (pp. 77–104). New York: Plenum Press.

Golledge, R. G., Ruggles, A. J., Pellegrino, J. W., & Gale, N. D. (1993). Integrating route knowledge in an unfamiliar neighborhood: Along and across route experiments. *Journal of Environmental Psychology,* 13(4), 293–307.

Golledge, R. G., Smith, T., Pellegrino, J. W., Doherty, S., & Marshall, S. (1985). A conceptual model and empirical analysis of children's acquisition of spatial knowledge. *Journal of Environmental Psychology,* 5, 125–152.

Golledge, R. G., & Spector, A. (1978). Comprehending the urban environment: Theory and practice. *Geographical Analysis,* 10, 403–426.

Golledge, R. G., and Stimson, R. J. (1997). *Spatial behavior: A geographic perspective.* New York: Guilford Press.

Goodridge, J. P., & Taube, J. S. (1995). Preferential use of the landmark navigational system by head direction cells. *Behavioral Neuroscience,* 109, 49–61.

Gopal, S., Klatzky, R. L., & Smith, T. R. (1989). NAVIGATOR: A psychologically based model of environmental learning through navigation. *Journal of Environmental Psychology,* 9(4), 309–331.

Gothard, K., Skaggs, W. E., Moore, K. M., & McNaughton, B. L. (1996). Binding of hippocampal CA1 neural activity to multiple reference frames in a landmark-based navigation task. *Journal of the Neurological Sciences,* 16, 823–835.

Gould, J. L. (1985). Absence of human homing ability as measured by displacement experiments. In J. L. Kirschvink, D. S. Jones, & B. J. McFadden (eds.), *Magnetite biomineralization and magnetoreception in organisms* (pp. 595–599). New York: Plenum Press.

Gould, J. L. (1986). The locale map of honey bees: Do insects have cognitive maps? *Science,* 232, 861–863.

Gould, P. (1966). *On mental maps.* Discussion Paper No. 9. Ann Arbor: Community of Mathematical Geographers, University of Michigan.

Gould, P., & White, R. (1974). *Mental maps.* Baltimore: Penguin.

Grasso, R., Glasauer, S., Takei, Y., & Berthöz, A. (1996). The predictive brain: Anticipatory control of head direction for the steering of locomotion. *Neuroreport,* 7, 1170–1174.

Graue, L. C. (1963). The effect of phase shifts in the day-night cycle on pigeon homing at distances of less than one mile. *Ohio Journal of Science,* 63, 214–217.

Griffin, A. S., & Etienne, A. S. (in press). Updating the path integrator through a visual fix. *Psychobiology.*

Griffin, D. (1948). *Topographical orientation.* In E. Boring, M. Langfeld, & H. Wald (eds.), Foundations of psychology. New York: Wiley. (Reprinted in part in R. Downs & D. Stea (eds.), 1973. *Image and environment: Cognitive mapping and spatial behavior* (pp. 296–329). Chicago: Aldine.

Griffin, D. R. (1955). Bird navigation. In A. Wolfson (ed.), *Recent studies in avian biology* (pp. 154–197). Urbana: University of Illinois Press.

Grüsser, O.-J., Guldin, W., Harris, L., Lefèbre, J. C., & Pause, M. (1991). Cortical representation of head-in-space movement and some psychophysical experiments on head movement. In A. Berthöz, W. Graf, & P. P. Vidal (eds.), *The head-neck sensorimotor system* (pp. 497–509). Oxford: Oxford University Press.

Grüsser, O.-J., & Landis, T. (1991). Getting lost in the world: Topographical disorientation and topographical agnosia. In J. Cronly-Dillon (ed.), *Vision and visual dysfunction,* Vol. 12: *Visual agnosias and other disturbances of visual perception and cognition* (pp. 411–430). London: Macmillan.

Grüsser, O.-J., Pause, M., & Schreiter, U. (1990a). Localisation and responses of neurones in the parieto-insular vestibular cortex of awake monkeys (*Macaca fascicularis*). *Journal of Physiology,* 430, 537–557.

Grüsser, O.-J., Pause, M., & Schreiter, U. (1990b). Vestibular neurons in the parieto-insular cortex of monkeys (*Macaca fascicularis*): Visual and neck receptor responses. *Journal of Physiology,* 430, 559–583.

Grüter, M., & Wiltschko, R. (1990). Pigeon homing: The effect of local experience on initial orientation and homing success. *Ethology,* 84, 239–255.

Guariglia, C., Padovani, A., Pantano, P., & Pizzamiglio, L. (1993). Unilateral neglect restricted to visual imagery. *Nature,* 364, 235–237.

Guilford, J. P. (1967). *The nature of human intelligence.* New York: McGraw-Hill.

Guilford, J. P., & Zimmerman, W. S. (1941). *Guilford-Zimmerman aptitude survey,* Part V: *Spatial orientation.* Beverly Hills: Sheridan Supply.

Guldin, W. O., Mirring, S., & Grüsser, O.-J. (1993). Connections from the neocortex to the vestibular brain stem nuclei in the common marmoset. *Neuroreport,* 5, 113–116.

Gulyás, B., & Roland, P. E. (1994). Binocular disparity discrimination in human cerebral cortex: Functional anatomy by positron emission tomography. *Proceedings of the National Academy of Sciences USA,* 91, 1239–1243.

Guth, D. A., & Rieser, J. J. (1998). Perception and the control of locomotion by blind and visually impaired pedestrians. In B. Blasch, W. Wiener, & R. Welch (eds.), *Handbook of orientation and mobility* (pp. 9–39). New York: American Foundation for the Blind.

Gwinner, E. (1968). Artspezifische Muster der Zugunruhe bei Laubsängern und ihre mögliche Bedeutung für die Beendigung des Zuges ins Winterquatier. *Zeitschrift für Tierpsychologie,* 25, 843–853.

Gwinner, E., & Wiltschko, W. (1978). Circannual changes in migratory orientation in the Garden Warbler (*Sylvia borin*). *Behavioral Ecology and Sociobiology,* 7, 73–78.

Gwinner, E., & Wiltschko, W. (1980). Endogenously controlled changes in the migratory direction of the Garden Warbler (*Sylvia borin*). *Journal of Comparative Physiology,* 125, 267–273.

Haber, L. R., Haber, R. N., Penningroth, S., Novak, K., & Radgowski, H. (1993). Comparison of nine methods of indicating the direction to objects: Data from blind adults. *Perception,* 22, 35–47.

Habib, M. (1987). Pure topographical disorientation: A definition and anatomical basis. *Cortex,* 23, 73–85.

Haken, H. (1983). *Synergetics: An introduction* (3rd ed.). Berlin: Springer-Verlag.

Haken, H. (1991). *Synergetic computers and cognition.* Berlin: Springer-Verlag.

Haken, H., & Portugali, J. (1996). Synergetics, inter-representation networks, and cognitive maps. In J. Portugali (ed.), *The construction of cognitive maps* (pp. 45–67). Dordrecht: Kluwer Academic.

Hanley, G. L., & Levine, M. (1983). Spatial problem solving: The integration of independently learned cognitive maps. *Memory and Cognition,* 11(4), 415–422.

Hanson, S. (1977). Measuring cognitive levels of urban residents. *Geographiska Annaler,* B59, 67–81.

Hanson, S. (1984). Environmental cognition and travel behavior. In T. D. Herbert & R. J. Johnston (eds.), *Geography and the urban environment: Progress in research and application* (pp. 95–126). London: John Wiley & Sons.

Hardwick, D. A., McIntyre, C. W., & Pick, H. L, Jr. (1976). The content and manipulation of cognitive maps in children and adults. *Monographs of the Society for Research in Child Development,* 41(3), serial no. 166.

Hart, R. A., & Moore, G. T. (1973). The development of spatial cognition, A review. In R. M. Downs & D. Stea (eds.), *Image and environment: Cognitive mapping and spatial behavior* (pp. 246–288). Chicago: Aldine.

Harvey, I., Husbands, P., & Cliff, D. (1994). Seeing the light: Artificial evolution, real vision. In D. Cliff, P. Husbands, J.-A. Meyer, & S. W. Wilson (eds.) *From animals to animats 3: Proceedings of the Third International Conference on Simulation of Adaptive Behaviour* (pp. 392–400). Cambridge, MA: MIT Press.

Hasher, L., & Zacks, R. (1979). Automatic versus effortful processes in memory. *Journal of Experimental Psychology: General,* 105, 356–388.

Haxby, J. V., Grady, C. L., Horwitz, B., Ungerleider, L. G., Mishkin, M., Carson, R. E., Herscovitch, P., Shapiro, M. B., & Rapoport, S. I. (1991). Dissociation of object and spatial visual processing pathways in human extrastriate cortex. *Proceedings of the National Academy of Sciences USA,* 88, 1621–1625.

Hayes-Roth, B., & Hayes-Roth, F. (1979). A cognitive model of planning. *Cognitive Science,* 3, 275–310.

Hebb, D. O. (1938). Studies of the organization of behavior. I. Behavior of the rat in a field orientation. *Journal of Comparative Psychology,* 25, 333–352.

Hebb, D. O. (1949). *The Organization of Behavior.* New York: Wiley.

Hécaen, H., Tzortzis, C., & Rondot, P. (1980). Loss of topographic memory with learning deficits. *Cortex,* 16, 525–542.

Heft, H. (1983). Wayfinding as the perception of information over time. *Population and Environment,* 6, 133–150.

Helbig, A. J. (1991). Inheritance of migratory direction in a bird species: A cross-breeding experiment with SE- and SW-migrating blackcaps (*Sylvia atricapilla*). *Behavioral Ecology and Sociobiology,* 28, 9–12.

Herman, J. F. (1980). Children's cognitive maps of large-scale spaces: Effects of exploration, direction, and repeated experience. *Journal of Experimental Child Psychology,* 29, 126–143.

Herman, J. F., Chatman, S. P., & Roth, S. F. (1983). Cognitive mapping in blind people: Acquisition of spatial relationships in a large-scale environment. *Journal of Visual Impairment and Blindness,* 77, 161–166.

Herman, J. F., & Coyne, A. C. (1980). Mental manipulation of spatial information in young and elderly adults. *Developmental Psychology,* 16, 537–538.

Hernandez, D. (1991). Relative representation of spatial knowledge: The 2-D case. In D. M. Mark & A. U. Frank (eds.), *Cognitive and linguistic aspects of geographic space* (pp. 373–385). Dordrecht: Kluwer Academic.

Hill, A. J. (1978). First occurrence of hippocampal spatial firing in a new environment. *Experimental Neurology,* 62, 282–297.

Hill, D. E. (1979). Orientation by jumping spiders of the genus Phidippus (Araneae: Salticidae) during the pursuit of prey. *Behavioral Ecology and Sociobiology,* 5, 301–322.

Hill, E., & Ponder, P. (1976). *Orientation and mobility techniques: A guide for the practitioner.* New York: American Foundation for the Blind.

Hill, E. W., Rieser, J. J., Hill, M. M., Hill, M., Halpin, J., & Halpin, R. (1993). How persons with visual impairments explore novel spaces: Strategies of good and poor performers. *Journal of Visual Impairment and Blindness,* October, 295–301.

Hintzman, D. L., O'Dell, C. S., & Arndt, D. R. (1981). Orientation in cognitive maps. *Cognitive Psychology,* 13, 149–206.

Hirtle, S. C., & Gärling, T. (1992). Heuristic rules for sequential spatial decisions. *Geoforum,* 23(2), 227–238.

Hirtle, S. C., & Jonides, J. (1985). Evidence of hierarchies in cognitive maps. *Memory and Cognition,* 13, 208–217.

Hoffman, A. J., & Wolfe, P. (1985). History. In E. L. Lawler, J. K. Lenstra, A.H.G. Rinnooy Kan, & D. B. Shmoys (eds.), *The traveling salesman problem* (pp. 1–15). New York: Wiley.

Holland, J. H. (1975). *Adaptation in natural and artificial systems.* Ann Arbor: University of Michigan Press.

Hong, J., Tan, X., Pinette, B., Weiss, R., & Riseman, E. M. (1991). Image based homing. In *Proceedings of the IEEE International Conference on Robotics and Automation 1991,* 620–625.

Horn, D. L. (1996). *The effects of spatial field on performance for a visually directed task.* Unpublished bachelor of arts honors thesis, Department of Psychology, University of California, Santa Barbara.

Hudson, J. A., Shapiro, L. R., & Sosa, B. B. (1994). *Planning in the real world: Preschool children's scripts and plans for familiar events.* Unpublished manuscript.

Hughey, D. J., & Koppenaal, R. J. (1987). Hippocampal lesions in rats alter learning about intramaze cues. *Behavioral Neuroscience, 101,* 634–643.

Hunt, E., Pellegrino, J. W., Frick, R. W., Farr, S. A., & Alderton, D. (1988). The ability to reason about movement in the visual field. *Intelligence, 12,* 77–100.

Hutchins, E. (1995). *Cognition in the wild.* Cambridge, MA: MIT Press.

Huttenlocher, J., Hedges, L. V., & Duncan, S. (1991). Categories and particulars: Prototype effects in estimating spatial location. *Psychological Review, 98,* 352–376.

Huttenlocher, J., Newcombe, N., & Sandberg, E. H. (1994). The coding of spatial location in young children. *Cognitive Psychology, 27,* 115–148.

Huttenlocher, J., & Presson, C. C. (1973). Mental rotation and the perspective change problem. *Cognitive Psychology, 4,* 277–299.

Huttenlocher, J., & Presson, C. C. (1979). The coding and transformation of spatial information. *Cognitive Psychology, 11,* 375–394.

Incisa della Rocchetta, A., Cipolotti, L., & Warrington, K. (1996). Topographical disorientation: Selective impairment of locomotor space? *Cortex, 32,* 727–735.

Intons-Peterson, M. J. (1996). Integrating the components of imagery. In M. de Vega, M. Intons-Peterson, P. Johnson-Laird, M. Denis, & M. Marschark (eds.), *Models of visuospatial cognition* (pp. 20–89). New York: Oxford University Press.

Ioalè, P., Wallraff, H. G., Papi, F., & Foa, A. (1983). Long-distance releases to determine the spatial range of pigeon navigation. *Comparative Biochemistry and Physiology, 76A,* 733–743.

Israël, I., & Berthöz, A. (1989). Contribution of the otoliths to the calculation of linear deplacement. *Journal of Neurophysiology, 62,* 247–263.

Israël, I., & Berthöz, A. (1992). Representations of space and motion in man. In G. E. Stelmach & J. Requin (eds.), Tutorials in motor behavior (pp. 195–209). Amsterdam: Elsevier.

Israël, I., Chapuis, N., Glasauer, S., Charade, O., & Berthöz, A. (1993). Estimation of passive horizontal linear whole-body displacement in humans. *Journal of Neurophysiology, 70*(3), 1270–1273.

Israël, I., Fetter, M., & Koenig, E. (1993). Vestibular perception of passive whole body rotation about the horizontal and vertical axes in humans: Goal directed vestibuloocular reflex and vestibular memory contingent saccades. *Experimental Brain Research, 96*(2), 335–346.

Israël, I., Rivaud, S., Berthöz, A., & Pierrot-Deseilligny, C. (1992). Cortical control of vestibular memory-guided saccades. *Annals of the New York Academy of Science, 656,* 472–484.

Israël, I., Rivaud, S., Gaymard, B., Berthöz, A., & Pierrot-Deseilligny, C. (1995). Cortical control of vestibular-guided saccades in man. *Brain, 118,* 1169–1183.

Israël, I., Rivaud, S., Pierrot-Deseilligny, P., & Berthöz, A. (1991). "Delayed VOR": An assessment of vestibular memory for self motion. In J. Requin & J. Stelmach (eds.), *Tutorials in motor neuroscience* (pp. 599–607). Dordrecht: Kluwer Academic.

Ivanenko, Y. P., Grasso, R., Israël, I., & Berthöz, A. (1997). The contribution of otoliths and semicircular canals to the perception of two-dimensional passive whole-body motion in humans. *Journal of Physiology, 502,* 223–233.

Jacobs, R. G., Moeller, M. B., Turner, S. M., & Wall, C. (1985). Otoneurological examination in panic disorder and agoraphobia with panic attacks: A pilot study. *American Journal of Psychiatry, 142,* 715–720.

Jain, A. K., & Dubes, R. C. (1988). *Algorithms for clustering data.* Englewood Cliffs, NJ: Prentice-Hall.

James, W. (1892). *Psychology: The briefer course* (1961 ed.). New York: Harper and Row.

Jammer, M. (1993). *Concepts of space: The history of theories of space in physics* (3rd ed.). New York: Dover Publications.

Jamon, M. (1994). An analysis of trail-following behavior in the wood mouse, *Apodemus sylvaticus. Animal Behavior, 47,* 1127–1134.

Jeannerod, M. (1994). The representing brain: Neural correlates of motor imagery and intention. *Behavioral and Brain Sciences, 17,* 187–245.

Judd, S. P. D., & Collett, T. S. (1998). Multiple stored views and landmark guidance in ants. *Nature,* 392, 710–714.

Junger, W. (1991). Waterstriders (*Gerris paludum* F.) compensate for drift with a discontinuously working visual position servo. *Journal of Comprehensive Physiology,* A169, 633–639.

Just, M. A., & Carpenter, P. A. (1985). Cognitive coordinate systems: Accounts of mental rotation and individual differences in spatial ability. *Psychological Review,* 92, 137–172.

Juurmaa, J., & Lehtinen-Railo, S. (1994). Visual experience and access to spatial knowledge. *Journal of Visual Impairment and Blindness,* March–April, 157–170.

Juurmaa, J., & Suonio, K. (1975). The role of audition and motion in the spatial orientation of the blind and sighted. *Scandinavian Journal of Psychology,* 16, 209–216.

Kahneman, D., Slovic, P. L., & Tversky, A. (1982). *Judgement under uncertainty: Heuristics and biases.* Cambridge: Cambridge University Press.

Kail, R. (1991). Developmental change in speed of processing during childhood and adolescence. *Psychological Bulletin,* 109, 490–501.

Kaplan, S. (1970). The role of location processing in the perception of the environment. In J. Archea & C. Eastman (eds.), *Proceedings of the Second Annual Environmental Design Research Association Conference* (pp. 131–134). Pittsburgh: Carnegie-Mellon University.

Kaplan, S. (1973). Cognitive maps in perception and thought. In R. M. Downs & D. Stea (eds.), *Image and environment: Cognitive mapping and spatial behavior* (pp. 63–78). Chicago: Aldine.

Kaplan, S., & Kaplan, R. (1982). *Cognition and Environment* (1983 ed.). Ann Arbor, MI: Ulrichs.

Kaplan, S., Sonntag, M., & Chown, E. (1991). Tracing recurrent activity in cognitive elements (TRACE): A model of temporal dynamics in a cell assembly. *Connection Science,* 3, 176–206.

Kearns, M., & Valiant, L. G. (1994). Cryptographic limitations on learning Boolean formulae and finite automata. *Journal of the ACM,* 41(1), 67–95.

Keeney, R. L., & Raifa, H. (1976). *Decisions with multiple objectives: Preferences and trade-offs.* New York: Wiley.

Keeton, W. T. (1969). Orientation by pigeons: Is the sun necessary? *Science,* 165, 922–928.

Keeton, W. T. (1973). Release-site bias as a possible guide to the "map" component in pigeon homing. *Journal of Comparative Physiology,* 86, 1–16.

Keeton, W. T. (1974). The orientational and navigational basis of homing in birds. *Advances in the Study of Behavior,* 5, 47–132.

Kendler, H. H., & Gasser, W. P. (1948). Variables in spatial learning: I. Number of reinforcements during training. *Journal of Comparative Physiology,* 41, 178–187.

Kershner, J. (1974). Relationship of motor development to visuospatial cognitive growth. *Journal of Special Education,* 8, 91–102.

Khattak, J. A., Schofer, L. J., & Koppelman, S. F. (1992). Commuters' enroute diversion and return decisions: Analysis and implications for advanced traveler information systems. *Transportation Research,* A27(2), 101–111.

Kiepenheuer, J. (1978). Pigeon navigation and magnetic field. *Naturwissenschaften,* 65, 113.

Kirasic, K. C. (1991). Spatial cognition and behavior in young and elderly adults: Implications for learning new environments. *Psychology and Aging,* 6(1), 10–18.

Kirasic, K. C. (1996). *Spatial and propositional representations of novel environments.* Paper presented at the Cognitive Aging Conference, Atlanta.

Kirasic, K. C. (1997). *Predictors of spatial performance and spatial behavior in young and elderly adults.* Paper presented at meetings of the American Association of Geographers, Ft. Worth.

Kirasic, K. C., Allen, G. L., & Haggerty, D. (1992). Age-related differences in adults' macrospatial cognitive processes. *Experimental Aging Research,* 18, 33–39.

Kirasic, K. C., Allen, G. L., & Siegel, A. W. (1984). Expression of configurational knowledge of large-scale environments. *Environment and Behavior,* 16, 687–712.

Kirschvink, J. L., Jones, D. S., & McFadden, B. J. (eds.). (1985). *Magnetite biomineralization and magnetoreception in organisms*. New York: Plenum Press.

Kitchin, R. M. (1994). Cognitive maps: What are they and why study them? *Journal of Environmental Psychology*, 14(1), 1–19.

Klatzky, R. L., Beall, A. C., Loomis, J. M., Golledge, R. G., & Philbeck, J. W. (unpublished). *Human navigation ability: Tests of the encoding-error model of path integration*. Unpublished manuscript. Department of Psychology, Carnegie-Mellon University, Pittsburgh.

Klatzky, R. L., Loomis, J. M., Beall, A. C., Chance, S. S., & Golledge, R. G. (1998). Spatial updating of self-position and orientation during real, imagined, and virtual locomotion. *Psychological Science*, 9(4), 293–298.

Klatzky, R. L., Loomis, J. M., & Golledge, R. G. (1997). Encoding spatial representations through nonvisually guided locomotion: Tests of human path integration. In D. Medin (ed.), *The psychology of learning and motivation* (Vol. 37, pp. 41–84). San Diego: Academic Press.

Klatzky, R. L., Loomis, J. M., Golledge, R. G., Cicinelli, J. G., Doherty, S., & Pellegrino, J. W. (1990). Acquisition of route and survey knowledge in the absence of vision. *Journal of Motor Behavior*, 22(1), 19–43.

Knierim, J. J., Kudrimoti, H. S., & McNaughton, B. L. (1995). Place cells, head direction cells, and the learning of landmark stability. *Journal of Neuroscience*, 15, 1648–1659.

Knierim, J. J., Kudrimoti, H. S., Skaggs, W. E., & McNaughton, B. L. (1996). The interaction between vestibular cues and visual landmark learning in spatial navigation. In T. Ono, B. McNaughton, S. Molotchnikoff, E. Rolls, & H. Nishijo (eds.), *Perception, memory and emotion: Frontiers in neuroscience* (pp. 343–357). Oxford: Pergamon.

Knierim, J. J., McNaughton, B. L., Duffield, C., & Bliss, J. (1993). On the binding of hippocampal place fields to the inertial orientation system. *Society for Neuroscience*, 19, 795.

Kortenkamp, D. (1993). *Cognitive maps for mobile robots: A representation for mapping and navigation*. Unpublished Ph.D. dissertation, Computer Science Department, University of Michigan, Ann Arbor.

Kortenkamp, D., & Chown, E. (1993). A directional spreading activation network for mobile robot navigation. In J.-A. Meyer, H. Roitblat, & S.W. Wilson (eds.), *From animals to animats 2: Proceedings of the Second International Conference on Simulation of Adaptive Behavior* (pp. 218–224). Cambridge, MA: MIT Press.

Kortenkamp, D., & Weymouth, T. (1994). Topological mapping for mobile robots using a combination of sonar and vision sensing. In *Proceedings of the Twelfth National Conference on Artificial Intelligence* (AAAI-94). Seattle: University of Washington Press.

Kosslyn, S. M. (1978). Measuring the visual angle of the mind's eye. *Cognitive Psychology*, 10, 356–389.

Kosslyn, S. M. (1980). *Image and Mind*. Cambridge, MA: Harvard University Press.

Kosslyn, S. M. (1981). The medium and the message in mental imagery: A theory. *Psychological Review*, 88, 46–66.

Kosslyn, S. M. (1991). A cognitive neuroscience of visual cognition: Further developments. In R. H. Logie & M. Denis (eds.), *Mental Images in Human Cognition* (pp. 351–381). Amsterdam: Elsevier Science.

Kosslyn, S. M., Ball, T. M., & Reiser, B. J. (1978). Visual images preserve metric spatial information: Evidence from studies of image scanning. *Journal of Experimental Psychology: Human Perception and Performance*, 4, 47–60.

Kosslyn, S. M., Brunn, J., Cave, K. R., & Wallach, R. W. (1984). Individual difference in mental image ability: A computational analysis. *Cognition*, 18, 195–243.

Kosslyn, S. M., & Koenig, O. (1992). *Wet mind: The new cognitive neuroscience*. New York: Free Press.

Kosslyn, S. M., Van Kleek, M. H., & Kirby, K. N. (1990). A neurologically plausible model of individual differences in visual mental imagery. In P. J. Hampson, D. F. Marks, & J. T. E. Richardson (eds.), *Imagery: Current developments* (pp. 39–77). New York: Routledge.

Kozlowski, L. T., & Bryant, K. J. (1977). Sense of direction, spatial orientation, and cognitive maps. *Journal of Experimental Psychology: Human Perception and Performance,* 3, 590–598.

Kramer, G. (1959). Recent experiments on bird orientation. *Ibis,* 101, 399–416.

Kritchevsky, M. (1988). The elementary spatial functions of the brain. In J. Stiles-Davis, M. Kritchevsky, & U. Bellugi (eds.), *Spatial cognition: Brain bases and development* (pp. 111–140). Hillsdale, NJ: Lawrence Erlbaum.

Kubie, J. L., & Ranck, J. B., Jr. (1983). Sensory-behavioral correlates in individual hippocampal neurons in three situations: Space and context. In W. Seifert (ed.), *Neurobiology of the Hippocampus* (pp. 433–447). London: Academic Press.

Kuipers, B. J. (1978). Modeling spatial knowledge, *Cognitive Science,* 2, 129–153.

Kuipers, B. J. (1982). The "map in the head" metaphor. *Environment and Behavior,* 14, 202–220.

Kuipers, B. J. (1983). The cognitive map: Could it have been any other way? In H. Pick & L. Acredolo (eds.), *Spatial orientation: Theory, research, and applications* (pp. 345–359). New York: Plenum Press.

Kuipers, B. J., & Byun, Y. (1991). A robot exploration and mapping strategy based on a semantic hierarchy of spatial representations. *Journal of Robotics and Autonomous Systems,* 8, 47–63.

Kuipers, B. J., & Levitt, T. (1988). Navigation and mapping in large-scale space. *AI Magazine,* 9, 24–46.

Kyllonen, P. C. (1993). Aptitude testing inspired by information processing: A test of the four-source model. *Journal of General Psychology,* 120, 375–405.

Lackner, J. R., & DiZio, P. (1988). Visual stimulation affects the perception of voluntary leg movements during walking. *Perception,* 17, 71–80.

Lakoff, G. (1987). *Women, fire, and dangerous things: What categories reveal about the mind.* Chicago: University of Chicago Press.

Landau, B. (1986). Early map use as an unlearned ability. *Cognition,* 22, 210–223.

Landau, B., Gleitman, H., & Spelke, E. (1981). Spatial knowledge and geometric representation in a child blind from birth. *Science,* 213, 1275–1278.

Landau, B., & Jackendoff, R. (1993) "What" and "where" in spatial language and spatial cognition. *Behavioral and Brain Sciences,* 16, 217–238.

Landau, B., Spelke, E., & Gleitman, H. (1984). Spatial knowledge in a young blind child. *Cognition,* 16, 225–260.

Lauer, J., & Lindauer, M. (1971). Genetisch fixierte Lerndispositionen bei der Honigbiene. In *Informationsaufnahme und Informationsverarbeitung im lebenden Organismus* (Vol. 1, pp. 1–87). Wiesbaden: Franz Steiner Verlag.

Laurent, M., & Cavallo, V. (1985). Role des modalités de prise d'informations visuelles dans un pointage locomoteur. *L'Annee Psychologique,* 85, 41–48.

Lavenex, P., & Schenk, F. (1995). Influence of local environmental olfactory cues on place learning in rats. *Physiology and Behavior,* 58, 1059–1066.

Lavenex, P., & Schenk, F. (1996). Integration of olfactory information in a spatial representation enabling accurate arm choice in the radial arm maze. *Learning and Memory,* 2, 299–319.

Lavenex, P., & Schenk, F. (1998). Olfactory cues potentiate learning of distant visuospatial information. *Neurobiology of Learning and Memory,* 68, 140–153.

Lawton, C. A. (1996). Strategies for indoor wayfinding: The role of orientation. *Journal of Environmental Psychology,* 16(2), 137–145.

Lea, G. (1975). Chronometric analysis of the method of loci. *Journal of Experimental Psychology: Human Perception and Performance,* 1, 95–104.

Lee, D. B., Jr. (1968). *Models and techniques for urban planning.* Ithaca, NY: Cornell University Press.

Lee, D. N. (1978). The functions of vision. In H. Pick & E. Saltzman (eds.), *Modes of perceiving and processing information.* Hillsdale, NJ: Lawrence Erlbaum.

Lee, D. N. (1980). The optic flow field: The foundation of vision. *Philosophical Transactions of the Royal Society of London, Series B*, 290, 169–179.

Lee, T. (1970). Perceived distance as a function of direction in the city. *Environment and Behavior*, 2, 40–51.

Lehrer, M., Srinivasan, M. V., Zhang, S. W., & Horridge, G. A. (1988). Motion cues provide the bee's visual world with a third dimension. *Nature*, 332, 356–357.

Lehtinen-Railo, S., & Juurmaa, J. (1994). Effect of visual experience on locational judgments after perspective change in small-scale space. *Scandinavian Journal of Psychology*, 35, 175–183.

Leiser, D., & Zilbershatz, A. (1989). The traveller: A computational model of spatial network learning. *Environment and Behavior*, 21(4), 435–463.

Lepecq, J.-C. (1989). The early development of position constancy in a no-landmark environment. *British Journal of Developmental Psychology*, 7, 289–306.

Lesperance, R. M. (1990). *The location system: Using approximate location and size information for scene segmentation.* Unpublished Ph.D. dissertation, Department of Computer Science, University of Michigan, Ann Arbor.

Levenick, J. R. (1991). NAPS: A connectionist implementation of cognitive maps. *Connection Science*, 3, 107–126.

Levesque, H. J., & Brachman, R. J. (1985). A fundamental tradeoff in knowledge representation and reasoning. In *Readings in Knowledge Representation* (pp. 41–70). Los Altos, CA: Morgan Kaufman.

Levin, I. P., Louviere, J. J., & Schepanski, A. A. (1983). Validation test and applications of laboratory studies of information integration. *Organizational Behavior and Human Performance*, 31, 173–193.

Levine, D. N., Warach, J., & Farah, M. (1985). Two visual systems in mental imagery: Dissociation of "what" and "where" in imagery disorders due to bilateral posterior cerebral lesions. *Neurology*, 35, 1010–1018.

Levine, M. (1982). You-are-here maps: Psychological considerations. *Environment and Behavior*, 14(2), 221–237.

Levine, M., Jankovic, I. N., & Palij, M. (1982). Principles of spatial problem solving. *Journal of Experimental Psychology: General*, 111(2), 157–175.

Levine, M., Marchon, I., & Hanley, G. (1984). The placement and misplacement of you-are-here maps. *Environment and Behavior*, 16(2), 139–157.

Lewis, D. (1976). Route finding by desert aborigines in Australia. *Journal of Navigation*, 29, 21–38.

Ley, D. (1974). *The black inner city as a frontier outpost: Images and behavior of a North Philadelphia neighborhood.* Washington, DC: Association of American Geographers.

Liben, L. S. (1981). Spatial representation and behavior: Multiple perspectives. In L. S. Liben, A. H. Patterson, & N. Newcombe (eds.), *Spatial representation and behavior across the lifespan* (pp. 3–36). New York: Academic Press.

Liben, L. S. (1982). Children's large-scale spatial cognition: Is the measure the message? In R. Cohen (ed.), *New directions for child development: Children's conceptions of spatial relationships* (pp. 51–64). San Francisco: Jossey-Bass.

Liben, L. S., & Downs, R. M. (1986). *Children's production and comprehension of maps: Increasing graphic literacy.* Final report for Grant 6–83–0025, National Institute of Education, Washington, DC.

Lieblich, I., & Arbib, M. (1982). Multiple representation of space underlying behavior and associated commentaries. *The Behavior and Brain Sciences*, 5(4), 627–660.

Light, L., & Zelinski, E. M. (1983). Memory for spatial information in young and old adults. *Developmental Psychology*, 19, 901–906.

Lindauer, M. (1963). Kompassorientierung. *Ergebnisse der Biologie*, 26, 15–181.

Lindberg, E., & Gärling, T. (1981a). Acquisition of locational information about reference points during blindfolded and sighted locomotion: Effects of a concurrent task and locomotion paths. *Scandinavian Journal of Psychology*, 22, 101–108.

Lindberg, E., & Gärling, T. (1981b). Acquisition of locational information about reference points during locomotion with and without a concurrent task: Effects of number of reference points. *Scandinavian Journal of Psychology*, 22, 109–115.

Lindberg, E., & Gärling, T. (1982). Acquisition of locational information about reference points during locomotion: The role of central information processing. *Scandinavian Journal of Psychology*, 23, 207–218.

Lindberg, E., & Gärling, T. (1983). Acquisition of different types of locational information in cognitive maps: Automatic or effortful processing? *Psychological Research*, 45, 19–38.

Lloyd, R. E. (1982). A look at images. *Annals of the Association of American Geographers*, 72(4), 532–548.

Lloyd, R. E. (1989). Cognitive maps: Encoding and decoding information. *Annals of the American Association of Geographers*, 79(1), 101–124.

Lloyd, R. E. (1993). Cognitive processes and cartographic maps. In T. Garling & R. G. Golledge (eds.), *Behavior and environment: Psychological and geographical approaches* (pp. 141–169). Amsterdam: Elsevier.

Lloyd, R. E., Cammack, R., & Holliday, W. (1996). Learning environments and switching perspectives. *Cartographica*, April, 5–17.

Lloyd, R. E., & Heivly, C. (1987). Systematic distortion in urban cognitive maps. *Annals of the Association of American Geographers*, 77, 191–207.

Loarer, E., & Savoyant, A. (1991). Visual imagery in locomotor movement without vision. In R. H. Logie & M. Denis (eds.), *Mental images in human cognition* (pp. 35–46). Amsterdam: Elsevier Science.

Lobel, E., Le Bihan, D., Leroy-Willig, A., & Berthöz, A. (1996). Searching for the vestibular cortex with functional MRI. *Neuroimage*, 3, S351 (Abstract).

Loewenstein, G., & Prelec, D. (1993). Preferences for sequences of outcomes. *Psychological Review*, 100, 91–108.

Loftus, G. (1978). Comprehending compass directions. *Memory and Cognition*, 6, 416–422.

Logothetis, N. K., & Sheinberg, D. L. (1996). Visual object recognition. In W. Cowan, E. Shooter, C. Stevens, & R. Thompson (eds.), *Annual review of neuroscience* (Vol. 19, pp. 577–621). Palo Alto, CA: Annual Reviews.

Lohman, D. F. (1979). *Spatial ability: A review and reanalysis of the correlational literature*. Technical Report No. 8, Aptitude Research Project. Stanford: Stanford University School of Education.

Lohman, D. F. (1988). Spatial abilities as traits, processes, and knowledge. In R. J. Sternberg (ed.), *Advances in the psychology of human intelligence* (Vol. 4, pp. 181–248). Hillsdale, NJ: Lawrence Erlbaum.

Löhrl, H. (1959). Zur Frage des Zeitpunkts einer Prägung auf die Heimatregion beim Halsbandschnäpper (Ficedula albicollis). *Journal für Ornithologie*, 100, 132–140.

Loomis, J. M. (1996). Visual space and other internal representations involved in the control of spatial behavior. Paper presented at ENP School Conference on the Representation of Space, San Feliu, Spain.

Loomis, J. M. (in press). Locomotion and navigation: Control by stimulation and internal representation. In M. McBeath (ed.), *Spatial navigational principles used by humans, animals, and machines*. London: Sage.

Loomis, J. M., Beall, A. C., Klatzky, R. L., Golledge, R. G., & Philbeck, J. W. (1995a). *Evaluating the sensory inputs to path integration*. Paper presented at the meeting of the Psychonomic Society, Los Angeles.

Loomis, J. M., Da Silva, J. A., Fujita, N., & Fukusima, S. S. (1992). Visual space perception and visually directed action. *Journal of Experimental Psychology: Human Perception and Performance*, 18(4), 906–921.

Loomis, J. M., Da Silva, J. A., Philbeck, J. W., & Fukusima, S. S. (1996). Visual perception of location and distance. *Current Directions in Psychological Science*, 5(3), 72–77.

Loomis, J. M., Golledge, R. G., & Klatzky, R. L. (1995b). *Personal guidance system for blind persons.* In *Proceedings of the Conference on Orientation and Navigation Systems for Blind Persons,* Hatfield, England, February 1–2.

Loomis, J. M., Golledge, R. G., & Klatzky, R. L. (1998). Navigation system for the blind: Auditory display modes and guidance. *Presence, 7,* 193–203.

Loomis, J. M., Golledge, R. G., Klatzky, R. L., Speigle, J. M., & Tietz, J. (1994). Personal guidance system for the visually impaired. In *Proceedings of the First Annual ACM Conference on Assistive Technologies* (pp. 85–91). New York: Association for Computing Machinery.

Loomis, J. M., Klatzky, R. L., Golledge, R. G., Cicinelli, J. G., Pellegrino, J. W., & Fry, P. A. (1993). Nonvisual navigation by blind and sighted: Assessment of path integration ability. *Journal of Experimental Psychology: General,* 122(1), 73–91.

Loomis, J. M., Klatzky, R. L., Philbeck, J. W., & Golledge, R. G. (1998). Assessing auditory distance perception using perceptually directed action. *Perception & Psychophysics,* 60, 966–980.

Lorenz, C. A., & Neisser, U. (1986). *Ecological and psychometric dimensions of spatial ability.* Technical Report No. 10. Atlanta, GA: Emory Cognition Project, Emory University.

Lorenz, K. (1973). *Die Rückseite des Spiegels.* Munich: R. Piper.

Lovelace, K. L., & Montello, D. R. (1997). Spatial cognition: Are there sex-related differences, and what do they mean? Paper presented at meetings of the American Association of Geographers, Ft. Worth.

Luce, R. D. (1986). *Response times: Their role in inferring elementary mental organization.* Oxford: Oxford University Press.

Luschi, P., & Dall'Antonia, P. (1993). Anosmic pigeons orient from familiar sites by relying on the map-and-compass mechanism. *Animal Behavior,* 46, 1195–1203.

Lynch, K. (1960). *The image of the city.* Cambridge, MA: MIT Press.

McClelland, J. L., McNaughton, B. L., & O'Reilly, R. C. (1995). Why there are complementary learning systems in the hippocampus and neocortex: Insights from the successes and failures of connectionist modes of learning and memory. *Psychological Review,* 102, 419–457.

MacCorquodale, K., & Meehl, P. (1953). Hypothetical constructs and intervening variables. In H. Feigl & M. Brodbeck (eds.), *Readings in the philosophy of science* (pp. 596–611). New York: Appleton-Century-Crofts.

McDonald, T. P., & Pellegrino, J. W. (1993). Psychological perspectives on cognitive mapping. In T. Gärling & R. G. Golledge (eds.), *Behavior and environment: Psychological and geographical approaches* (pp. 47–82). Amsterdam: Elsevier Science.

MacEachren, A. M. (1986). A linear view of the world: Strip maps as a unique form of cartographic representation. *American Cartographer,* 13(1), 7–25.

MacEachren, A. M. (1991). The role of maps in spatial knowledge acquisition. *Cartographic Journal,* 28(2), 152–162.

MacEachren, A. M. (1992a). Application of environmental learning theory to spatial knowledge acquisition from maps. *Annals of the Association of American Geographers,* 82(2), 245–274.

MacEachren, A. M. (1992b). Learning spatial information from maps: Can orientation-specificity be overcome? *Professional Geographer,* 44(4), 431–443.

MacEachren, A. M. (1995). *How maps work: Representation, visualization, and design.* New York: Guilford.

McGee, M. G. (1979). Human spatial abilities: Psychometric studies and environmental, genetic, hormonal, and neurological influences. *Psychological Bulletin,* 86(5), 889–918.

McKenzie, B. E., Day, R. H., & Ihsen, E. (1984). Localization of events in space: Infants are not always egocentric. *British Journal of Developmental Psychology,* 2, 1–9.

McNamara, T. P. (1986). Mental representations of spatial relations. *Cognitive Psychology,* 18, 87–121.

McNamara, T. P. (1992). Spatial representation. *Geoforum*, 23(2), 139–150.

McNamara, T. P., Halpin, J. A., & Hardy, J. K. (1992). The representation and integration in memory of spatial and nonspatial information. *Memory and Cognition*, 20, 519–532.

McNamara, T. P., Hardy, J. K., & Hirtle, S. (1989). Subjective hierarchies in spatial memory. *Journal of Experimental Psychology: Learning, Memory, and Cognition*, 15, 211–227.

McNamara, T. P., Ratcliff, R., & McKoon, G. (1984). The mental representation of knowledge acquired from maps. *Journal of Experimental Psychology: Learning, Memory, and Cognition*, 10, 723–732.

McNaughton, B. L. (1989). Neuronal mechanisms for spatial computation and information storage. In L. Nadel, L. A. Cooper, P. Culicover, & R. M. Harnish (eds.), *Neural connections, mental computations* (pp. 285–350). Cambridge, MA: MIT Press.

McNaughton, B. L. (1996). Cognitive cartography. *Nature*, 381, 368–369.

McNaughton, B. L., Barnes, C. A., Gerrard, J. L., Gothard, K., Jung, M. W., Knierim, J. J., Kudrimoti, H., Qin, Y., Skaggs, W. E., Suster, M., & Weaver, K. L. (1996). Deciphering the hippocampal polyglot: The hippocampus as a path integration system. *Journal of Experimental Biology*, 199(1), 173–185.

McNaughton, B. L., Barnes, C. A., & O'Keefe, J. (1983). The contributions of position, direction, and velocity to single unit activity in the hippocampus of freely-moving rats. *Experimental Brain Research*, 52, 41–49.

McNaughton, B. L., Chen, L. L., & Markus, E. J. (1991). "Dead reckoning," landmark learning, and the sense of direction: A neurophysiological and computational hypothesis. *Journal of Cognitive Neuroscience*, 3, 190–202.

McNaughton, B. L., Knierim, J. L., & Wilson, M. A. (1994). Vector encoding and the vestibular foundations of spatial cognition: Neurophysiological and computational mechanisms. In M. Gazzaniga (ed.), *The cognitive neurosciences* (pp. 585–595). Cambridge, MA: MIT Press.

Magliano, J. P., Cohen, R., Allen, G. L., & Rodrigue, J. R. (1995). The impact of a wayfinder's goal on learning a new environment: Different types of spatial knowledge of goals. *Journal of Environmental Psychology*, 15, 65–75.

Maguire, E. (1997). Hippocampal function in human topographical memory. Paper presented at the Royal Society Discussion Meeting, "What Are the Parietal and Hippocampal Contributions to Spatial Cognition," London.

Mandler, J. M., Seegmiller, D., & Day, J. (1977). On the coding of spatial information. *Memory and Cognition*, 5, 10–16.

Mataric, M. J. (1990). *A distributed model for mobile robot environment-learning and navigation*. AI Technical Report No. 1228. Cambridge, MA: Massachusetts Institute of Technology.

Mataric, M. J. (1992). Integration of representation into goal-driven behavior-based robots. *IEEE Transactions on Robotics and Automation*, 8(3), 304–312.

Mather, J. G., & Baker, R. R. (1980). Magnetic sense of direction in woodmice for route-based navigation. *Nature*, 291, 152–155.

Matthews, G. V. T. (1951). The experimental investigation of navigation in homing pigeons. *Journal of Experimental Biology*, 28, 508–536.

Matthews, M. H. (1992). *Making sense of place: Children's understanding of large-scale environments*. Savage, MD: Barnes and Noble.

Maule, J. A., & Svenson, O. (1993). Theoretical and empirical approaches to behavioral decision making and their relation to time constraints. In O. Svenson & J. A. Maule (eds.), *Time pressure and stress in human judgment and decision making* (pp. 3–25). New York: Plenum Press.

Maurer, R. (1998). A connectionist model of path integration with and without a representation of distance. *Psychobiology*, 26, 21–35.

Maurer, R., & Séguinot, V. (1995). What is modelling for? A critical review of the models of path integration. *Journal of Theoretical Biology*, 175, 457–475.

May, M. (1996). Cognitive and embodied modes of spatial imagery. *Psychologische Beiträge*, 38, 418–434.

Mecklinger, A., & Pfeifer, E. (1996). Event-related potentials reveal topographical and temporal distinct neuronal activation patterns for spatial and object working memory. *Cognitive Brain Research*, 4, 211–224.

Mellet, E., Tzourio, N., Denis, M., & Mazoyer, B. (1995). A positron emission tomography study of visual and mental exploration. *Journal of Cognitive Neuroscience*, 7, 433–445.

Menzel, E. W. (1973). Chimpanzee spatial memory organization. *Science*, 182, 943–945.

Menzel, E. W. (1979). Cognitive mapping in chimpanzees. In S. H. Hulse, H. F. Fowler, & W. K. Honig (eds.), *Cognitive aspects of animal behaviour* (pp. 375–422). Hillsdale, NJ: Lawrence Erlbaum.

Menzel, R., Geiger, K., Chittka, L., Joerges, J., Kunze J., & Müller, U. (1996). The knowledge base of bee navigation. *Journal of Experimental Biology*, 199, 141–146.

Michener, M. C., & Walcott, C. (1967). Homing in single pigeons—Analysis of tracks. *Journal of Experimental Biology*, 47, 99–131.

Millar, S. (1994). *Understanding and representing space. Theory and evidence from studies with blind and sighted children.* Oxford: Clarendon Press.

Miller, S., Potegal, M., & Abraham, L. (1983).Vestibular involvement in a passive transport and return task. *Physiological Psychology*, 11, 1–10.

Mishkin, M., Ungerleider, L. G., & Macko, K. A. (1983). Object vision and spatial vision: Two cortical pathways. *Trends in Neurosciences*, 6, 414–417.

Mittelstaedt, H. (1985). Analytical cybernetics of spider navigation. In F. G. Barth (ed.), *Neurobiology of arachnids* (pp. 298–316). Berlin: Springer-Verlag.

Mittelstaedt, H., & Mittelstaedt, M. L. (1973). Mechanismen der Orientierung ohne richtende Aussenreize. *Fortschritte der Zoologie*, 21, 46–58.

Mittelstaedt, H., & Mittelstaedt, M. L. (1982). Homing by path integration. In F. Papi & H. G. Walraff (eds.), *Avian navigation* (pp. 290–297). New York: Springer-Verlag.

Mittelstaedt, M. L., & Glasauer, S. (1991). Idiothetic navigation in gerbils and humans. *Zoologische, Jahrbücher: Abteilungen für allgemeine Zoologie und Physiologie der Tiere*, 95, 427–435.

Mittelstaedt, M. L., & Mittelstaedt, H. (1980). Homing by path integration in a mammal. *Naturwissenschaften*, 67, 566–567.

Mizumori, S.J.Y. (1994). Neural representations during spatial navigation. *Current Directions in Psychological Science*, 3, 125–129.

Mizumori, S.J.Y., Miya, D. Y., & Ward, K. E. (1994). Reversible inactivation of the lateral dorsal thalamus disrupts hippocampal place representation and impairs spatial learning. *Brain Research*, 644, 168–174.

Mizumori, S.J.Y., & Williams, J. D. (1991). Mnemonic properties of visual-sensitive head direction cells in lateral dorsal thalamus. *Society for Neuroscience*, 17, 482.

Mizumori, S.J.Y., & Williams, J. D. (1993). Directionally selective mnemonic properties of neurons in the lateral dorsal nucleus of the thalamus of rats. *Journal of Neuroscience*, 13, 4015–4028.

Moar, I., & Carleton, L. R. (1982). Memory for routes. *Quarterly Journal of Experimental Psychology*, 34A, 381–394.

Montello, D. R. (1989). The geometry of environmental knowledge. In A. U. Frank, I. Campari, & V. Formentini (eds.), *Theories and methods of spatio-temporal reasoning in geographic space* (pp. 136–152). Lecture Notes in Computer Science No. 639. Berlin: Springer-Verlag.

Montello, D. R. (1991a). The measurement of cognitive distance: Methods and construct validity. *Journal of Environmental Psychology*, 11(2), 101–122.

Montello, D. R. (1991b). Spatial orientation and the angularity of urban routes: A field study. *Environment and Behavior*, 23(1), 47–69.

Montello, D. R., & Lemberg, D. S. (1995) The Minotaur's revenge: Geographic disorientation in caves. Paper presented at the International Conference on Spatial Analysis in Environment-Behaviour Studies, Eindhoven, The Netherlands.

Moore, G. T., & Golledge, R. G. (eds.). (1976a). *Environmental knowing*. Stroudsburg, PA: Dowden, Hutchinson, and Ross.

Moore, G. T., & Golledge, R. G. (1976b). Environmental knowing: Concepts and theories. In G. T. Moore & R. G. Golledge (eds.), *Environmental knowing* (pp. 3–24). Stroudsburg, PA: Dowden, Hutchinson, and Ross.

Morrow, L., & Ratcliff, G. (1988). Neuropsychology of spatial cognition: Evidence from cerebral lesions. In J. Stiles-Davis, M. Kritchevsky, & U. Bellugi (eds.), *Spatial cognition: Brain bases and development* (pp. 5–32). Hillsdale, NJ: Lawrence Erlbaum.

Müller, M., & Wehner, R. (1988). Path integration in desert ants, *Cataglyphis fortis. Proceedings of the National Academy of Sciences USA*, 85, 5287–5290.

Müller, M., & Wehner, R. (1994). The hidden spiral: Systematic search and path integration in desert ants, *Cataglyphus fortis. Journal of Comparative Physiology*, A175, 525–530.

Muller, R. (1996). A quarter of a century of place cells. *Neuron*, 17, 813–822.

Muller, R. U., Ranck, J. B., & Taube, J. S. (1996). Head direction cells: Properties and functional significance. *Current Opinion in Neurobiology*, 6, 196–206.

Mumaw, R. J., & Pellegrino, J. W. (1984). Individual differences in complex spatial processing. *Journal of Educational Psychology*, 76, 920–939.

Mumaw, R. J., Pellegrino, J. W., Kail, R. V., & Carter P. (1984). Different slopes for different folks: Process analysis of spatial aptitude. *Memory and Cognition*, 12, 515–521.

Munro, U., & Wiltschko, R. (1993). Clock-shift experiments with migratory Yellow-faced Honeyeaters, *Lichenostomus chrysops* (Meliphagidae), an Australian day migrating bird. *Journal of Experimental Biology*, 181, 233–244.

Myers, C. T. (1957). *Some observations and opinions concerning spatial relations tests.* Research Memorandum No. RM-57-7. Princeton, NJ: Educational Testing Service.

Nadel, L. (1991). The hippocampus and space revisited. *Hippocampus*, 1, 221–229.

Nadel, L. (1992). Multiple memory systems: What and why. *Journal of Cognitive Neuroscience*, 4, 179–188.

Nadel, L. (1994). Multiple memory systems: What and why. An update. In D. Schacter and E. Tulving (eds.), *Memory systems* (pp. 39–63). Cambridge, MA: MIT Press.

Nadel, L., & MacDonald, L. (1980). Hippocampus: Cognitive map or working memory? *Behavioral and Neural Biology*, 29, 405–449.

Nadel, L., & Moscovitch, M. (1997) Memory consolidation, retrograde amnesia and the hippocampal formation. *Current Opinion in Neurobiology*, 7, 217–227.

Nadel, L., & O'Keefe, J. (1974). The hippocampus in pieces and patches: An essay on modes of explanation in physiological psychology. In R. Bellairs & E. G. Gray (eds.), *Essays on the nervous system: A festschrift for Prof. J. Z. Young* (pp. 367–390). Oxford: Clarendon Press.

Naveh-Benjamin, M. (1987). Coding of spatial locations: An automatic process? *Journal of Experimental Psychology: Learning, Memory, and Cognition*, 13, 595–605.

Neisser, U. (1976). *Cognition and reality: Principles and implications of cognitive psychology.* San Francisco: W. H. Freeman.

Neisser, U. (1985). A sense of where you are: Functions of the spatial module. In P. Ellen & C. Thinus-Blanc (eds.), *Cognitive processes and spatial orientation in animal and man,* Vol. 2: *Neurophysiology and developmental aspects* (pp. 293–310). Dordrecht: Martinus Nijhoff.

Newcombe, N., & Huttenlocher, J. (1992). Children's early ability to solve perspective-taking problems. *Developmental Psychology*, 28, 635–643.

Newell, A., & Simon, H. A. (1972). *Human problem solving*. Englewood Cliffs, NJ: Prentice-Hall.

Nisbett, R., & Ross, L. (1980). *Human inference: Strategies and shortcomings of social judgment.* Englewood Cliffs, NJ: Prentice-Hall.

Ochaita, E., & Huertas, J. A. (1993). Spatial representation by persons who are blind: A study of the effects of learning and development. *Journal of Visual Impairment and Blindness*, 87, 37–41.

Ohta, R. J., Walsh, D. A., & Krauss, I. K. (1981). Spatial perspective-taking ability in young and elderly adults. *Experimental Aging Research*, 7, 45–63.

O'Keefe, J. (1976). Place units in the hippocampus of the freely moving rat. *Experimental Neurology*, 51, 78–109.

O'Keefe, J. (1979). A review of the hippocampal place cells. *Progress in Neurobiology*, 13, 419–439.

O'Keefe, J. (1989). Computations the hippocampus might perform. In L. Nadel., L. A. Cooper, P. Culicover, & R. M. Harnish (eds.), *Neural connections, mental computations* (pp. 225–284). Cambridge, MA: MIT Press.

O'Keefe J. (1991a). An allocentric spatial model for the hippocampal cognitive map. *Hippocampus*, 1, 230–235.

O'Keefe, J. (1991b). The hippocampal cognitive map and navigational strategies. In J. Paillard (ed.), *Brain and space* (pp. 273–295). Oxford: Oxford University Press.

O'Keefe, J., & Burgess, N. (1996). Geometrical determinants of the space field of hippocampal neurons. *Nature*, 381, 425–428.

O'Keefe, J., & Conway, D. H. (1980). On the trail of the hippocampal engram. *Physiological Psychology*, 8, 229–238.

O'Keefe, J., & Dostrovsky, J. (1971). The hippocampus as a spatial map. Preliminary evidence from unit activity in the freely-moving rat. *Brain Research*, 34, 171–175.

O'Keefe, J., & Nadel, L. (1978). *The hippocampus as a cognitive map*. Oxford: Oxford University Press.

O'Keefe, J., Nadel, L., Keightley, S., & Kill, D. (1975). Fornix lesions selectively abolish place learning in the rat. *Experimental Neurology*, 48, 152–166.

O'Keefe, J., & Speakman, A. (1987). Single unit activity in the rat hippocampus during a spatial memory task. *Experimental Brain Research*, 12, 1–27.

Olson, D. R., & Bialystok, E. (1983). *Spatial cognition*. Hillsdale, NJ: Lawrence Erlbaum.

O'Mara, S. M., Rolls, E. T., Berthöz, A., & Kesner, R. P. (1997). Neurons responding to whole body motion in the primate hippocampus. *Journal of Neuroscience*, 14, 6511–6523.

Ondracek, P. J. (1995). *Children's development of spatial mental models from verbal information: Listening to or reading environmental descriptions*. Unpublished Ph.D. dissertation, Department of Psychology, University of South Carolina, Columbia.

O'Neill, M. J. (1990). Computer simulation of the cognitive map: A validation study. In R. I. Selby, K. H. Anthony, J. Choi, & B. Orland (eds.), *Coming of age: Proceedings of the twenty-first annual conference of the Environmental Design Research Association* (pp. 353–360). Oklahoma City: EDRA.

O'Neill, M. J. (1991). A biologically based model of spatial cognition and wayfinding. *Journal of Environmental Psychology*, 11, 299–320.

Pailhouse, J. (1969). Representation de l'espace urbain at cheminements. *Le Travail Humain*, 32, 239–250.

Pallis, C. A. (1955). Impaired identification of faces and places with agnosia for colors. *Journal of Neurology, Neurosurgery, and Psychiatry*, 18, 218–224.

Palmer, S. E. (1978). Fundamental aspects of cognitive representation. In E. Rosch & B. B. Lloyd (eds.), *Cognition and categorization* (pp. 259–303). Hillsdale, NJ: Lawrence Erlbaum.

Papi, F. (1982). Olfaction and homing in pigeons: Ten years of experiments. In F. Papi & H. G. Wallraff (eds.), *Avian navigation* (pp. 149–159). Berlin: Springer-Verlag.

Papi, F. (1986). Pigeon navigation: Solved problems and open questions. *Monitore Zoologico Italiano* (n.s.), 20, 471–517.

Pardi, L., & Ercolini, A. (1986). Zonal recovery mechanisms in talitrid crustaceans. *Bolletino di Zoologia*, 53, 139–160.

Partridge, D. (1996). Representation of knowledge. In M. A. Boden (ed.), *Artificial intelligence. Handbook of perception and cognition* (2nd ed.) (pp. 55–87). San Diego: Academic Press.

Passini, R. (1980a). Wayfinding: A conceptual framework. *Man-Environment Systems*, 10, 22–30.

Passini, R. (1980b). Wayfinding in complex buildings: An environmental analysis. *Man-Environment Systems*, 10, 31–40.

Passini, R. (1984a). Spatial representations: A wayfinding perspective. *Journal of Environmental Psychology*, 4, 153–164.

Passini, R. (1984b). *Wayfinding in architecture*. New York: Van Nostrand Rheinhold.

Passini, R., Proulx, G., & Rainville, C. (1990). The spatio-cognitive abilities of the visually impaired population. *Environment and Behavior*, 22, 91–116.

Paterson, A. (1994). A case of topographical disorientation associated with a unilateral cerebral lesion. *Brain*, 68, 188–212.

Payne, J. W. (1976). Task complexity and contingent processing in decision making: An information search and protocol analysis. *Organizational Behavior and Human Performance*, 16, 366–387.

Payne, J. W. (1982). Contingent decision behavior. *Psychological Bulletin*, 92, 382–402.

Payne, J. W., Bettman, J. R., Coupey, E., & Johnson, E. J. (1992). A constructive process view of decision making: Multiple strategies in judgment and choice. *Acta Psychologica*, 80, 107–141.

Payne, J. W., Bettman, J. R., & Johnson, E. J. (1988). Adaptive strategy selection in decision making. *Journal of Experimental Psychology: Learning, Memory, and Cognition*, 14, 534–552.

Payne, J. W., Bettman, J. R., & Johnson, E. J. (1992). Behavioral decision research: A constructive processing perspective. *Annual Review of Psychology*, 43, 87–131.

Payne, J. W., Bettman, J. R., & Johnson, E. J. (1993). *The adaptive decision maker*. Cambridge: Cambridge University Press.

Payne, J. W., & Braunstein, M. L. (1978). Risky choice: An examination of information acquisition behavior. *Memory and Cognition*, 5, 554–561.

Pellegrino, J. W., Alderton, D. L., & Shute, V. J. (1984). Understanding spatial ability. *Educational Psychologist*, 19, 239–253.

Pellegrino, J. W., & Kail, R. V. (1982). Process analyses of spatial aptitude. In R. Sternberg (ed.), *Advances in the psychology of human intelligence* (Vol. 1, pp. 311–365). Hillsdale, NJ: Lawrence Erlbaum.

Pellegrino, J. W., Mumaw, R. J., & Shute, V. J. (1985). Analyses of spatial aptitude and expertise. In S. Whitely (ed.), *Test design: Contributions from psychology, education, and psychometrics* (pp. 45–76). New York: Academic Press.

Perdeck, A. C. (1958). Two types of orientation in migrating *Sturnus vulgaris* and *Fringilla coelebs* as revealed by displacement experiments. *Ardea*, 46, 1–37.

Perdeck, A. C. (1967). Orientation of starlings after displacement to Spain. *Ardea*, 55, 194–202.

Perdeck, A. C. (1983). An experiment of the orientation of juvenile starlings during spring migration: An addendum. *Ardea*, 71, 255.

Perrig, W., & Kintch, W. (1985). Propositional and situational representations of text. *Journal of Memory and Language*, 24, 503–518.

Péruch, P., Giraudo, M.-D., & Gärling, T. (1989). Distance cognition of taxi drivers and lay people. *Journal of Environmental Psychology*, 9, 233–239.

Péruch, P., & Lapin, E. A. (1993). Route knowledge in different spatial frames of reference. *Acta Psychologica*, 84(3), 253–269.

Peterson, M. A., Nadel, L., Bloom, P., & Garrett, M. F. (1996). Space and language. In P. Bloom, M. A. Peterson, L. Nadel, & M. Garrett (eds.), *Language and space* (pp. 553–557). Cambridge, MA: MIT Press.

Pezdek, K. (1983). Memory for items and their spatial locations by young and elderly adults. *Developmental Psychology*, 19, 895–900.

Pezdek, K., & Evans, G. W. (1979). Visual and verbal memory for objects and their spatial locations. *Journal of Experimental Psychology: Human Learning and Memory*, 5, 360–373.

Philbeck, J. W., & Loomis, J. M. (1997). Comparison of two indicators of perceived egocentric distance under full-cue and reduced-cue conditions. *Journal of Experimental Psychology: Human Perception and Performance*, 23, 72–85.

Philbeck, J. W., Loomis, J. M., & Beall, A. C. (1997). Visually perceived location is an invariant in the control of action. *Perception and Psychophysics,* 59, 601–612.

Piaget, J. (1937). *La construction du réel chez l'enfant.* Paris: Delachaux et Niestlé.

Piaget, J. (1950). *The psychology of intelligence* (M. Piercy & D. Berlyne, trans.). London: Routledge and Kegan Paul.

Piaget, J. (1951). *Plays, dramas, and imitation in childhood* [La Formation du symbole; original French ed., 1946]. New York: W. W. Norton.

Piaget, J. (1956). *La Psychologie de l'Intelligence.* Paris: Armand Colin.

Piaget, J., & Inhelder, B. (1967). *The child's conception of space.* New York: W. W. Norton.

Piaget. J., Inhelder, B., & Szeminska, A. (1960). *The child's conception of geometry.* New York: Basic Books.

Pick, H. L., Jr. (1990). Issues in the development of mobility. In H. Bloch & B. Bertenthal (eds.), *Sensory-motor organizations and development in infancy and early childhood* (pp. 419–436). Dordrecht: Kluwer Academic.

Pick, H. L., Jr. (1993). Organization of spatial knowledge in children. In N. Eilan, R. McCarthy, & B. Brewer (eds.), *Spatial representation* (pp. 31–42). Oxford: Blackwell.

Pick, H. L., Jr., Montello, D. R., & Somerville, S. C. (1988). Landmarks and the coordination and integration of spatial information. *British Journal of Developmental Psychology,* 6, 372–375.

Pick, H. L., Jr., & Rieser, J. J. (1982). Children's cognitive mapping. In M. Potegal (ed.), *Spatial abilities: Development and physiological foundations* (pp. 107–128). San Diego: Academic Press.

Pick, H. L., Jr., Rieser, J. J., Wagner, M., & Garing, A. E. (in press). The recalibration of rotational locomotion. *Journal of Experimental Psychology: Human Perception and Performance.*

Pierrot-Deseilligny, C., Israël, I., Berthöz, A., Rivaud, S., & Gaymard, B. (1993). Role of the different frontal lobe areas in the control of the horizontal component of memory-guided saccades in man. *Experimental Brain Research,* 95, 166–171.

Pigott, S., & Milner, B. (1993). Memory for different aspects of complex visual scenes after unilateral temporal- or frontal-lobe resection. *Neuropsychologia,* 31, 1–15.

Pinker, S. (1980). Mental imagery and the third dimension. *Journal of Experimental Psychology: General,* 109, 354–371.

Pinker, S. (1985). Visual cognition: An introduction. In S. Pinker (ed.), *Visual cognition* (pp. 1–63). Cambridge, MA: MIT Press.

Plude, D. J., & Hoyer, W. J. (1985). Attention and performance: Identifying and localizing age deficits. In N. Charness (ed.), *Aging and human performance* (pp. 47–99). New York: John Wiley and Sons.

Pocock, D. C. D. (1973). Urban environmental perception and behaviour: A review. *Tijdschrift voor Economische en Sociale Geografie,* 64, 251–257.

Poincaré, H. (1904). *La Valeur de la Science.* Paris: Flammarion.

Poitevineau, J., & Lecoutre, B. (1986). PIF: Un programme d'inférence fiducio-bayésienne. *Informatique et Sciences Humaines,* 68–69, 77–78.

Poltrock, S. E., & Agnoli, F. (1986). Are spatial visualization ability and visual imagery ability equivalent? In R. Sternberg (ed.), *Advances in the psychology of human intelligence* (Vol. 3, pp. 255–296). Hillsdale, NJ: Lawrence Erlbaum.

Poltrock, S. E., & Brown, P. (1984). Individual differences in visual imagery and spatial ability. *Intelligence,* 8, 93–138.

Portugali, J. (1990). Social synergetics, cognitive maps and environmental recognition. In H. Haken & M. Stadler (eds.), *Synergetics of cognition* (pp. 379–392). Berlin: Springer-Verlag.

Portugali, J. (1996). Inter-representation networks and cognitive maps. In J. Portugali (ed.), *The construction of cognitive maps* (pp. 11–43). Dordrecht: Kluwer Academic.

Portugali, J., & Haken, H. (1992). Synergetics and cognitive maps. *Geoforum,* 23(2), 111–130.

Potegal, M. (1982). Vestibular and neostriatal contribution to spatial orientation. In M. Potegal (ed.), *Spatial abilities. Development and physiological foundations* (pp. 361–387). New York: Academic Press.

Potegal, M. (1987). The vestibular navigation hypothesis: A progress report. In P. Ellen & C. Thinus-Blanc (eds.), *Cognitive processes and spatial orientation in animal and man,* Vol. 2: *Neurophysiology and developmental aspects* (pp. 28–34). Dordrecht: Martinus Nijhoff.

Poucet, B. (1993). Spatial cognitive maps in animals: New hypothesis on their structure and neural mechanisms. *Psychological Review,* 100, 163–182.

Poucet, B. (1996). Hippocampal contribution to spatial processing: Lesion and electrophysiological evidence. Paper presented at ENP School Conference on the Representation of Space, San Feliu, Spain.

Poucet, B., Chapuis, N., Durup, M., & Thinus-Blanc, C. (1986). Exploratory behavior as an index of spatial knowledge in hamsters. *Animal Learning and Behavior,* 14, 93–100.

Pozzo, T., Berthöz, A., & Lefort, L. (1990). Head stabilisation during various locomotor tasks in humans. I. Normal subjects. *Experimental Brain Research,* 82, 97–106.

Pozzo, T., Berthöz, A., Lefort, L., & Vitte, E. (1991). Head stabilisation during various locomotory tasks in humans. II. Patients with peripheral vestibular deficits. *Experimental Brain Research,* 85, 1–10.

Pratt, J. G., & Thouless, R. H. (1955). Homing orientation in pigeons in relation to opportunity to observe the sun before release. *Journal of Experimental Biology,* 32, 140–157.

Presson, C. C., DeLange, N., & Hazelrigg, M. D. (1987). Orientation-specificity in kinesthetic spatial learning: The role of multiple orientations. *Memory and Cognition,* 15, 225–229.

Presson, C. C., DeLange, N., & Hazelrigg, M. D. (1989). Orientation specificity in spatial memory: What makes a path different from a map of the path? *Journal of Experimental Psychology: Learning, Memory, and Cognition,* 15, 887–897.

Presson, C. C., & Hazelrigg, M. D. (1984). Building spatial representations through primary and secondary learning. *Journal of Experimental Psychology: Learning, Memory, and Cognition,* 10, 716–722.

Presson, C. C., & Montello, D. R. (1994). Updating after rotational and translational body movements: Coordinate structure of perspective space. *Perception,* 23, 1447–1455.

Presson, C. C., & Somerville, S. C. (1985). Beyond egocentrism: A new look at the beginnings of spatial representation. In H. Wellman (ed.), *Children's searching: The development of search skill and spatial representation* (pp. 1–26). Hillsdale, NJ: Lawrence Erlbaum.

Prinz, K., & Wiltschko, W. (1992). Migratory orientation of Pied Flycatchers: Interaction of stellar and magnetic information during ontogeny. *Animal Behavior,* 44, 539–545.

Radvansky, G. A., Carlson-Radvansky, L. A., & Irwin, D. E. (1995). Uncertainty in estimating distances from memory. *Memory and Cognition,* 23(5), 596–606.

Ranck, J. B., Jr. (1973). Studies on single neurons in dorsal hippocampal formation and septum in unrestrained rats. *Experimental Neurology,* 41, 461–555.

Ranck, J. B., Jr. (1984). Head direction cells in the deep cell layer of dorsal presubiculum in freely moving rats. *Society for Neuroscience,* 10, 599.

Rasmussen, M., Barnes, C. A., & McNaughton, B. L. (1989). A systematic test of cognitive mapping, working-memory and temporal discontiguity theories of hippocampal function. *Psychobiology,* 17, 335–348.

Ratcliff, G. (1982). Disturbances of spatial orientation associated with cerebral lesions. In M. Potegal (ed.), *Spatial abilities: Development and physiological foundations* (pp. 301–331). New York: Academic Press.

Ratcliff, G., & Newcombe, F. (1973). Spatial orientation in man: Effects of left, right, and bilateral posterior cerebral lesions. *Journal of Neurology, Neurosurgery, and Psychiatry,* 36, 448–454.

Recce, M. L. (1994). *The representation of space in the rat hippocampus as revealed using new computer-based methods.* Unpublished Ph.D. dissertation, University College, London.

Redish, A. D., & Touretzky, D. S. (1997). Cognitive maps beyond the hippocampus. *Hippocampus,* 7, 15–53.

Regolin, L., Vallortigara, G., & Zanforlin, M. (1995). Object and spatial representations in detour problems by chicks. *Animal Behavior,* 49, 195–199.

Restle, F. (1957). Discrimination of cues in mazes: A resolution of the "place-versus-response" question. *Psychological Review,* 64, 217–228.

Richardson, G. D. (1982). *Spatial cognition.* Unpublished Ph.D. dissertation, Department of Geography, University of California, Santa Barbara.

Rider, E. A., & Rieser, J. J. (1988). Pointing at objects in other rooms: Young children's sensitivity to perspective after walking with and without vision. *Child Development,* 59, 480–494.

Rieser, J. J. (1989). Access to knowledge of spatial structure at novel points of observation. *Journal of Experimental Psychology: Human Learning, Memory, and Cognition,* 15, 1157–1165.

Rieser, J. J. (1990). Development of perceptual-motor control while walking without vision: The calibration of perception and action. In H. Bloch & B. I. Bertenthal (eds.), *Sensory-motor organizations and development in infancy and early childhood* (pp. 379–408). Netherlands: Kluwer Academic.

Rieser, J. J. (1993). Imagery and action: Knowing how objects look when they are moved by hand without vision. Paper presented to Psychonomics Conference, Washington, DC.

Rieser, J. J. (1994). Ego-centered and environment-centered perceptions of self-movement: A commentary on A. Wertheim's article "Motion perception during self-movement." *Behavioral Sciences,* 17, 328–329.

Rieser, J. J., Ashmead, D. H., Talor, C. R., & Youngquist, G. A. (1990). Visual perception and the guidance of locomotion without vision to previously seen targets. *Perception,* 19, 675–689.

Rieser, J. J., & Frymire, M. (1995). Locomotion with vision is coupled with knowledge of real and imagined surroundings. Paper presented at the meeting of the Psychonomic Society, Los Angeles.

Rieser, J. J., Frymire, M., & Berry, D. (1997). Geometrical constraints on imagery and action when walking without vision. Paper presented to the Psychonomic Society, Philadelphia.

Rieser, J. J., Garing, A. E., & Young, M. F. (1994). Imagery, action, and young children's spatial orientation: Its not being there that counts, its what one has in mind. *Child Development,* 65, 1262–1278.

Rieser, J. J., Guth, D. A., & Hill, E. W. (1982). Mental processes mediating independent travel: Implications for orientation and mobility. *Journal of Visual Impairment and Blindness,* 76(6), 213–218.

Rieser, J. J., Guth, D. A., & Hill, E. W. (1986). Sensitivity to perspective structure while walking without vision. *Perception,* 15, 173–188.

Rieser, J. J., & Heiman, M. L. (1982). Spatial self-reference systems and shortest-route behavior in toddlers. *Child Development,* 53, 524–533.

Rieser, J. J., Hill, E. W., Talor, C. R., Bradfield, A., & Rosen, S. (1992). Visual experience, visual field size, and the development of nonvisual sensitivity to the spatial structure of outdoor neighborhoods explored by walking. *Journal of Experimental Psychology: General,* 121, 210–222.

Rieser, J. J., Pick, H. L., Ashmead, D. H., & Garing, A. E. (1995). The calibration of human locomotion and models of perceptual-motor organization. *Journal of Experimental Psychology: Human Perception and Performance,* 21, 480–497.

Rieser, J. J., & Rider, E. A. (1991). Young children's spatial orientation with respect to multiple targets when walking without vision. *Developmental Psychology,* 27, 97–107.

Rivizzigno, V. L. (1976). *Cognitive representations of an urban area.* Unpublished Ph.D. dissertation, Department of Geography, Ohio State University, Columbus.

Röfer, T. (1995). Controlling a robot with image-based homing. In B. Krieg-Brückner & C. Herwig (eds.), *Kognitive robotik*. ZKW Bericht 3. Bremen: Zentrum für Kognitionswissenschaften, Universität Bremen.

Rogers, D. (1970). *The role of search and learning in consumer space behavior: The case of urban migrants*. Unpublished master's thesis, Department of Geography, University of Wisconsin, Madison.

Roitblat, H. L. (1982). The meaning of representation in animal memory. *Behavioral Brain Science*, 5, 353–406.

Rolls, E. T. (1989). Functions of neuronal networks in the hippocampus and neocortex in memory. In J. H. Byrne & W. O. Berry (eds.), *Neural models of plasticity: Theoretical and empirical approaches* (pp. 240–265). New York: Academic Press.

Rolls, E. T., & O'Mara, S. (1995). View-responsive neurons in the primate hippocampal complex. *Hippocampus*, 5, 409–424.

Ronacher, B., & Wehner, R. (1995). Desert ants (*Cataglyphis fortis*) use self-induced optic flow to measure distances traveled. *Journal of Comparative Physiology*, A177, 21–27.

Rosch, R., & Mervis, C. B. (1975). Family resemblances: Studies in the internal structure of categories. *Cognitive Psychology*, 7, 573–605.

Rosch, E., Mervis, C. B., Gray, W., Johnson, D., & Boyes-Braem, P. (1976). Basic objects in natural categories. *Cognitive Psychology*, 8, 382–439.

Rossano, M. J., & Warren, D. H. (1989). Misaligned maps lead to predictable errors. *Perception*, 18, 215–229.

Rossel, S., & Wehner, R. (1982). The bee's map of the E-vector pattern in the sky. *Proceedings of the National Academy of Sciences USA*, 79, 4451–4455.

Rossier, J., Grobéty, M.-C., & Schenk, F. (1996). Place learning under discontinuous access to a limited number of visual cues. *European Journal of Neuroscience Supplement*, 9, 119.

Rotenberg, A., Kubie, J., & Muller, R. (1993). Variable coupling between a stimulus object and place cell firing fields. *Society for Neuroscience Abstracts*, 19, 357.

Rouanet, H. (1996). Bayesian methods for assessing importance of effects. *Psychological Bulletin*, 119, 149–158.

Rouanet, H., & Lecoutre, B. (1983). Specific inference in ANOVA: From significance tests to Bayesian procedures. *British Journal of Mathematical and Statistical Psychology*, 36, 252–268.

Royden, C. S. (in press). Human perception of heading and object motion: Computational and psychophysical studies. In M. McBeath (ed.), *Spatial navigational principles used by humans, animals, and machines*. London: Sage.

Rudy, J. W., Stadler-Morris, S., & Albert, P. (1987). Ontogeny of spatial navigation behaviors in the rat: Dissociation of "proximal"- and "distal"-cue based behaviors. *Behavioral Neuroscience*, 101, 62–73.

Rueckl, J., Cave, K., & Kosslyn, S. (1988). Why are "what" and "where" processed by separate cortical visual systems? A computational investigation. *Journal of Cognitive Neuroscience*, 1(2), 171–186.

Rumelhart, D. E., Smolensky, P., McClelland, J. L., & Hinton, G. E. (1986). Schemata and sequential thought processes in PDP models. In J. L. McClelland & D. E. Rumelhart (eds.), *Parallel distributed processing, explorations in the microstructure of cognition*, Vol. 2: *Psychological and biological models* (pp. 7–57). Cambridge, MA: MIT Press.

Sadalla, E. K., Burroughs, W. J., & Staplin, L. J. (1980). Reference points in spatial cognition. *Journal of Experimental Psychology: Human Learning and Memory*, 6(5), 516–528.

Sadalla, E. K., & Montello, D. R. (1989). Remembering changes in direction. *Environment and Behavior*, 21, 346–363.

Säisä, J., & Gärling, T. (1987). Sequential spatial choices in the large-scale environment. *Environment and Behavior*, 19, 614–635.

Säisä, J., Svensson-Gärling, A., Gärling, T., & Lindberg, E. (1986). Intra-urban cognitive distance: The relationship between judgments of straight-line distances, travel distances, and travel times. *Geographical Analysis*, 18(2), 167–174.

Salthouse, T. A. (1992). Reasoning and spatial abilities. In F. Craik & T. Salthouse (eds.), *The handbook of aging and cognition* (pp. 167–211). Hillsdale, NJ: Lawrence Erlbaum.

Samsonovich, A., & McNaughton, B. L. (1997). Path integration and cognitive mapping in a continuous attractor neural network model. *Journal of Neuroscience*, 17, 5900–5920.

Sauve, J. P. (1989). *L'orientation spatiale: Formalization d'un modèle de mémorisation egocentrée et expérimentation chez l'homme*. Ph.D. thesis, Department of Neurosciences, Université d'Aix-Marseille II, Marseille, France.

Savage, L. J. (1954). *The foundations of statistics*. New York: John Wiley & Sons.

Schear, J. M., & Nebes, R. D. (1980). Memory for verbal and spatial information as a function of age. *Experimental Aging Research*, 6, 271–282.

Schenk, F. (1985). Development of place navigation in rats from weaning to puberty. *Behavioral and Neural Biology*, 43, 69–85.

Schenk, F., Grobéty, M. C., & Gafner, M. (1997). Spatial learning by rats across visually disconnected environments. *Quarterly Journal of Experimental Psychology*, 50B, 54–78.

Schiff, W., & Oldak, R. (1990). Accuracy of judging time to arrival: Effects of modality, trajectory, and gender. *Journal of Experimental Psychology: Human Perception and Performance*, 16, 303–316.

Schlichte, H.-J. (1973). Untersuchungen über die Bedeutung optischer Parameter für das Heimkehrverhalten der Brieftauben. *Zeitschrift für Tierpsychologie*, 32, 257–280.

Schmidt-Koenig, K. (1958). Experimentelle Einflußnahme auf die 24-Stunden-Periodik bei Brieftauben und deren Auswirkung unter besonderer Berücksichtigung des Heimfindevermögens. *Zeitschrift für Tierpsychologie*, 15, 301–331.

Schmidt-Koenig, K. (1963). On the role of the loft, the distance and site of release in pigeon homing (the "cross loft experiment"). *Biological Bulletin*, 125, 154–164.

Schmidt-Koenig, K. (1965). Current problems in bird orientation. *Advances in the Study of Behavior*, 1, 217–276.

Schmidt-Koenig, K. (1975). *Migration and homing in animals*. Berlin: Springer-Verlag.

Schmidt-Koenig, K., & Walcott, C. (1973). Flugwege und Verbleib von Brieftauben mit getrübten Haftschalen. *Naturwissenschaften*, 60, 108–109.

Schober, M. F. (1993). Spatial perspective-taking in conversation. *Cognition*, 47, 1–24.

Scoville, W. B., & Milner, B. (1957). Loss of recent memory after bilateral hippocampal lesions. *Journal of Neurology, Neurosurgery and Psychiatry*, 20, 11–21.

Séguinot, V., Maurer, R., & Etienne, A. S. (1993). Dead reckoning in a small mammal: The evaluation of distance. *Journal of Comparative Physiology*, A173, 103–113.

Self, C. M., Gopal, S., Golledge, R. G., & Fenstermaker, S. (1992). Gender-related differences in spatial abilities. *Progress in Human Geography*, 16, 315–342.

Sharp, P. E. (1991). Computer simulation of hippocampal place cells. *Psychobiology*, 19, 103–115.

Sharp, P. E., Blair, H. T., Etkin, D., & Tzanetos, D. B. (1995). Influences of vestibular and visual motion information on the spatial firing patterns of hippocampal place cells. *Journal of Neuroscience*, 15, 173–189.

Sheffi, J., Mahmassani, H., & Powell, B. W. (1982). A transportation network evacuation model. *Transportation Research*, A16, 209–218.

Shemyakin, F. N. (1962). General problems of orientation in space and space representations. In B. G. Anan'yev et al. (eds.), *Psychological science in the U.S.S.R.* (Vol. 1, pp. 186–255). NTIS Report TT62–11083. Washington, DC: Office of Technical Services.

Shepard, R. N., & Cooper, L. A. (1982). *Mental images and their transformations*. Cambridge, MA: MIT Press.

Shepard, R. N., & Hurwitz, S. (1984). Upward direction, mental rotation, and discrimination of left and right turns in maps. *Cognition*, 18, 161–193.

Shepard, R. N., & Metzler, J. (1971). Mental rotation of three-dimensional objects. *Science*, 171, 701–703.

Sholl, M. J. (1987). Cognitive maps as orienting schemata. *Journal of Experimental Psychology: Learning, Memory, and Cognition,* 13(4), 615–628.

Sholl, M. J. (1988). The relation between sense of direction and mental geographic updating. *Intelligence,* 12, 299–314.

Sholl, M. J. (1989). The relation between horizontality and rod-and-frame and vestibular navigational performance. *Journal of Experimental Psychology: Learning, Memory, and Cognition,* 15, 110–125.

Sholl, M. J. (1996). From visual information to cognitive maps. In J. Portugali (ed.), *The construction of cognitive maps* (pp. 157–186). Dordrecht: Kluwer Academic.

Shute, V. J. (1984). *Characteristics of cognitive cartography.* Unpublished Ph.D. dissertation, Graduate School of Education, University of California, Santa Barbara.

Siegel, A. W. (1981). The externalization of cognitive maps by children and adults: In search of better ways to ask better questions. In L. S. Liben, A. Patterson, & N. Newcombe (eds.), *Spatial representation and behavior across the life span: Theory and application* (pp. 167–194). New York: Academic Press.

Siegel, A. W., & Cousins, J. H. (1985). The symbolizing and symbolized child in the enterprise of cognitive mapping. In R. Cohen (ed.), *The development of spatial cognition* (pp. 347–368). Hillsdale, NJ: Lawrence Erlbaum.

Siegel, A. W., Herman, J. F., Allen, G. L., & Kirasic, K. C. (1979). The development of cognitive maps in large- and small-scale spaces. *Child Development,* 50, 582–585.

Siegel, A. W., Kirasic, K. C., & Kail, R. V. (1978). Stalking the elusive cognitive map: The development of children's representations of geographic space. In I. Altman & J. Wohlwill (eds.), *Human behavior and environment,* Vol. 3: *Children and the environment* (pp. 223–258). New York: Plenum Press.

Siegel, A. W., & White, S. H. (1975). The development of spatial representations of large-scale environments. In H. W. Reese (ed.), *Advances in child development and behavior* (Vol. 10, pp. 9–55). New York: Academic Press.

Simon, H. A. (1956). Rational choice and the structure of the environment. *Psychological Review,* 63, 129–138.

Simon, H. A. (1957). *Models of man.* New York: John Wiley & Sons.

Simon, H. A. (1969). The architecture of complexity. In H. Simon (ed.) *The sciences of the artificial* (pp. 192–229). Cambridge: Cambridge University Press. [Reprinted from *Proceedings of the American Philosophical Society,* 106, 467–482 (1962).]

Simon, H. A. (1990). Invariants of human behavior. *Annual Review of Psychology,* 41, 1–19.

Skaggs, W. E., Knierim, J. J., Kudrimoti, H. S., & McNaughton, B. D. (1995). A model of the neural basis of the rat's sense of direction. In D. S. Touretzky, G. D. Tesauro, & T. Leen (eds.), *Advances in neural information processing systems* (Vol. 7, pp. 173–180). Cambridge, MA: MIT Press.

Skaggs, W. E., & McNaughton, B. L. (1995). Replay of neuronal firing sequences in rat hippocampus during sleep following spatial experience. *Science,* 271, 1870–1873.

Smythe, M. M., & Kennedy, J. (1982). Orientation in spatial representations within multiple frames of reference. *British Journal of Psychology,* 73, 527–535.

Snow, R. E., Kyllonen, P. C., & Marshalek, B. (1984). The topography of ability and learning correlations. In R. Sternberg (ed.), *Advances in the psychology of human intelligence* (Vol. 2, pp. 47–103). Hillsdale, NJ: Lawrence Erlbaum.

Snow, R. E., Lohman, D. F., Marshalek, B., Yalow, E., & Webb, N. (1977). *Correlational analyses of reference aptitude constructs.* Technical Report No. 5. Stanford, CA: Aptitude Research Project, Stanford University School of Education.

Sokolov, L. V., Bolshakov, K. V., Vinogradova, N. V., Dolnik, T. V., Lyuleeva, D. S., Payevsky, V. A., Shumakov, M. E., & Yablonkevich, M. L. (1984). The testing of the ability for imprinting and finding the site of future nesting in young chaffinches. *Zoological Journal (Moscow),* 43, 1671–1681 (in Russian).

Solomon, D., & Cohen, B. (1992a). Stabilization of gaze during circular locomotion in light. I. Compensatory head and eye nystagmus in the running monkey. *Journal of Neurophysiology,* 67, 1146–1157.

Solomon, D., and Cohen, B. (1992b). Stabilization of gaze during circular locomotion in light. II. Contribution of velocity storage to compensatory eye and head nystagmus in the running monkey. *Journal of Neurophysiology*, 67, 1158–1170.

Spector, A. N., & Rivizzigno, V. (1982). Sampling designs and recovering cognitive representations of an urban area. In R. G. Golledge & J. Rayner (eds.), *Proximity and preference* (pp. 47–79). Minneapolis,: University of Minnesota Press.

Spencer, C. P., Blades, M., & Morsley, K. (1989). *The child in the physical environment.* Chichester, UK: John Wiley & Sons.

Spencer, C. P., & Darvizeh, Z. (1981). The case for developing a cognitive environmental psychology that does not underestimate the abilities of young children. *Journal of Environmental Psychology*, 1, 21–31.

Srinivasan, M. V., Zhang, S. W., Lehrer, M., & Collett, T. S. (1996). Honeybee navigation en route to the goal: Visual flight control and odometry. *Journal of Experimental Biology*, 199(1), 237–244.

Stea, D. (1969). The measurement of mental maps: An experimental model for studying conceptual spaces. In K. R. Cox & R. G. Golledge (eds.), *Behavioral problems in geography: A symposium* (pp. 228–253). Studies in Geography No. 17. Evanston, IL: Northwestern University.

Stea, D. (1997). A diachronic perspective on universal mapping. Paper presented at the 93rd annual meeting of the Association of American Geographers, Ft. Worth.

Stea, D., & Downs, R. (eds.) (1970). Cognitive representations of man's spatial environment. *Environment and Behavior*, 2(1).

Steenhuis, R. E., & Goodale, M. A. (1988). The effects of time and distance on accuracy of target-directed locomotion: Does an accurate short-term memory for spatial location exist? *Journal of Motor Behavior*, 20, 399–415.

Stern, E. (1983). Are geography students more spatially oriented than others? *South African Geographer*, 11(2), 149–160.

Stern, E. (1996). Congestion response behavior in Tel Aviv metropolitan area. In Y. Gradus & G. Lipshitz (eds.), *The mosaic of Israeli geography* (pp. 147–158). Beer Sheva: Ben-Gurion University Press.

Stern, E. (1998). Travel choice in congestions: Modeling and research needs. In T. Gärling, K. Westin, & T. Laitila (eds.), *Theoretical foundations of travel choice modeling* (pp. 173–200). Oxford: Pergamon.

Stern, E., & Azrieli, A. (1995). *Tourists in a strange city—Visitation patterns.* Research paper. Beer Sheva, Israel: Ben Gurion University, Department of Geography (in Hebrew).

Stern, E., & Leiser, D. (1988). Levels of spatial knowledge and urban travel modeling. *Geographical Analysis*, 20(2), 140–155.

Stevens, A., & Coupe, P. (1978). Distortions in judged spatial relations. *Cognitive Psychology*, 10, 422–437.

Strauss, A. (1961). *Images of the American city.* New York: Free Press.

Strong, G. W. (1994). Separability of reference frame distinctions from motor and visual images. *Behavioral and Brain Sciences*, 17, 224–225.

Suzuki, S., Augerinos, G., & Black, A. H. (1980). Stimulus control of spatial behavior on the eight-arm maze in rats. *Learning and Motivation*, 11, 1–15.

Svenson, O. (1979). Process descriptions of decision making. *Organizational Behavior and Human Performance*, 23, 86–112.

Svenson, O. (1992). Differentiation and consolidation theory of human decision making: A frame of reference for the study of pre- and post-decision processes. *Acta Psychologica*, 80, 143–168.

Swinson, R. P., Cox, B. J., Rutka, J., Mai, M., Kerr, S., & Kuch, K. (1993). Otoneurological functioning in panic disorder patients with prominent dizziness. *Comprehensive Psychiatry*, 2, 127–129.

Takei, Y., Grasso, R., Amorim, M. A., & Berthöz, A. (1997). Circular trajectory formation during blind locomotion. A test for path integration and motor memory. *Experimental Brain Research*, 115, 361–368.

Taube, J. S. (1992). Qualitative analysis of head-direction cells recorded in the rat anterior thalamus. *Society for Neuroscience,* 18, 708.

Taube, J. S. (1995). Head direction cells recorded in the anterior thalamic nuclei of freely moving rats. *Journal of Neuroscience,* 15, 70–86.

Taube, J. S., & Burton, H. L. (1995). Head direction cell activity monitored in a novel environment and during a cue conflict situation. *Journal of Neurophysics,* 74(5), 1953–1971.

Taube, J. S., Goodridge, J. P., Golob, E. J., Dudchenko, P. A., & Stackman, R. W. (1996). Processing the head direction cell signal: A review and commentary. *Brain Research Bulletin,* 40, 477–486.

Taube, J. S., Muller, R. U., & Ranck, J. B., Jr. (1990a). Head-direction cells recorded from the postsubiculum in freely moving rats, I. Description and quantitative analysis. *Journal of Neuroscience,* 10, 420–435.

Taube, J. S., Muller, R. U., & Ranck, J. B., Jr. (1990b). Head-direction cells recorded from the postsubiculum in freely moving rats, II. Effects of environmental manipulations. *Journal of Neuroscience,* 10, 436–447.

Taylor, H. A., & Tversky, B. (1992a). Descriptions and depictions of environments. *Memory and Cognition,* 20, 483–496.

Taylor, H. A., & Tversky, B. (1992b). Spatial mental models derived from survey and route descriptions. *Journal of Memory and Language,* 31, 261–292.

Tellevik, J. M. (1992). Influence of spatial exploration patterns on cognitive mapping by blindfolded sighted persons. *Journal of Visual Impairment and Blindness,* 86, 221–224.

Teroni, E., Portenier, V., & Etienne, A. S. (1987). Spatial orientation of the golden hamster in conditions of conflicting location-based and route-based information. *Behavioral Ecology and Sociobiology,* 20, 389–397.

Teroni, E., Portenier, V., & Etienne, A. S. (1990). Utilisation par le hamster doré d'indices visuels dans un environment visuellement symétrique. *Biology of Behavior,* 15, 74–92.

Thinus-Blanc, C. (1978). Discrimination entre deux modalités (ouvert/fermé) d'une propriété topologique chez le hamster doré; effet des modifications métriques. *L'Année Psychologique,* 78, 7–27.

Thinus-Blanc, C. (1987). The concept of cognitive maps and its consequences. In P. Ellen and C. Thinus-Blanc (eds.), *Cognitive processes and spatial orientation in animal and man* (Vol. 1, pp. 1–19). Dordrecht: Martinus Nijhoff.

Thinus-Blanc, C. (1988). Animal spatial cognition. In L. Weiskrantz (ed.), *Thought without language* (pp. 371–395). London: Oxford University Press.

Thinus-Blanc, C. (1996). *Animal spatial cognition. Behavioral and neural approaches.* Singapore: World Scientific.

Thinus-Blanc, C., Bouzouba, L., Chaix, K., Chapuis, N., Durup, M., & Poucet, B. (1987). A study of spatial parameters encoded during exploration in hamsters. *Journal of Experimental Psychology: Animal Behavior Processes,* 13, 418–427.

Thinus-Blanc, C., Durup, M., & Poucet, B. (1992). The spatial parameters encoded by hamsters during exploration: A further study. *Behavioural Processes,* 26, 43–57.

Thinus-Blanc, C., & Gaunet, F. (1997). Representation of space in blind persons: Vision as a spatial sense? *Psychological Bulletin,* 121(1), 20–41.

Thinus-Blanc, C., Gaunet, F., & Péruch, P. (1996). Des cartes mentales pour nous orienter. *Sciences et Vie Hors, Série: A Quoi Sert le Cerveau?* 195, 18–27.

Thomson, J. A. (1980). How do we use visual information to control locomotion? *Trends in Neuroscience,* 3, 247–250.

Thomson, J. A. (1983). Is continuous visual monitoring necessary in visually guided locomotion? *Journal of Experimental Psychology: Human Perception and Performance,* 9(3), 427–443.

Thorndyke, P. W. (1981). Distance estimation from cognitive maps. *Cognitive Psychology,* 13, 526–550.

Thorndyke, P. W., & Hayes-Roth, B. (1982). Differences in spatial knowledge acquired from maps and navigation. *Cognitive Psychology,* 14, 560–589.

Thorndyke, P. W., & Stasz, C. (1980). Individual differences in knowledge acquisition from maps. *Cognitive Psychology*, 12, 137–175.

Thurstone, L. L. (1938). *Primary mental abilities*. Chicago: University of Chicago Press.

Timmermans, H.J.P., & Golledge, R. G. (1990). Applications of behavioral research on spatial problems II: Preference and choice. *Progress in Human Geography*, 14, 311–354.

Tinbergen, N. (1972). *The Animal in its World*, Vol 1: *Field Studies*. London: George Allen and Unwin.

Toates, F. (1994). Hierarchies of control-changing weightings of levels. In M.-A. Rodrigues & M.-H. Lee (eds.), *Perceptual control theory* (pp. 71–86). Aberystwyth: University of Wales.

Tobler, W. R. (1978). DARCY (Discussion Paper). Department of Geography, University of California, Santa Barbara.

Tobler, W. R. (1994). Bidimensional regression. *Geographical Analysis*, 26(3), 187–212.

Tolman, E. C. (1948). Cognitive maps in rats and men. *Psychological Review*, 55, 189–208.

Tolman, E. C. (1949). There is more than one kind of learning. *Psychological Review*, 56, 144–155.

Tolman, E. C., & Honzik, C. H. (1930). "Insight" in rats. *University of California Publication in Psychology*, 4, 215–232.

Tolman, E. C., Ritchie, B. F., & Kalish, D. (1946). Studies in spatial learning. 1. Orientation and the short-cut. *Journal of Experimental Psychology*, 36, 13–24.

Tomlinson, W. T., & Johnston, T. D. (1991). Hamsters remember spatial information derived from olfactory cues. *Animal Learning Behavior*, 19, 185–190.

Touretzky, D. S., & Redish, A. D. (1996). Theory of rodent navigation based on interacting representations of space. *Hippocampus*, 6, 247–270.

Trowbridge, C. C. (1913). On fundamental methods of orientation and imaginary maps. *Science*, 38, 888–897.

Tsuzuku, T., Vitte, E., Sémont, A., & Berthöz, A. (1995). Modification of parameters in vertical optokinetic nystagmus after repeated vertical optokinetic stimulation in patients with vestibular lesions. *Acta Otolaryngologica* (Stockholm), 115, Supplement 520, Part 2, pp. 419–422.

Tuan, Y.-F. (1977). *Space and place: The perspective of experience*. Minneapolis: University of Minnesota Press.

Tversky, B. (1981). Distortions in memory for maps. *Cognitive Psychology*, 13, 407–433.

Tversky, B. (1992). Distortions in cognitive maps. *Geoforum*, 23(2), 131–138.

Tversky, B., & Schiano, D. J. (1989). Perceptual and conceptual factors in distortions in memory for graphs and maps. *Journal of Experimental Psychology: General*, 118, 387–398.

Ungar, S. J., Blades, M., Spencer, C., & Morsley, K. (1994). Can visually impaired children use tactile maps to estimate directions? *Journal of Visual Impairment and Blindness*, 88, 221–223.

Ungerleider, I. G., & Mishkin, M. (1982). Two cortical visual systems. In D. J. Ingle, M. A. Goodale, & R. J. W. Mansfield (eds.), *Analysis of Visual Behavior* (pp. 549–586). Cambridge, MA: MIT Press.

Uttal, D. H. (1994). Preschoolers' and adults' scale translation and reconstruction of spatial information acquired from maps. *British Journal of Developmental Psychology*, 12, 259–275.

Uttal, D. H. (1997). Seeing the big picture: Children's mental representation of spatial information acquired from maps. Paper presented at the 93rd annual meeting of the Association of American Geographers, Ft. Worth.

Uttal, D. H., & Wellman, H. M. (1989). Young children's representation of spatial information acquired from maps. *Developmental Psychology*, 25, 128–138.

Vallar, G., Sterzi, R., Bottini, G., Cappa, S., & Rusconi, M. L. (1990). Temporary remisssion of left hemianesthesia after vestibular stimulation: A sensory neglect phenomenon. *Cortex*, 26, 123–131.

Vauclair, J. (1987). A comparative approach to cognitive mapping. In P. Ellen & C. Thinus-Blanc (eds.), *Cognitive processes and spatial orientation in animal and man* (Vol. 1, pp. 89–96). Dordrecht: Martinus Nijhoff.

Ventre, J., & Faugier-Grimaud, S. (1988). Projections of the temporo-parietal cortex on vestibular complex in the macaque monkey. *Experimental Brain Research, 72,* 653–658.

Viguier, C. (1882). Le sens de l'orientation et ses organes chez les animaux et chez l'homme. *Revue Philosophique de la France et de l'Étranger, 14,* 1–36.

Vischer, K., & Seeley, T. D. (1982). Foraging strategy of honeybee colonies in a temperate deciduous forest. *Ecology, 63,* 1790–1801.

Vishton, P. M., & Cutting, J. E. (1995). Wayfinding, displacements, and mental maps: Velocity fields are not typically used to determine one's aimpoint. *Journal of Experimental Psychology: Human Perception and Performance, 21,* 978–995.

Vitte, E., Derosier, C., Caritu, Y., Berthöz, A., Hasboun, D., & Soulié, D. (1996). Activation of the hippocampal formation by vestibular stimulation: A functional magnetic resonance imaging study. *Experimental Brain Research, 112,* 523–526.

von Frisch, K. (1965). *Tanzsprache und Orientierung der Bienen.* Berlin: Springer-Verlag.

von Frisch, K. (1967). *The dance language and orientation of bees.* London: Oxford University Press.

von Frisch, K., & Lindauer, M. (1954). Himmel und Erde in Konkurrenz bei der Orientierung der Bienen. *Naturwissenschaften, 41,* 245–253.

von Saint Paul, U. (1982). Do geese use path integration for walking home? In F. Papi & H. G. Wallraff (eds.), *Avian navigation* (pp. 296–307). New York: Springer-Verlag.

von Uexküll, J. (1909). *Umwelt und Innenleben der Tiere.* Berlin: Springer-Verlag.

von Uexküll, J. (1934). *Streifzüge durch die Umwelten von Tieren und Menschen.* Berlin: Springer-Verlag.

von Wright, J. M., Gebhard, P., & Kartunnen, M. (1975). A developmental study of the recall of spatial locations. *Journal of Experimental Child Psychology, 2,* 431–445.

Vygotsky, L. S. (1978). *Mind in society.* Cambridge, MA: Cambridge University Press.

Wagener, M., Wender, K. F., & Wagner, V. (1990). Role of routes in spatial memory. Paper presented at the meeting of the Psychonomic Society, New Orleans.

Walcott, C., & Schmidt-Koenig, K. (1973). The effect on homing of anesthesia during displacement. *Auk, 90,* 281–286.

Wallraff, H. G. (1959). Örtlich und zeitlich bedingte Variabilität des Heimkehrverhaltens von Brieftauben. *Zeitschrift für Tierpsychologie, 16,* 513–544.

Wallraff, H. G. (1970). Über die Flugrichtungen verfrachteter Brieftauben in Abhängigkeit vom Heimatort und vom Ort der Freilassung. *Zeitschrift für Tierpsychologie, 27,* 303–351.

Wallraff, H. G. (1974). *Das Navigationssystem der Vögel: Ein theoretischer Beitrag zur Analyse ungeklärter Orientierungsleistungen.* Schriftenreihe "Kybernetik." Munich: R. Oldenbourg Verlag.

Wallraff, H. G. (1978). Preferred compass direction in initial orientation of homing pigeons. In K. Schmidt-Koenig & W. T. Keeton (eds.), *Animal migration, navigation, and homing* (pp. 171–183). Berlin: Springer-Verlag.

Wallraff, H. G. (1980). Does pigeon homing depend on stimuli perceived during displacement? I. Experiments in Germany. *Journal of Comparative Physiology, 139,* 193–201.

Wallraff, H. G. (1991). Conceptual approaches to avian navigation systems. In P. Berthold (ed.), *Orientation in birds* (pp. 128–165). Basel: Birkhäuser.

Wallraff, H. G., & Graue, L. C. (1973). Orientation of pigeons after transatlantic displacement. *Behaviour, 64,* 1–35.

Walsh, S. E., & Martland, J. R. (1996). Maintaining orientation within route following tasks. *Cartographica,* April, 30–37.

Wan, H. S., Touretzky, D. S., & Redish, A. D. (1994a). A rodent navigation model that combines place code, head direction, and path integration information. *Society for Neuroscience Abstracts, 20,* 1205.

Wan, H. S., Touretzky, D. S., & Redish, A. D. (1994b). Towards a computational theory of rat navigation. In M. Mozer, P. Smolensky, D. Touretzky, J. Elman, & A. Weigend (eds.), *Proceedings of the 1993 connectionist models summer school* (pp. 11–19). Hillsdale, NJ: Lawrence Erlbaum.

Warren, D. H., & Scott, T. E. (1993). Map alignment in traveling multisegment routes. *Environment and Behavior,* 25(5), 643–666.

Warren, W. H., & Hannon, D. J. (1990). Eye movements and optical flow. *Journal of the Optical Society of America,* A7, 160–169.

Webster (1995). *Webster's Collegiate Dictionary.* Springfield, MA: Merriam-Webster.

Wehner, R. (1981). Spatial vision in arthropods. In H. Autrum (ed.), *Handbook of sensory physiology* (Vol. VII/6C, pp. 287–616). New York: Springer-Verlag.

Wehner, R. (1987). *Spatial organization of foraging in individually searching desert ants,* Cataglyphis *(Sahara desert) and* Ocymyrmex *(Namib desert).* Experientia Supplementum 54, Behavior in social insects. Basel: Birkhäeuser Verlag.

Wehner, R. (1991). Visuelle navigation: Kleinstgehirn-Strategien. *Verhandlungen der Deutschen Zoologischen Gesellschaft,* 84, 89–104.

Wehner, R. (1992). Homing in arthropods. In F. Papi (ed.), *Animal homing* (pp. 45–144). London: Chapman and Hall.

Wehner, R. (1994). The polarization-vision project: Championing organismic biology. In K. Schildberger & N. Elsner (eds.), *Neural basis of behavioral adaptation* (pp. 103–143). New York: G. Fischer.

Wehner, R. (1996). Middle-scale navigation: The insect case. *Journal of Experimental Biology,* 199(1), 125–127.

Wehner, R., Bleuler, S., Nievergelt, C., & Shah, D. (1990). Bees navigate by using vectors and routes rather than maps. *Naturwissenschaften,* 77, 479–482.

Wehner, R., & Flatt, I. (1977). Visual fixation in freely flying bees. *Zeitschrift für Naturforschung,* 32c, 469–471.

Wehner, R., & Menzel, R. (1990). Do insects have cognitive maps? In W. Cowan, E. Shooter, C. Stevens, & R. Thompson (eds.), *Annual review of neuroscience* (Vol. 13, pp. 403–414). Palo Alto, CA: Annual Reviews.

Wehner, R., Michel, B., & Antonsen, P. (1996). Visual navigation in insects: Coupling of egocentric and geocentric information. *Journal of Experimental Biology,* 199(1), 129–140.

Wehner, R., & Müller, M. (1993). How do insects acquire their celestial ephemeris function? *Naturwissenschaften,* 80, 331–333.

Wehner, R., & Räber, F. (1979). Visual spatial memory in desert ants, *Cataglyphis fortis* (Hymenoptera, Formicidae). *Experientia,* 35, 1569–1571.

Wehner, R., & Rossel, S. (1985). The bee's celestial compass—A case study in behavioral neurobiology. *Fortschritte der Zoologie,* 31, 11–53.

Wehner, R., & Srinivasan, M. V. (1981). Searching behavior of desert ants, genus *Cataglyphis* (Formicidae, Hymenoptera). *Journal of Comparative Physiology,* 142, 315–338.

Wehner, R., & Wehner, S. (1986). Path integration in desert ants. Approaching a long-standing puzzle in insect navigation. *Monitore Zoologico Italiano,* 20, 309–331.

Wehner, R., & Wehner, S. (1990). Insect navigation: Use of maps or Ariadne's thread? *Ethology, Ecology, and Evolution,* 2, 27–48.

Weindler, P., Wiltschko, R., & Wiltschko, W. (1996). Magnetic information affects the stellar orientation of young bird migrants. *Nature,* 383, 158–160.

Welsh, R. L., & Blasch, B. B. (eds.). (1980). *Foundation of orientation and mobility.* New York: American Foundation for the Blind.

Wertheim, A. H., & Warren, R. (eds.). (1990). *Perception and control of self-motion.* Hillsdale, NJ: Lawrence Erlbaum.

Whiteley, A. M., & Warrington, E. K. (1978). Selective impairment of topographical memory: A single case study. *Journal of Neurology, Neurosurgery, and Psychiatry,* 41, 575–578.

Wiener, S. I. (1993). Spatial and behavioral correlates of striatal neurons in rats performing a self-initiated navigation task. *Journal of Neuroscience,* 13, 3802–3817.

Wiener, S. I., & Berthöz, A. (1993). Forebrain structures mediating the vestibular con-tribution during navigation. In A. Berthöz (ed.), *Multisensory control of movement* (pp. 427–455). Oxford: Oxford University Press.

Wiener, S. I., Korshunov, V. A., Garcia, R., & Berthöz, A. (1995). Inertial, substratal and landmark cue control of hippocampal CA1 place cell activity. *European Journal of Neuroscience, 7*, 2206–2219.

Wiest, G., Baumgartner, C., Deecke, L., Olbrich, A., Steinhof, N., & Müller, C. (1996). Effects of hippocampal lesions on vestibular memory in whole body rotation. *Journal of Vestibular Research, 6*(4S), S17.

Willshaw, D. J., Buneman, O. P., & Longuet-Higgins, H. C. (1969). Non-holographic associative memory. *Nature, 222*, 960–962.

Wilson, F.A.W., Scalaidhe, S. P. O., & Goldman-Rakic, P. S. (1992). Dissociation of object and spatial processing domains in primate prefrontal cortex. *Science, 260*, 1955–1958.

Wilson, M. A., & McNaughton, B. L. (1993). Dynamics of the hippocampal ensemble code for space. *Science, 261*, 1055–1058.

Wilson, M. A., & McNaughton, B. L. (1994). Reactivation of hippocampal ensemble mem-ories during sleep. *Science, 265*, 676–679.

Wilton, R. N. (1979). Knowledge of spatial relations: The specification of the informa-tion used in making inferences. *Quarterly Journal of Experimental Psychology, 31*, 133–146.

Wiltschko, R. (1991). The role of experience in avian navigation and homing. In P. Berthold (ed.), *Orientation in birds* (pp. 250–269). Berlin: Birkhäuser Verlag.

Wiltschko, R. (1992a). Das Verhalten verfrachteter Vögel. *Vogelwarte, 36*, 249–310.

Wiltschko, R. (1992b). The flying behavior of homing pigeons, *Columba livia*, immedi-ately after release. *Ethology, 91*, 279–290.

Wiltschko, R. (1993). Release site biases and their interpretation. In Orientation and navi-gation—Birds, humans and other animals. Paper No. 15. *Proceedings of the Conference of the Royal Institute of Navigation*. Oxford: Royal Institute of Navigation.

Wiltschko, R., Kumpfmüller, R., Muth, R., & Wiltschko, W. (1994). Pigeon homing: The effect of clock-shift is often smaller than predicted. *Behavioral Ecology and Sociobiology, 35*, 63–73.

Wiltschko, R., Munro, U., Ford, H., & Wiltschko, W. (in press). After-effects of expo-sure to conflicting celestial and magnetic cues at sunset in migratory Silvereyes, *Zosterops l. lateralis. Journal of Avian Biology.*

Wiltschko, R., & Wiltschko, W. (1978a). Evidence for the use of magnetic outward-journey information in homing pigeons. *Naturwissenschaften, 65*, 112.

Wiltschko, R., & Wiltschko, W. (1985). Pigeon homing: Change in navigational strat-egy during ontogeny. *Animal Behavior, 33*, 583–590.

Wiltschko, R., & Wiltschko, W. (1995). *Magnetic orientation in animals.* Berlin: Springer-Verlag.

Wiltschko, R., Wiltschko, W., & Munro, U. (1997). Migratory orientation in birds: The effects and after-effects of exposure to conflicting celestial and magnetic cues. In *Proceedings of the Conference of the Royal Institute of Navigation* (pp. 6-1–6-14). Oxford: Royal Institute of Navigation.

Wiltschko, W., Beason, R., & Wiltschko, R. (1991). Introduction and concluding remarks to the symposium "Sensory Basis of Orientation." In M. Williamsson (ed.), *Acta XX, Congressus Internationalis Ornithologici* (pp. 1803–1850). Christchurch, New Zealand: Christchurch.

Wiltschko, W., Daum, P., Fergenbauer-Kimmel, A., & Wiltschko, R. (1987). The develop-ment of the star compass in Garden Warblers (*Sylvia borin*). *Ethology, 74*, 285–292.

Wiltschko, W., & Gwinner, E. (1974). Evidence for an innate magnetic compass in Garden Warblers. *Naturwissenschaften, 61*, 406.

Wiltschko, W., & Wiltschko, R. (1978b). A theoretical model for migratory orientation and homing in birds. *Oikos, 30*, 177–187.

Wiltschko, W., & Wiltschko, R. (1982). The role of outward journey information in the orientation of homing pigeons. In F. Papi & H. G. Wallraff (eds.), *Avian navigation* (pp. 239–252). Berlin: Springer-Verlag.

Wiltschko, W., & Wiltschko, R. (1987). Cognitive maps and navigaton in homing pigeons. In P. Ellen & C. Thinus-Blanc (eds.), *Cognitive processes and spatial orientation in animal and man* (pp. 201–216). Dordrecht: Martinus Nijhoff.

Wiltschko, W., & Wiltschko, R. (in press). The navigation system of birds and its development. In I. M. Pepperberg, A. C. Kamil, & R. P. Balda (eds.), *Animal cognition in nature*. New York: Academic Press.

Wiltschko, W., Wiltschko, R., & Keeton, W. T. (1984). The effect of a "permanent" clockshift on the orientation of experienced homing pigeons. I. Experiments in Ithaca, New York. *Behavioral Ecology and Sociobiology*, 15, 263–272.

Wittman, T. (1995). Modelling landmark navigation. In B. Krieg-Brückner & C. Herwig (eds.), *Kognitive Robotik* (ZKW Bericht Nr. 3.). Bremen: Zentrum für Kognitionswissenschaften, Universität Bremen.

Worchel, P. (1951). Space perception and orientation in the blind. *Psychological Monographs: General and Applied*, 65, 1–27.

Worchel, P. (1952). The role of the vestibular organs in space orientation. *Journal of Experimental Psychology*, 44, 4–10.

Yamamoto, T. (1991). A longitudinal study of the development of spatial problem solving ability in the early blind. *Japanese Journal of Psychology*, 61, 413–417 (in Japanese).

Yardley, L. Britton, J., Lear, S., Bird, J., & Luxon, L. (1995). Relationship between balance system function and agoraphobic avoidance. *Behavioral Research and Therapy*, 33(4), 435–439.

Yardley, L., & Putnam, J. (1992). Quantitative analysis of factors contributing to handicap and distress in vertiginous patients: A questionnaire study. *Clinical Otolaryngology*, 17(3), 231–236.

Young, M. F. (1989). *Cognitive repositioning: A constraint on access to spatial knowledge*. Ph.D. dissertation, Vanderbilt University, Nashville.

Zannaras, G. (1976). The relation between cognitive structure and urban form. In G. Moore & R. G. Golledge (eds.), *Environmental knowing* (pp. 336–350). Stroudsburg, PA: Dowden, Hutchinson, and Ross.

Zeil, J. (1993a). Orientation flights of solitary wasps (*Cerceris*; Sphecidae; Hymenoptera), I. Description of flight. *Journal of Comparative Physiology*, A172, 189–205.

Zeil, J. (1993b). Orientation flights of solitary wasps (*Cerceris*; Sphecidae; Hymenoptera), II. Similarities between orientation and return flights and the use of motion parallax. *Journal of Comparative Physiology*, A172, 207–222.

Zeil, J., Kelber, A., & Voss, R. (1996). Structure and function of learning flights in bees and wasps. *Journal of Experimental Biology*, 199, 245–252.

Zhang, K. (1996). Representation of spatial orientation by the intrinsic dynamics of the head-direction cell ensemble: A theory. *Journal of Neuroscience*, 16, 2112–2126.

Zhang, W., & Dietterich, T. G. (1995). A reinforcement learning approach to job shop scheduling. In *Proceedings of the International Joint Conference on Artificial Intelligence, Montreal, Canada* (pp. 1114–112). San Francisco: Morgan Kaufmann.

Zola-Morgan, S., & Kritchevsky, M. (1988). Spatial cognition in adults. In J. Stiles-Davis, M. Kritchevsky, & U. Bellugi (eds.), *Spatial cognition: Brain bases and development* (pp. 415–422). Hillsdale, NJ: Lawrence Erlbaum.

Zola-Morgan, S., Squire, L. R., & Amaral, D. G. (1986). Human amnesia and the medial temporal region: Enduring memory impairment following a bilateral lesion limited to field CA1 of the hippocampus. *Journal of Neuroscience*, 6, 2950–2967.

Contributors

GARY L. ALLEN
Department of Psychology, University of South Carolina at Columbia,
   Columbia, South Carolina 29208, USA. E-mail: allen@garnet.cla.sc.edu

MICHEL-ANGE AMORIM
Laboratoire de Physiologie de la Perception et de l'Action (LPPA),
   CNRS Collège de France, 11 Place Marcelin Berthelot, 75005 Paris,
   France. E-mail: amorim@ccr.jussieu.fr

ALAIN BERTHÖZ
Laboratoire de Physiologie de la Perception et de l'Action (LPPA),
   CNRS-Collège de France, 11 Place Marcelin Berthelot, 75005 Paris,
   France. E-mail: aber@ccr.jussieu.fr

ERIC CHOWN
Department of Computer Science, Bowdoin College, Brunswick,
   Maine 04011. E-mail: echown@polar.bowdoin.edu

THOMAS S. COLLETT
Sussex Centre for Neuroscience, School of Biological Sciences,
   University of Sussex, Brighton BN1 9QG, United Kingdom.
   E-mail: t.s.collett@sussex.ac.uk

KYRAN DALE
Sussex Centre for Neuroscience, School of Biological Sciences,
    University of Sussex, Brighton BN1 9QG, United Kingdom.
    E-mail: kyrand@cogs.sussex.ac.uk

ARIANE S. ETIENNE
Laboratory for Ethology, University of Geneva, 54 rte. des Acacias,
    1227 Geneva, Switzerland. E-mail: etienne@uni2a.unige.ch

TOMMY GÄRLING
Department of Psychology, Göteborg University, P.O. Box 500,
    SE-40530 Göteborg, Sweden. E-mail: tommy.garling@psy.gu.se

FLORENCE GAUNET
Laboratoire de Neurosciences Cognitives, CNRS,
    31 Chemin Joseph Aiguier, 13402 Marsellie cedex 20, France

JOSEPHINE GEORGAKOPOULOS
Laboratory for Ethology, University of Geneva, 54 rte. des Acacias,
    1227 Geneva, Switzerland. E-mail: georgako@uni2a.unige.ch

STEPHAN GLASAUER
Laboratoire de Physiologie de la Perception et de l'Action (LPPA),
    CNRS Collège de France, 11 Place Marcelin Berthelot, 75005 Paris,
    France

REGINALD G. GOLLEDGE
Department of Geography, University of California Santa Barbara,
    Santa Barbara, California 93106, USA. E-mail: golledge@geog.ucsb.edu

RENATO GRASSO
Laboratoire de Physiologie de la Perception et de l'Action (LPPA),
    CNRS Collège de France, 11 Place Marcelin Berthelot, 75005 Paris,
    France

ANDREA GRIFFIN
Laboratory for Ethology, University of Geneva, 54 rte. des Acacias,
    1227 Geneva, Switzerland

SIMON P. D. JUDD
Sussex Centre for Neuroscience, School of Biological Sciences,
    University of Sussex, Brighton BN1 9QG, United Kingdom.
    E-mail: s.p.d.judd@susscx.ac.uk

ROBERTA L. KLATZKY
Department of Psychology, Carnegie Mellon University, Pittsburgh,
    Pennsylvania 15213, USA. E-mail: klatzky@cmu.edu

JACK M. LOOMIS
Department of Psychology, University of California Santa Barbara,
    Santa Barbara, California 93106, USA. E-mail: loomis@psych.ucsb.edu

ROLAND MAURER
Laboratory for Ethology, University of Geneva, 54 rte. des Acacias,
1227 Geneva, Switzerland. E-mail: maurerr@uni2a.unige.ch

LYNN NADEL
Department of Psychology, University of Arizona, Tucson,
Arizona 85721, USA. E-mail: nadel@u.arizona.edu

JOHN W. PHILBECK
Department of Psychology, Carnegie Mellon University, Pittsburgh,
Pennsylvania 15213, USA. E-mail: philbeck+@andrew.cmu.edu

JUVAL PORTUGALI
Department of Geography, Tel Aviv University, P.O. Box 39040,
RAMAT Aviv 69978, Tel Aviv, Israel. E-mail: juval@ccsg.tau.ac.il

JOHN J. RIESER
Department of Psychology and Human Development, Vanderbilt
University, Box 512 Peabody, Nashville, Tennessee 37203, USA.
E-mail: rieserjj@ctrvax.vanderbilt.edu

ELIAHU STERN
Department of Geography and Environmental Development,
Ben Gurion University of the Negev, Beer Sheva 84105, Israel.
E-mail: eli@river.bgu.ac.il

YASUIKO TAKEI
Laboratoire de Physiologie de la Perception et de l'Action (LPPA),
CNRS Collège de France, 11 Place Marcelin Berthelot, 75005 Paris,
France

CATHERINE THINUS-BLANC
Centre de Recherche en Neurosciences Cognitives, CNRS,
31 Chemin Joseph Aiguier, 13402 Marsellie cedex 20, France.
E-mail: thinus@lnf.cnrs-mrs.fr

ISABELLE VIAUD-DELMON
Laboratoire de Physiologie de la Perception et de l'Action (LPPA),
CNRS Collège de France, 11 Place Marcelin Berthelot, 75005 Paris,
France

ROSWITHA WILTSCHKO
Fachbereich Biologie der Universität, Zoologie, Siesmayerstrasse 70,
D-60054 Frankfurt a.M., Germany.
E-mail: wiltschko@zoology.uni-frankfurt.d400.de

WOLFGANG WILTSCHKO
Fachbereich Biologie der Universität, Zoologie, Siesmayerstrasse 70,
D-60054 Frankfurt a.M., Germany.
E-mail: wiltschko@zoology.uni-frankfurt.d400.de

Index

Page numbers for entries in figures are followed by an *f*; those for entries in tables are followed by a *t*.